THE NATURE

*Earth, Plants, Animals, Man
and Their Effect On Each Other*

A Chanticleer Press Edition

OF LIFE

Lorus and Margery Milne

Photographs by

Emil Javorsky
Clem Haagner
J. A. L. Cooke
Fred Bruemmer
Lorus and Margery Milne
Willis Peterson
Manuel V. Rubio
Wolf Suschitzky
Rolf Blomberg
Martin Litton

and others

CROWN PUBLISHERS, INC. · NEW YORK

Photographs on the half-title, title, and contents
pages are by, respectively, Helen Buttfield,
Lisa Gensetter, and Bill and Mary Lou Stackhouse

Published in the United States by
Crown Publishers, Inc., New York

Planned, prepared and produced by Chanticleer Press,
New York

Manufactured by Amilcare Pizzi S.p.A. in Milan,
Italy

Library of Congress Catalog Card Number 72-130316

Contents

Foreword

In the 1740's and 1750's, a Swedish physician-naturalist, who signed himself Carolus Linnaeus, attempted to give a short scientific name and description for every kind of plant and animal on earth. He succeeded because, two centuries ago, no one had had a reliable international guidebook to follow in appreciating the truly immense variety among living things.

Today about one-and-a-half million species have been named, two-thirds of them animals. The total grows every year, but no good estimate can be made as to whether another century of exploration and research will raise the number closest to two million, or three, or four. No longer does the total matter, for no human mind or lifetime is adequate for getting acquainted with more than a fraction of this array. The names of inconspicuous plants and animals have importance only to a few scientists. Yet everyone must be impressed that each of these many kinds of life has individual peculiarities of geological past history, of inheritance, of combinations in requirements for survival and of geographic range.

Perhaps the greatest surprise comes from realizing that each kind of life occupies only a small number among the many places in the world that would be suitable for it. The lack of gorillas in Central America, of polar bears in the Antarctic, of giraffes in Australia, is due to barriers of various types that have prevented the ancestors of these creatures from spreading so far beyond their modern boundaries. Nor is there any indication that the geological disturbances that produced mountains and coastlines had as marked effects on living things as the climatic conditions that follow from prevailing winds and ocean currents.

In our overview of life on the continents and islands, in fresh waters and salt, and of the steps shown by the fossil record toward modern times, we have simplified —oversimplified—to keep manageable the number of kinds for which historical patterns might be sought. We have emphasized these patterns among the familiar, the large, the conspicuous, the common, the colorful and the scientifically meaningful. Most kinds of living things are quite otherwise—unfamiliar, small, inconspicuous, uncommon, drab, or too little known to provide broad interest. We can continue to overlook them while gaining a global perspective from which to think about plants and animals that are our neighbors, wherever we live or travel.

Our heartfelt thanks go to the many people in every walk of life who have helped us become acquainted with native plants and animals on the six continents we have visited, and to the uncounted careful scientists who have made available to every reader of scientific journals their discoveries about the world. A number of these skilled men and women have been generous with their time and knowledge in answering specific questions about plants and animals whose names or geographic range eluded us.

We are grateful to Dr. Milton Rugoff of Chanticleer Press for his editorial advice, and to his associate, Miss Susan Weiley, for her resourcefulness in gathering the illustrations that accompany our chapters.

April 1970 *L. J. M. and M. M.*

1

The Dynamic Earth and its Mobile Life

Every twenty-four hours, on the average, the earth trembles in three hundred places with a vigor strong enough to be felt and heard. Three of these quakes are generally severe, causing obvious damage or loss of life. Yet the newscasts concentrate on the few that create major disasters: in both Morocco and Chile in 1960, Iran in 1962, Yugoslavia in 1963, Alaska in 1964, Chile again in 1965, Turkey in 1966, Venezuela in 1967, Iran and Japan in 1968, Turkey again in 1970.

In the Alaskan quake, the land shifted down six feet or up more than thirty feet in just a few minutes. Yet seismologists, whose instruments record more than 100,000 tremors each year, tell us that we live in an age of a "quiet earth." These scientists have had no chance to measure the continental convulsions that produce mountains. No mountains, other than cones of volcanic debris, have been formed anywhere on the earth in recent times.

Each year about half a dozen volcanos become active, often after centuries of complete quiet. South of Iceland, where fishermen knew the sea to have been 425 feet deep, a volcano broke through the water surface for the first time on November 14, 1963 and built a new island. Named Surtsey, after the awesome, fire-carrying giant Surtur in Norse mythology, it continued to grow. By May 1965, the highest crest of its cinder cones stood 568 feet above sea level. The island had an area of nearly one and one-quarter square miles, and was adding about an acre more every day.

From all over the world, vulcanologists came promptly to study the phenomenon from aircraft. They knew that unless Surtsey produced lava and shielded the particles of loose volcanic matter with solid rock, erosion would soon cause the island to disappear beneath the waves. Lava flow did begin on April 4, 1965 and it continued until June 1967, giving Surtsey a possibility of permanence. Wherever the island cooled, biologists stepped ashore from boats to discover how life would colonize the site. Which kinds would come first, and how would their presence change the conditions under which still other forms would find a place for roots or feet?

Within a few months, seaweeds took hold where the ebbing tide did not expose them for more than a few hours to the dry air and sun. A film of molds and ferns began to grow over the bare rock from spores brought to the island by the winds. Other vegetation washed ashore. Seeds arrived by air, sometimes on the muddy feet of birds, such as gulls, which settled on Surtsey to rest, and some lifted by gales out of the northeast, from Iceland 18 miles away and the Westman Islands some 10 miles across open stormy water. Young seals came to bask on the shores. Soon the new cliffs thronged with sea birds. And in June of 1967 the first flowers opened. They were magenta, cross-shaped blossoms of sea rocket, a fleshy annual common along coasts of Iceland, eastern America and the remote Azores.

Islands rising from the sea and mountain tops thrusting upward from level land are similar in offering to plants and animals an assortment of places to live that differ markedly from any in the immediate surroundings. The farther the island or the mountain is from the next bit of land at similar elevation, the greater is the test of mobility. The steeper the slope and the higher the summit, the more the environments challenge the living things that manage to arrive. Only those kinds that can adapt in some way to the conditions on the slopes and crests can continue to survive.

We have visited many islands and also a considerable

Rising steam shows that Bromo volcano is still active in eastern Java. (Victor Englebert) Overleaf: After a heavy rain, erosion marks the fields of cinder and ash on the slopes of Irazú, an 11,260-foot volcano in Costa Rica. (Emil Javorsky)

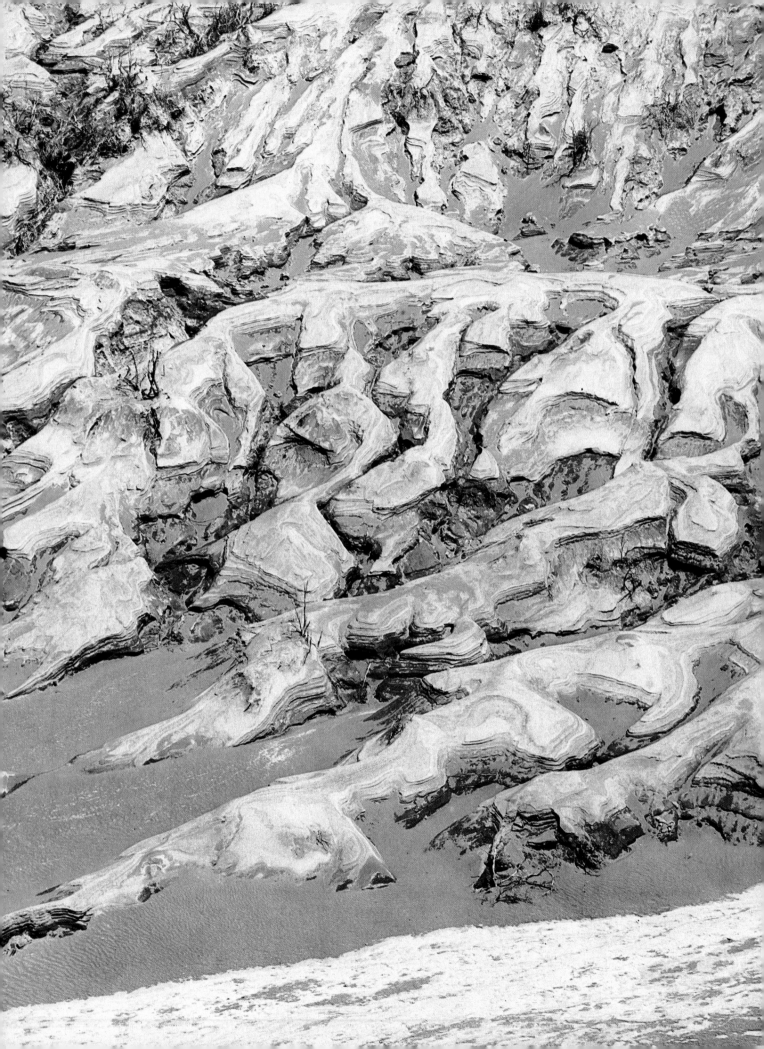

number of mountain tops in order to understand how the plants and animals got there and how they fare. Often we are impressed by the impermanence of the places to which these creatures cling. The islands may be clearly the shrinking remains of a long peninsula—a headland that storm waves isolated by cutting into it from both sides. A bar of gravel or sand may still show the connection when the tide is fully out, as it does between Percé Rock and the Gaspé coast of Canada's Quebec province. It suggests, in this case, that Bonaventure Island originated in the same way, although it is farther from the shore. Percé Rock is small and treeless, but the forest on Bonaventure Island seems no different from the pine woods that formerly covered all of this coastal area. Only the sea birds on Bonaventure seem special. They return each year to nest along the lofty cliffs, raising their young where the saltwater barrier keeps out four-legged predators.

Storms carve away at the eastern flank of Bonaventure and all sides of Percé Rock, although probably with far less vigor than they erode the coast of Norway or the inhospitable islands of Tierra del Fuego. Yet without a series of measurements extending over many centuries, we can scarcely estimate the time scale for the shrinking of the land. Surely it would be much slower than the four feet a year at which Niagara Falls wears away the rocky lip over which the river pours. We suspect it would be far faster than the rate of one foot each thousand years that the winds and scanty rains carry away the sides of the White Mountains in California, where the oldest known living things—the bristlecone pines—are as old as 4700 years.

Scenes take on new meaning if we think of the changes that produced them during millions of years. The high, pointed rocky cones of The Pitons on the island of St. Lucia in the West Indies, and Mitre Peak in Milford Sound near the southwest corner of South Island, New Zealand, are obviously the resistant remains of lava plugs that solidified in the throats of

Above left: Mallard ducks on migration are silhouetted against sunset skies over much of the Northern Hemisphere. (Joe Van Wormer) Far left: Extra moisture and protection from both wind and intense sun reward plants growing in a rock crevice. (Martin Litton) Left: The large land snail (Achatina fulica) native to coasts of the western Indian Ocean has been introduced in India, Malaysia, the East Indies, Hawaii and islands of the South Pacific with disastrous effect. (Lorus and Margery Milne)

volcanos that now have eroded away completely. Geologists know that glaciers carved Milford Sound during the last expansion of the Ice Ages, and that Mitre Peak emerged from this grinding action only 10,000 to 12,000 years ago. However, no erosion of this kind affected St. Lucia, which probably means that the changes took longer. Correspondingly, plants and animals have had a short time to colonize the fiordland of New Zealand, and a long while to spread in the West Indies.

On the island of Trinidad, we get no feeling of change in recent times. Trinidad may be isolated from northern South America mostly through subsidence of the intervening land or a rise in sea level. It is an outpost of the continent, sharing many of the kinds of life. Even the salt in the sea barrier between Trinidad and Venezuela is washed away once or twice a year when the Orinoco River is in flood. Fresh water from South America then almost surrounds the island. Great rafts of vegetation, some more than an acre in area, come floating along. Some carry with them South American animals as large as pumas. A few of these creatures swim ashore on Trinidad before being swept out to the open sea. Others are cast on the beaches by storms, which may drive the rafts into shallow water.

The strange matamata turtle seems to arrive in this way; it enters Trinidad's rivers almost every year without finding a suitable place to breed. It is one of the world's few kinds of side-necked turtles, given this name because it withdraws its head under the edge of its shell by bending its neck in a horizontal S-curve instead of a vertical one. A Trinidad fisherman introduced us to a matamata fully sixteen inches long in the shell, and laughed at us when we feared that the animal would suddenly cease feigning death and bite. It let its big head hang limply and made no attempt to defend itself. A matamata's lower jaw, in fact, is supported feebly by only two slender rods of bone, which extend almost to its ears and can exert no significant force.

The matamata gets its prey by subterfuge. In shallow water it waits, inconspicuous and motionless on the bottom, using its sharp, pointed snout as a snorkel through which to breathe. Along the sides of its extended neck, irregular flaps wave gently, like tags of vegetation fluttering in a current. If a fish swims under the turtle's head, as though it were just part of an old stick slanting upward from the bottom, the matamata suddenly dilates its broad mouth. Sucking in the fish along with a gulp of water, it swallows its prey whole.

The replenishment of rivers in Trinidad with matamatas is part of a continual flow of life from South

Born on November 14, 1963, the island of Surtsey off the coast of Iceland achieved permanence in 1965 when lava spread over its surface, protecting it from erosion. (Willard Parsons: Cranbrook Institute of Science)

America to the island. Contact with corresponding populations on the mainland is too great for Trinidad to have evolved any peculiar kinds of animals and plants. Instead, within its less than 2000 square miles, living things afford a particularly convenient sample of the flora and fauna from the northern part of the great continent. These survive close to good roads and other forms of human transportation. Opportunities for growth are varied; life thrives along fresh waters and in swamps, in forests and caves and sea grottos, around the famous asphalt lake and sandy and rocky shores, as well as on the three low mountains for which the island was named. More than most islands, Trinidad seems an ideal place to measure the mobility of living things.

Mobile Animals and Plants

We could easily think of Trinidad as a kind of way station or steppingstone for plants and animals of many kinds, spreading from South American land across open water to one after another of the islands in the West Indies. Around the eastern fringe of the Caribbean Sea these bits of land are almost as regular as beads on a necklace: from Trinidad to Tobago, Grenada, St. Vincent, St. Lucia, Martinique, Guadeloupe, Antigua, across the Virgin Islands to Puerto Rico, then westward to Hispaniola (the Dominican Republic and Haiti), to Jamaica or Cuba, and finally to the Bahamas and Florida. Twice each year the blue-winged teals become the commonest ducks in the West Indies, as they fly this route around the Caribbean between their wintering grounds in northern South America and their breeding grounds in the northern United States and Canada.

Actually only a small assortment of life has spread northward past Tobago. Still fewer kinds have come south from Florida and the Bahamas to Cuba. Instead, the roster of native plants and animals on West Indian islands reads more like a list from Central America. Apparently the water barrier was formerly far less extensive than it is now between Yucatán (Mexico) and Cuba, and between the angle of Honduras and Jamaica. Long before the peninsula of Florida formed and the Gulf Stream became an ocean river flowing diagonally across the North Atlantic, the barriers were least where shallow water still extends under the Gulf of Mexico. These may never have been land bridges, but they almost certainly formed the principal routes along which the wild colonists moved into the island havens of the West Indies.

Most living things may be readier than we generally realize to move for considerable distances, at least during some part of their lives. The exceptions may impress us most: the storks that migrate from Scandinavia to South Africa, or the salmon between their natal streams and the open seas. We recognize that many kinds of birds fly for hundreds (or thousands) of miles along seasonal migration routes. We may know that some kinds of bats, such as the hoary bat of North America, travel on a comparable schedule. Yet we think of insects, whose history of flying extends much farther back into the past, as flitting about pretty much at random.

If we need a reminder that some insects do fly for enormous distances without mishap, we can notice one particular kind of butterfly on Trinidad—one that is likely to be familiar no matter whether we come from North America or South, from Hawaii or Tahiti, Fiji or Australia. In the United States it is known as the monarch; in some parts of Canada as the King Billy (in honor of William III of England, whose heraldic colors the butterfly bears); and in Australia, logically, as the wanderer. Its caterpillar, encircled by narrow stripes of yellow alternating with black and with a pair of black thread-like "feelers" at each end, feeds on milkweed. It pupates in a pale green chrysalis studded with golden dots. Flying adults of this insect, marked with numbered tags while on autumn migration, have been recaptured many times after they had flown 1300 miles or more, for example from Ontario, Canada, to southern Texas. One recovered at San Luis Potosi, Mexico, was 1870 miles in a direct line from the place where it had received its distinctive marker four months previously.

Monarchs generally fly southwesterly in autumn, apparently orienting their course in relation to the time of day and the polarization pattern they can detect in the blue sky. Pregnant females make the return trip northeasterly in early spring, back to the general areas in which these insects spent their caterpillar stages. Their eggs start off new generations in the old sites.

Thousands of years ago, monarch butterflies spread into the West Indies and northern South America, staying there instead of returning. They evolved into a slightly different, non-migratory subspecies, *Danaus*

plexippus megalippe, which shows white spots in the outer corner of the fore wings where the North American insects are brownish orange. Still another contingent traveled onward into South America south of the Amazon basin, from southern Brazil to Patagonia and the Pacific coast of Chile. In the course of time, these most southern monarchs showed a different change in color pattern: the addition of white borders to the black veins on the underside of the hind wings, and loss of much of the black band along the rear border above on the fore wings. They are known as *Danaus plexippus erippe*, and may represent a separate species altogether.

Butterfly collectors in Britain and on the European continent have known the monarch for many years, for occasional specimens turn up on the European side of the North Atlantic Ocean. Almost all are of the North American kind. One or two have been the subspecies from the West Indies. Entomologists have never been completely convinced that monarchs make the trip under their own wing power, for the direction is wrong, the distance is great, the strong winds that might help are usually so chilly that a butterfly of this kind would stop flying; and although a monarch can alight and take off from calm water, it cannot rest there for more than an hour or two before becoming waterlogged. It seems more plausible that these insects are carried along on the ships of commerce. But they fail to colonize Britain and Europe because the food plant for their caterpillars—the milkweeds—do not grow there. Milkweeds are native to the Americas and to Africa south of the great deserts.

Milkweeds have been introduced in Pacific areas, however; and each time within a few years monarchs of the North American type have turned up to take advantage of the opportunity. They reached Hawaii about 1845, Ponape in the Caroline Islands of the South Pacific in 1857, Tonga in 1863, Samoa in 1869, and flew in large numbers around Brisbane, Australia starting in 1870. Their farthest south, on Tasmania, came in 1886—the same year that transatlantic travelers among the monarchs were caught in both Portugal and Gibraltar. Monarchs are now the commonest butterflies in the Carolines, just as they are on Bermuda. They are familiar today in the East Indies as far as Java, and on both coasts of Australia. With no deliberate help from man, these insects have demonstrated during the past two centuries a truly spectacular ability to spread as milkweeds allowed.

Scientists have often assumed that, if a plant such as milkweed or an animal such as a monarch butterfly had the physical ability to traverse a barrier such as an ocean in modern times, it would have made the move a long time ago, before any knowledgeable person was on hand to notice the change. This may be true. Yet there is often a world of difference between the arrival of a solitary individual and the actual colonization of a new land. For years New Zealanders knew the white herons from the east coast of Australia as strays 1500 miles farther eastward across open water. Winter census counts of bird life included totals ranging from fifty to two hundred individuals. Within the past decade, some twelve to twenty pairs nested on the west coast of the South Island, generally in the crowns of tree ferns. These birds naturalized themselves as members of the New Zealand community. Perhaps many came before any showed the versatility needed to accept unconventional nest sites, for in Australia the white herons nest high in dead eucalyptus trees.

New Zealanders were similarly familiar with the Australian white-eye, which is a partial migrant in its native states of New South Wales, Victoria and Tasmania. Scattered records began in 1832 at Milford Sound, but the real influx came in 1856 when flocks appeared in June around Wellington on the North Island, and a month later near Canterbury Bight, which is halfway down the east coast of the South Island. The white-eyes promptly settled in, nested, and became common throughout the country. We found them numerous as sparrows, confiding as chickadees, active as wood warblers, and always eager to supplement their varied diet of small fruits and insects with suet, granulated sugar, jam or bread crumbs. It seems incredible that colonization should have waited until ornithologists were present to take note, or that earlier populations of white-eyes on New Zealand could have become extinct or been overlooked. Apparently they had no help from anyone on their long journey across the Tasman Sea. The direction is right, though, for a high-speed tail wind to have helped them along. June is winter, a season of gales in those latitudes south of the Equator. It is also a month of short days, less than ten hours long, and the trip probably began or ended in the dark of night.

Even in a strong wind, a small bird in good health can maneuver skillfully. It is not carried along, tumbling uncontrollably, like a milkweed seed on its breeze-catching puff of radiating hairs or a dandelion seed on its glistening parachute. The bird has the advantage too in being unharmed by a dash of rain that would strike an airborne seed to the ground or the waves below. It is unaffected by chill that would paralyze a flying insect. In its larger body it may conceal a reserve of fat as fuel for a trip of many hours' duration, and sense organs that let it maintain its heading and altitude by day or night. We may not know how it manages, but we must recognize the bird's ability to travel.

In a true hurricane, the wild winds are too rough for the fragile bodies of birds, insects and most animals. However, branches torn from living trees are not so susceptible. The air currents carry them aloft, bang them around and drop them abruptly into the muddy chaos of storm debris, without doing them mortal harm. Many of them can still take root, far from the shattered trunk on which they originally grew. The distinctive character of the Florida Keys and of little elevations, called hammocks, in the great Everglades swampland near the tip of the low Florida peninsula seems due to transport by hurricanes. During the thousands (but not millions) of years since Florida arose from the sea, samples of vegetation have been ripped loose from Cuba, swirled high in the air over the Gulf Stream, and then deposited at random. Battered but alive, they grew to provide these southeastern parts of the United States with West Indian mahogany, gumbo limbo, poisonwood and coco plum, all of them native to the tropics. In its new location, the mahogany rarely grows large enough to yield the cabinet wood for which it is justly famous; the gumbo limbo is almost never tapped for sticky sap—a kind of birdlime that has been used to smear on branches and catch small birds by their feet; the poisonwood is cut out before people get hurt by its sap, which is similar to that of poison ivy; and the coco plums are seldom harvested as edible fruit, although they are delicious.

Along with living fragments of these hardwood trees, air-breathing snails (species of *Liguus*) made the trip from Cuba to Florida. Probably most of those that survived were, at the time, waiting out the dry season until rainy weather would again foster the growth of the bark fungi on which the snails feed. During drier times, these snails cement their shells by the rim to the smooth bark. Sealed away from the weather and stuck firmly in place, they could be whisked across the Gulf Stream and dropped helter-skelter. Those that emerged in small patches of tropical trees found conditions not

The two tall cones of volcanic rock called The Pitons on the island of St. Lucia in the West Indies are the eroded remains of lava that solidified in the throats of old volcanos. (J. Allan Cash)

quite the same as in their West Indian homes. Gradually they evolved distinctive colorations and bandings in the many Florida hammocks and on the various keys. When shell-fanciers arrived much later to collect trophies from these isolated samples of tropical vegetation, they found *Liguus* patterns a real challenge. Each shell has the shape of a tear drop, and measures about one and a half inches from tip to lip when the tan body of the animal inside attains full size.

We have found *Liguus* snails of one type safe in the Everglades National Park, and some with other color patterns surviving outside the sanctuary in hardwood hammocks, whose numbers are shrinking as real estate developers order the land bulldozed for human uses. Certainly many of the distinctive snails have been exterminated, and the inheritance that controlled their special patterns is now lost. Today more *Liguus* shells may be stored in zoological museums and private collections, usually segregated carefully by source and color pattern, than are worn by living snails.

People who become aware of *Liguus* and the past travels of these snails sometimes wonder whether the same or similar patterns might evolve again if, for ten or fifteen thousand years, Florida could once more be uninfluenced by mankind. We doubt that the same forces are at work now, even though the normal (if stormy and sporadic) redistribution of mobile plants and animals continues. History could scarcely repeat

The narrow, pointed snout and alert eyes of the soft-shell turtle apprise the animal of food or danger. (Hans W. Silvester)

itself unless changes in ground level, sea level, moisture and temperature also varied in approximately the same sequence they went through after the end of the last Ice Age. Changes in geological features, including soil, and in climate provide the selecting forces on the living things dropped in chance combinations on the various keys and hammocks, guiding the evolution of tree snails and their neighbors alike.

Living Dust

When we think of the spread of life from one region to another, we generally think in terms of fruits and seeds as the special packages for plant reproduction, and of animals that are fully grown, no smaller than a tiny insect. Yet a surprising wealth of both plant and animal life travels as minute particles no bigger than the mineral matter of dust. Hayfever sufferers know of the wind-borne pollen grains, particularly those of ragweed (*Ambrosia* species) and other seed plants, that bring them misery and that pollinate the flowers. Farmers recognize that corn pollen must be blown from tassel to silk if corn ears are to develop. But only the scientists who make pollen counts—the average number of pollen grains in a cubic foot of air, as measured in a standard way—realize how many kinds of spores are airlifted too. They come from algae and fungi, horsetails and ferns. Along with them in the living dust go some kinds of bacteria and the desiccated bodies of tiny animals from ponds that have dried up: single-celled protozoans, microscopic wheel animalcules and tardigrades. So great is the variety, in fact, that few of the technicians who make the published pollen counts take time to record more than the particular kinds that concern them.

Botanists at the University of Texas repeatedly sampled the air at a height of 82 feet above ground in April and May. On most days, algal spores outnumbered the pollen grains; but fungus spores were many times as numerous as the other two combined. When cultured and studied, the live algae proved to represent twenty-eight different genera of green algae (Chlorophyta), seventeen of blue greens, and seven of diatoms. Eleven of these identifiable kinds were collected also from an airplane flying at 3600 feet above the ground. The greatest concentration of airborne algae proved to be about 1600 feet up, and consisted largely of the small green alga *Chlorococcum*. This grows so commonly on the shaded side of tree trunks that Boy Scouts in the North Temperate Zone use it to determine which way is north.

Buff-backed herons, known in America as cattle egrets, watch for insects stirred by the hippo's movements. (Clem Haagner)

As might be predicted, the number of living particles in dust falls off at increasing altitude. At 11,000 feet —lower than the summit of Mount Everest—sampling devices operated from airplanes have proved the presence of between five and six microbes per cubic foot of air to be about average. Between 30,000 and 60,000 feet above the ground, the number is generally down to one spore or bacterium or animalcule per 400 cubic feet of air; from 60,000 to 90,000 feet, it diminishes to one per 2000 cubic feet. Yet a few motes of life go higher. Recently, scientists of the National Aeronautics and Space Administration reported that twice their rockets had captured in special samplers living spores of the mold *Penicillium* between 125,000 and 135,000 feet above the earth.

By being able to tolerate the severe conditions in the upper atmosphere—particularly the low temperature, intense ultraviolet light and extreme desiccation—certain of the bacteria, fungi, algae and microscopic animals have become truly cosmopolitan. They can be found in suitable situations everywhere—never as distinctively Australian or European or North American kinds. Most of the living dust, however, has less tolerance and often greater size and weight per particle. These kinds survive to reproduce only when carried along near the ground and for shorter distances. They include the mosses and liverworts, the horsetails and ferns, the clubmosses and a few types of seed plants (such as orchids) whose seeds are scarcely larger than specks of pepper.

Getting these motes airborne is no problem, for vertical air currents are fairly common. In the central areas of most continents, little "dust devils" whirl across the fields, often surrounding an upward current of air traveling at twenty-five miles per hour. For a

19

Monarch butterflies bearing numbered paper tags migrate from Canada sometimes far into Mexico, then return in spring to lay eggs on young milkweed plants. (Lorus and Margery Milne)

person they are only a momentarily unpleasant experience. A man can shut his eyes, hold his breath and wait for the swirl of dust to pass. But to a one-ounce mouse, a dust devil is formidable. Because of the mouse's small size, it presents fourteen times as great a surface in proportion to its weight. A wind of twenty-five miles per hour pulling at that surface can lift the mouse into the air and carry it for an amazing distance. In wetter places, similar winds sometimes lift the water out of a pond, carrying along small fishes and dumping them a few miles away, to the consternation of the human inhabitants. "Rains" of fishes, frogs and salamanders have been shown to occur in this way.

Once airborne, a lightweight particle is in no hurry to descend. A bacterium, perhaps enclosed in a microscopic droplet of mist barely bigger than itself, takes as much as 300 hours to settle 100 feet through the air. During that time, a steady wind of ten miles per hour can carry it a thousand miles. With an occasional updraft, a bacterial cell can stay aloft indefinitely.

Gain and Loss

Life runs its greatest risks at this time of propagation. The chance is infinitesimal that a spore or seed or microscopic animal will fall where it can become active and grow. Only a few succeed of the countless individuals that start out. The same is true for emigrants from one region that become immigrants in another. Most places they can reach by flying or swimming or walking are already occupied by plants and animals that are well adapted to the situation and able to repel invaders. The test becomes a fair one, however, because the majority of new arrivals represent kinds of life that are extraordinarily successful and numerous in the place from which they came. They are mostly pioneers—not refugees.

Occasionally an immigrant in a new land finds that its customary way of life brings it into no competition with any of the established residents. Apparently the buff-backed heron of Africa made such a beneficial move in 1930, when it somehow crossed the South Atlantic Ocean to the New World. Previously this bird had associated with elephants, rhinoceroses, antelopes of many kinds and domestic cattle over a wide area from southern Spain to Madagascar, from Ghana to the Caucasus. Its principal food was grasshoppers and frogs stirred into action by the feet of the grazing mammals. To a lesser extent the buff-backed heron scavenged the backs of the big mammals for engorged ticks, which explains why the grazers show no annoyance when a heron or two alight on their backs. We have seen these birds perched on hippopotamuses that bulged like big boulders from a water hole; the game ranger assured us that the herons cleaned off every leech they could see.

In British Guiana, where the buff-backed herons were first reported from the New World, they found domestic cattle and also some of the tropical American cowbirds. But the herons, now called "cattle egrets," ate food too big for a cowbird and found no conflict. They prospered. By 1943 they had spread west into Venezuela, and in 1946 east into Surinam. They invaded Colombia in 1951, nested in Florida in 1952 and reached Cape May, New Jersey, the same year. Yet they were not seen in Jamaica until 1956, three years after they were nesting in both Cuba and Puerto Rico. Near San Juan we saw them chasing after a disk harrow being pulled by a noisy tractor, ignoring the driver while catching all the insects and worms he exposed. For their purposes, the machine was every bit as useful as a grazing mammal—and as harmless. Now cattle egrets are known from Texas to Newfoundland, recognized by their reddish legs and feet, and a yellow beak that may be darker at the tip.

Prior to their crossing of the ocean, buff-backed

herons showed no remarkable increase in numbers over their African homeland. But the frequency with which vagrants of this kind reached Britain and Scandinavia did show a rise. And the population spread eastward almost simultaneously with an equally explosive production of new colonies. In the late 1950's they reached Australia, and by 1964 were breeding in New Zealand. Everywhere they went, the special role they took seemed vacant, waiting. No other carnivorous and insectivorous birds standing 18 inches tall had become so fearless of being kicked by grazing mammals, and so adept at catching food without being stepped on. Only in mating season do they show in their plumage the biscuit-brown color on crown and back for which they were given their Old-World name.

Colonization of a new area by plant life generally seems to be a fundamental gain, without compensatory loss. When a storm or a change in the pattern of oceanic currents produces a new sandy beach, where previously the water was deep enough for fishes and other marine creatures, we tend to forget the displaced aquatic life and think only of the opportunity for terrestrial colonists. If the new beach is in the tropics, it may soon receive among the debris thrown ashore by further storms a few coconuts which will sprout in the bare sand and soon provide food and shelter for other kinds of living things. In time the tree matures and begins dropping fruit of its own, often where the giant seeds in their waterproof buoyant husks will roll

Specialized grasses can maintain themselves despite drifting sand, being rippled by wind and tracked by animals (Helen Buttfield)

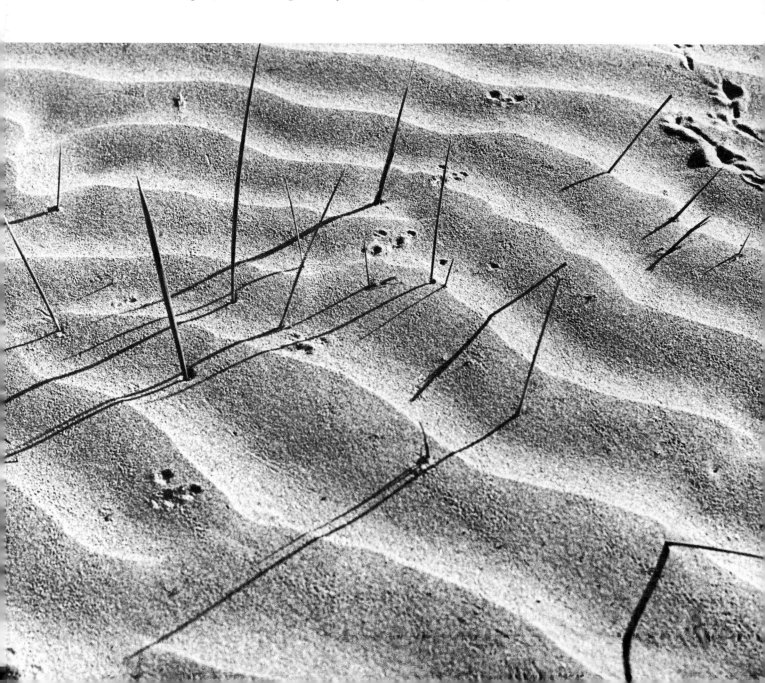

into the sea. Currents may carry them off and storms cast them ashore dozens or hundreds of miles away. Coconuts have been distributed so widely by natural means, in fact, that botanists cannot be sure where this kind of palm originated.

Mangroves colonize mudflats in the tropics and subtropics, and provide support for shellfish such as oysters and for nests of many kinds of birds. Gradually the many adventitious roots retain debris and reinforce it until it transforms into soil—perhaps more suitable for other kinds of vegetation than for mangroves. Similarly, the lichens and mosses convert bare rock and cooled lava into a substrate upon which larger and more permanent plants can grow. Animals gain from the new supply of food and shelter. What has seemed a wasteland is transformed into a rich community of life.

Replacement of one kind of life by another is a far commoner event. It may occur on a large scale at a slow pace, requiring thousands or millions of years, as did the disappearance of the vegetarian and carnivorous reptiles in Mesozoic times, making room for vegetarian and carnivorous birds and mammals of the

Cenozoic. Or it may be quick in a smaller area, perhaps through human intervention, as was the conversion of the Great Plains in North America from a home for native grasses, bison, prairie dogs, wolves and grizzly bears into agricultural country producing introduced grains, cattle, sheep and people.

All over the world, wherever the Eurasian house sparrow has been introduced, its numbers have increased only at the cost of a decrease in populations of native birds. Competition was less for food than for nest sites. In the United States and Canada we see this conflict continuing. House wrens, which previously contended mostly with tree swallows for suitable nest holes, have become restricted to places with a doorway too small for house sparrows. A house wren can squeeze through an opening an inch in diameter. The sparrow needs a full inch and a quarter, and will accept larger ones. But the introduced European starling can use a doorway an inch and a half in diameter, and seems able to oust either a house sparrow or the woodpecker that excavated the cavity in a dead tree.

Food is more critical where the rainfall is generally scanty, as in Australia. There the equivalence of one kind of grazer to another has become particularly obvious. Greenery within reach of the ground was food for kangaroos and emus until less than two centuries ago. Then colonists introduced sheep and cattle, and realized how much food could be saved for the domestic animals by reducing the populations of native grazers. But before much enthusiasm could be developed toward shooting kangaroos and emus for sport, a new game animal—the European rabbit—was introduced. Freed on an estate near Geelong, in the Australian state of Victoria on Christmas Day, 1859, the rabbits soon escaped to live wild. They prospered mightily. Ranchers calculated that ten rabbits ate as much as a cow, and five rabbits as much as a sheep. They also saw that rabbits reproduced so fast that, if unopposed, the progeny of each pair of rabbits could in three years number more than 13 million individuals!

Australians were soon supplying most of the world's demand for felt of rabbit fur, and also the ingredients

Left: Buoyant husks protect coconut seeds, letting storm waves throw them ashore to germinate on tropical beaches so widely that scientists have no clue to the origin of this useful tree. (Lorus and Margery Milne) Right: Wind has carved an elaborate series of terraces from the walls of the canyon cut by the Fish River in South West Africa. (G. G. Collins)

for countless rabbit pies and rabbit stews. In a single decade they exported 700 million rabbit skins and 157 million frozen carcasses without making a dent in the wild population. Yet the financial gain to rabbit hunters, furriers and meat handlers from five skins and five carcasses was less than a third as much as a rancher might have received for the wool from a single sheep eating the same amount of greenery. The sheep would still have yielded meat, tallow and a hide—all valuable commodities—while utilizing the same food. The enormous population of rabbits brought near disaster to production of sheep, cattle, kangaroos, emus and many other kinds of life that depended on Australian soil.

In equatorial Africa, the limiting feature on much of the highland country is water during the annual dry season. The native giraffes, antelopes and zebras manage for a week or two between visits to a water hole, and then satisfy themselves fully with about a twentieth as much water as an equal weight of domestic cattle would require in the same period. Wherever native people keep large numbers of cattle, principally as their preferred form of wealth, the land can support proportionately fewer of the wildlife for which the continent is famous. As the herds of cattle increase toward complete utilization of the waterholes during times of drought, the numbers of native wild mammals shrink toward zero.

The Time Scale of Change

When a person returns to the scenes of his youth, he is often shocked by the pace of change. We recall a pond with white water lilies that no one picked because the bottom was too deep for wading, and no passable road came near along which to bring a canoe or rowboat. Water beetles of many kinds, backswimmers and water boatmen swam among the sparse vegetation on the bottom near shore. Sunfish and stickleback nested in the shallows. Whirligig beetles and water striders scavenged over the surface film. Frogs and dragonflies perched on lily pads. Swallows and kingfishers found things to eat from the pond. Among the alders and willows along one side, phoebes perched and called between aerial forays to catch gnats and mosquitoes.

Twenty years later the pond was gone, its place taken by a wet meadow with sedges and grasses. One willow had become a big tree. The rest, like the alders, were gone. The little stream that once had fed the pond had gradually filled it with mineral matter and organic debris. Scarcely two feet wide, it still meandered through the meadow and flooded the low parts every spring.

But virtually all of the aquatic life had disappeared. Cottontail rabbits hid where water lilies formerly floated their leaves and blossoms in water six to ten feet deep.

Although many lives were affected by this modest and local change, no one recorded it in a history book. Nor, apparently, did the people of Pozzuoli see anything remarkable when their coastline across the bay from Naples, Italy, quietly sank beneath the waves of the Mediterranean. No published report tells when this area became dry land again. Yet between the fourth century A.D., when a temple was built near the shore as a place to worship the Egyptian goddess Serapis, and modern times, the ruins of that edifice were gently immersed and as gradually exposed to air. Today visitors can see conspicuous pits far up the standing columns where marine bivalves bored cavities into the marble. Some of the mussel shells are still in place, clear evidence of what happened long ago. They are visible, too, in fallen columns that are still far above sea level. The fall is usually attributed to the great earthquakes that took place in 1198 and 1538. To us these convulsions of the earth seem far less remarkable than the inconspicuous lowering and raising that allowed the sea to spread and then drained it off again.

Farther north in Europe, vertical movements of the earth have been at least as great in relatively recent times. Fishermen in the North Sea often snag their nets on the remains of human dwellings far from land, where the water is 60 to 100 feet deep. Fitted stones and hand tools have been hauled to the surface, proving that an island at least—if not a land bridge—existed between Britain and Scandinavia. But there is still no agreement as to whether the settlements, called "Doggerland" by people who knew the location of the Dogger Banks as fishing grounds, date from one of the three interglacial periods during the Ice Ages. Stone-Age men were in Britain during the second and third interglacials, but were driven out when the glaciers spread again. Geologists regard 7000 B.C. as the probable end of the latest extension of the ice, which scoured as far south as the estuaries of the Thames and Severn and presumably eliminated all life from Ireland. Between 6000 and 5000 B.C., the land level readjusted itself, letting the Rhine River drain southwest past the Low Countries and France into the Atlantic Ocean. This reestablished the Straits of Dover and isolated the British Isles from the Continent.

To anyone trying to account for the flora and fauna of England, Wales, Scotland and Ireland, these dates are highly significant. They provide only one or two thousand years during which Great Britain could receive terrestrial plants and animals across a land bridge. Arctic-alpine forms, such as peat moss, may have had a head start, by surviving in southernmost England during all of the glaciation. Certainly peat moss spread northward quickly in the wake of the melting ice. As the land drained, arctic birch colonized it, followed by pines and hazel, then oak and beech.

As the weather warmed and the plants became established, the corridor from the Continent to Britain grew narrower. Less than half of Europe's kinds of animals got across. Many animals for which the climate and food resources in the British Isles are now suitable are still absent. Although Britain during the interglacials had saber-toothed tigers and hyenas, horses and tapirs, rhinoceroses and hippopotamuses, mastodons and Old-World monkeys, none of these returned. Neither did hamsters, lemmings, beavers, bears, wolves or wild boars. No reindeer, ibexes or chamois crossed the bridge. Missing too are midwife toads, fire toads and yellow-bellied toads, although these are common today on the French side of the English Channel.

For a while after the melting of the last glaciers from the Ice Ages, a narrow neck of land linked Scotland to Ireland. Across it went a sampling of the plants and animals that had reached Great Britain from the Continent. But this second filtering action let through only one reptile (the viviparous brown lizard) out of six, and two amphibians (the common newt and the natterjack toad) from a possible half dozen. Aside from the hoary legend that St. Patrick drove the snakes out of Ireland, no evidence has been found for the existence in postglacial times of the five other reptiles found in Great Britain (the grass snake, the smooth snake, the poisonous adder, the sand lizard and the legless lizard known as the slow worm). Yet this account is consistent, for on the adjacent areas of continental Europe there are these same reptiles and six more, these same amphibians and another half-dozen kinds.

It is history of this type that the scientist seeks, linking the past to the present in accounting for the assortment of life he finds in the world today. Plants and animals are where they live as much because their ancestors were able to reach the place as because their adaptations fit them for the conditions they encounter. Time is needed to cross a bridge, and still more time for the improbable to happen—such as the arrival of a raft at a distant oceanic island, bringing a pregnant female or a pair of some kind of colonist. Time and isolation are both needed for evolution to provide fresh diversity and distinctive forms of life.

The pattern of life upon our planet has been building for at least 300 million years. When we think of this immensity of time, we might remember that since the beginning of the Christian Era, less than three-quarters of a million *days* have passed.

2

The Long Evolution of the World and its Life

In recent years, geologists have progressed beyond investigating the details of mountains and plains, rivers and lakes, shorelines and islands and toward examining the shapes of continents and their actual existence. These features, which show in truest proportion on any good global map of the earth, are products of the past, just as are fossils and living things. The world and its life have evolved together, and an understanding of the steps toward the present would go far toward accounting for the modern location of each feature.

We could scarcely think of continents, let alone of life, without water in the oceans and lakes. The amount of water has now been estimated with reasonable accuracy. Yet the total (about 322 million cubic miles) is almost meaningless until it is viewed in relation to the planet as a whole. On the scale of a globe twelve inches in diameter representing the earth, the total volume of salt water in the oceans would be shown by a cube one inch in each direction. All of the fresh water could be indicated by three smaller cubes, measuring respectively a quarter-inch, an eighth-inch, and a sixteenth-inch on a side. The biggest would show the relative volume of water congealed in ice caps and glaciers, the middle-sized cube the ground water within 2500 feet of the earth's surface, and the smallest, all of the water in lakes and rivers. On this scale, a cube to represent the water vapor in the atmosphere would be almost invisibly small.

To account for this much water on our planet, it is necessary only to assume that volcanos have been steaming out water vapor for the past 3000 million years at the rate that has been estimated for them during the last 140 million—since Jurassic times. Geologists calculate that the new water they continually contribute produces a rise in sea level of about three feet each million years. This seems extremely slow, and could easily be concealed by other changes. Oceanographers believe that a three-foot rise would occur merely by expansion if all the water now in the oceans were to increase in average temperature by 2° F.

The ocean basins slope upward so gently on each side that progressively more new water must be added to produce each extra foot in depth. Conversely, with smaller volume in the past, the oceans might be expected to have come less far up on the continents. With only about four-fifths as much water on the earth's surface 600 million years ago at the beginning of Cambrian times and of the first good fossil record, the oceans could have been at least 2000 feet shallower than at present.

The distribution of sedimentary rocks on the continents shows clearly that continental outlines, including the continental shelves, have remained essentially unchanged at least since Cambrian times. The shallow seas that spread in from the deep oceans and accumulated the sediments were no deeper than the modern Baltic Sea or Hudson Bay. Epicontinental seas of this dimension could not accommodate the extra water accumulating on the surface of the planet.

An alternative idea seems more plausible: that the extra weight of water on the ocean floors presses the crustal rocks there down into the fluid core of the earth, deepening the ocean basins without interfering with the continents. But a sag in the broad sea floors would be possible only through a simultaneous and greater rise in the relatively small areas of the continents. That matching movements of this magnitude have taken place all through geologic history was suggested first in 1889 and given the name isostatic change. It is now an accepted part of modern thinking about the earth.

The three-toed sloths of tropical America climb well and suspend themselves below the branches of the trees whose foliage they eat. They can swim, but only drag themselves across bare ground. (Rolf Blomberg: Full Hand)

The progressive rise in the continents as the sea floor sags would help explain why erosion never completes the work of leveling the continents. This process has been measured carefully in many ways. High on the slopes of the White Mountains in eastern California, where the bristlecone pines include patriarchs 4900 years old, wind and scanty rain (aided by frost) have been eroding the rocky soil around the tree roots at the rate of an inch each century. Many of the oldest ones thickened on their undersides and became vertical vanes of woody tissue, which now have been exposed to air. The American geologist James Gilluly calculates that about three and one-quarter cubic miles of the continents wash or blow into the oceans annually. If not for isostatic changes or some equivalent phenomenon, this would fill in the oceans enough to raise sea level by about 130 feet each million years. Eroding the mountains fastest and the lower land at a slower pace, it would flood and obliterate the continents—destroying all land life—in less than 19 million years. Quite evidently this has never happened, and is not likely to.

Accumulations of eroded matter in shallow epicontinental seas are best known because of their greater proximity to terrestrial man. The thickness of it, after consolidation into rock and elevation above sea level, is almost unbelievable. Erosive forces have left exposed the remains of more than 100,000 feet of Pre-Cambrian sedimentary rocks, another 100,000 feet of Paleozoic rocks in some parts of Europe, an additional 30,000 feet of Mesozoic deposits in North America, and nearly 20,000 feet of overlying remains from the Cenozoic in the Northern Hemisphere. During the last 600 million years alone the total rise of the continents is at least 150,000 feet—five times the height of Mount Everest.

Why Do Mountains Rise?

As the geologist studies the distribution of sedimentary rocks from each period in the past, he sees so often that the depression in which epicontinental seas stood for 100 million years or so suddenly became zones of major mountain building. Epicontinental troughs of Paleozoic age in North America rose to produce the Appalachian Mountains and the Rockies. Similarly the Himalayan region of Asia has risen in recent geologic time—perhaps the last 20 million years—from an epicontinental trough below sea level to above five miles higher than this reference line. In the same period the Tibetan plateaus, with an area of about 750,000 square miles, have been elevated to an average three miles above sea level.

The idea of isostatic change seems contradicted by this evidence. Why should it not level out a low place before so much new weight was added, or let any mountain chain sag of its own immense mass? How can the depression become progressively deeper, accumulating as much as 100,000 feet of sedimentary rock, before the process ends? Certainly the mass of this new rock is real enough. Wherever mountains of it have been raised along a coastline, the gravitational attraction from the rock above sea level pulls on the adjacent ocean. It produces a consistently higher water line, as measured from the center of the earth, than where the continental margins are nearly level.

Today, geologists know of a trough that is forming and filling along the coast of the Gulf of Mexico on each side of the Mississippi River's mouth. It deepens just fast enough to accommodate the sediments emptied by the great river. Annually these amount to about 750 million tons of solid matter, representing an average of about a foot of topsoil removed from the river basin each 8000 years. Based on past experience, geologists might predict that a range of east-west mountains will eventually rise where the trough is now being loaded with sediments. It is an area nearly 1000 miles long, 100 miles wide, but nowhere thick enough yet to expect a change in the process.

Until the 1950's no comprehensive scientific explanation was offered for slow changes in the earth's contours that seemed contrary to isostasy. Then came information from several different lines of research, supporting a new suggestion: that the vertical forces required in mountain building, and the previous sagging of the sea floor to accommodate sedimentary deposits, might both be due to changes in pressure and temperature deep in the earth. Intense pressure, of the order of seventy-five tons to the square inch, has strange effects on materials when the temperature is as high as 900° F. These are the pressures and temperatures where the less dense crust of the earth meets the underlying denser rocks of the mantle layer—the solid material outside the core.

What Happens at the Moho?

The location of this change in density of rocks was discovered in 1909 by the Yugoslav seismologist Andrija Mohorovičić. Known as "the M discontinuity" (or "the Moho"), it affects the rate at which earthquake waves travel. Under the ocean floors it is only about two extra miles down. Under the continents, it is twenty to thirty miles below sea level, and under the major mountain ranges as much as forty miles down. It is as though each continent were an island of lighter rock floating on the mantle, with only a thin skin of the same material as the crust under the oceans. The extra thickness under each big mountain might give extra buoyancy.

No one knows yet whether the M discontinuity is a zone where crustal rocks of one chemical constitution grade into mantle rocks of another chemical nature, or whether the boundary is merely between two different crystalline patterns in rock of a single chemical type. An attempt during the middle 1960's to drill a hole to the M discontinuity (the "Mohole") in the sea floor off California did not go far enough to get an answer.

The consequences of molecular rearrangement in a single chemical compound are familiar in parts of the world where water freezes or ice thaws. The same molecules occupy about 10 per cent more space in the form of ice than they do as water, which is why frost causes water pipes to burst and why ice floats. We thought about this recently while exploring along the east coast of Newfoundland, where some icebergs of medium and small size seemed anchored less than a mile offshore. They had run aground in a storm, each marked with a splash of colored dye dropped on it from an airplane to help scientists trace the travels of these hazards to navigation. The largest of the icebergs in view thrust skyward 100 feet or more, buoyed up above sea level by the immersed ice, which accounted for about 90 per cent of the total mass. Eventually, we knew, the late June sun would melt the top of each berg and transform exposed ice into water vapor, letting the berg rise progressively into air and freeing it from its contact with the bottom.

The effect of pressure on ice is evident when a person goes skating on a rink or frozen pond. The ice immediately under the hollow-ground blade of each skate changes to water with no alteration in temperature. This acts as a lubricant, helping the skater glide along. But as soon as the skate passes by, the pressure is reduced; the water freezes solid, filling the groove made by the skate and often showing hardly a mark to indicate that anything happened.

A similar phenomenon may affect the rocks at the M discontinuity, if the lighter crustal rocks are in one physical state and the denser mantle rocks merely the same chemical material in a different physical state. An increase in pressure or a decrease in temperature along the boundary would cause some crustal rock to become mantle rock; the converse would also be true. Where sediments accumulate on the floor of an epicontinental sea, the increase in pressure might cause the sea floor

The reproductive cone of the cycad Dion spinulosum *arises at the center of a whorl of evergreen fronds. (Emil Javorsky)*

to sag by isostatic change. But if the temperature is appropriate, the crustal rocks could transform under the pressure to become mantle rock. The M discontinuity would remain about the same distance below the surface, while the crustal layer grew thinner, accommodating the growing overburden of sedimentary material.

Geologists can imagine this process continuing for millions of years, until a difference in the heat conduction through the sedimentary rocks let temperatures below increase. Then a new reaction—mountain building—would begin because mantle rock at higher temperature was transforming into lighter crustal rock. Expanding, it would push up the overburden, buckling and folding the strata. As erosion carved at the new mountain peaks, reducing the pressure at the depth of the M discontinuity without lowering the temperature there, more mantle material would expand to become crust. The remains of the mountains would then be thrust even higher.

Similarly, the warming of the mantle rocks at the M discontinuity by a few tens of degrees from any other cause would start the uplift of whole continents or large plateaus from sea level. Progressively the change would shift to greater depths, increasing the volume of lighter rocks and floating the land mass isostatically to impressive heights.

Where Have the Continents Been?

Although there is good geological evidence that the continents have had approximately their present shape for at least 600 million years and that they are composed of lighter rocks than the underlying mantle which buoys them up, no proof can be found that their present locations on the globe date back that far. Most things that float can move unless anchored. If whole continents moved about in the past and were previously grouped in different ways, living things on land and along the shores in shallow water would have had quite a different world in which to spread. Where have the continents been?

Bracken ferns, with their wing-like fronds, grow in full sunlight across North America, Europe and eastern Asia. (Helen Buttfield)

Continental Drift

Fundamentally, the position of any continent is a matter of geology. Yet geologists ridiculed the idea of continental drift when it was proposed in 1912 by the German meteorologist Alfred Wegener. They chose to ignore the impressive list of similarities in geological features and fossils he presented as his reason for assuming that the South Atlantic Ocean is relatively new. The near-perfection of fit between the shoreline of eastern South America and western Africa was brushed aside as fortuitous. No force was then known that might shove continents apart. Until a reasonable explanation for such a phenomenon could be offered, the geologists preferred to ignore evidence for the phenomenon itself.

Biogeographers found Wegener's hypothesis fascinating. His map, which showed all of the continents

Fossilized fern fronds are common in sedimentary rocks from the Carboniferous age. (National Coal Assoc.)

fitted together like so many jigsaw-puzzle pieces into a single land mass named "Pangaea," brought into close proximity some widely-separated lands with similar floral and faunal features. Plants of the family Proteaceae, known only from southern South America, South Africa, temperate eastern Australia and New Zealand, might easily have spread over the tips of clustered continents and islands. So might the walking worms of phylum Onychophora, which still survive in these places and some of the West Indies. The fossils of the broad-leaved seed fern *Glossopteris*, which date from Permian times, become understandable in their distribution if the southern continents (including India and Antarctica) have all been connected.

By 1930, when Wegener died in Greenland, most biologists of the Northern Hemisphere had accepted the decision of virtually all geologists that continental drift was an illusion. For them the controversy died. But in the Southern Hemisphere, scientists kept on gathering evidence to support the idea. In 1937, Alexander L. du Toit of South Africa combined the new information with the old into a modification of Wegener's hypothesis. He interpreted the evidence as indicating that a cluster of continents just north of the Equator, to be called "Laurasia," separated into Eurasia (minus India), Greenland and North America, and moved northward into their present positions. A second cluster, to be named "Gondwanaland," broke up later into Antarctica, South America, Africa, India, Australia and New Zealand. In this way he tried to account for the widespread glaciation in the Southern Hemisphere about 200 million years ago, while coal deposits were simultaneously accumulating under sub-tropical conditions in continents now clearly northern.

Oceanographers who were mapping the sea floors found some strange features of geology in deep water. Down the middle of the Atlantic Ocean runs a peculiar ridge, which is strikingly parallel to the coasts of Africa and South America, of Europe and of North America plus Greenland. They traced it north between Greenland and Scandinavia, where it curves eastward through the Arctic Ocean and finally approaches the Siberian coast near the Lena River. Southward this ridge passes midway between Africa and Antarctica, then diagonals across the Indian Ocean toward the tip of the Malayan peninsula. Another mid-ocean ridge turned up in the Pacific Ocean. It begins just south of the Gulf of California, continues more or less parallel to the South American coast, then almost encircles Antarctica before turning northwest across the Indian Ocean in the general direction of Madagascar. It would be easy to interpret the great Rift Valley of eastern Africa as an extension of this mid-ocean ridge—one with no ocean above it.

The rock samples collected from the ocean floor, and from oceanic islands remote from continents, showed another peculiarity. Unlike rocks from continental areas, which date back into Pre-Cambrian and Paleozoic eras, the samples associated with the open ocean were never older than the Age of Reptiles—Mesozoic times. The oceanographers kept looking for older rocks, or an explanation for the absence of ancient sedimentary deposits.

These observations, and a whole array of measurements made by geophysicists with equipment developed during World War II, led to renewed interest in continental drift. After the middle of the twentieth century, even a geologist could consider the theory seriously, because a force sufficient to move continents had been suggested. Energy in the earth's interior might cause slow convection currents in the earth's mantle. Up-welling of rock from the planet's interior could be the cause of mid-ocean ridges, and would explain why heat flow outward through the crust there is two to eight times greater than normal for the sea floor. Such a slow upwelling could be compensated for by a down-turning of the sea floor to disappear below oceanic trenches. In these trenches the measurements of heat flow prove to be extraordinarily small—a mere tenth of average rate. The force of gravity too shows a strange lessening in these trenches, as though something peculiar is going on there.

Convection currents might create oceans, such as the Atlantic, where none had been before, if continental masses move apart. Otherwise, as in the Pacific, the sea floor would act like an endless belt. Forming at the mid-ocean ridge, it would move toward a continent at a speed of about an inch annually, only to vanish down a trench, carrying with it whatever sediments had accumulated. Presumably, oceanic islands such as Hawaii and the coral atolls of the South Pacific will disappear on the same schedule. Its rate is enough to move an area of sea floor 3400 miles in the years since the beginning of the Mesozoic era, which is more than halfway across the widest ocean.

The Hawaiian islands, in fact, show the probable progress of the sea floor. The easternmost and largest of the islands is Hawaii itself, with the most active volcanos. It must stand right above a volcanic pipe through which molten rock comes to the earth's surface. Maui, next to the west and with older craters, is followed by progressively older islands—each begun as a volcano but now eroding, and having permanence because of the work of coralline plants and animals. As the sea floor shifts westward, carrying all of these islands along, the volcanos on Hawaii are likely to cease; then another island, one not yet born, will appear in the chain and have its day to flare and rumble.

Fitting Continents Together

The knowledge of a force that can expand an ocean, shove continents apart, and cause the sea floor to renew itself 200 million years or less, has encouraged geophysicists to learn the latitude and longitude of the continents at various periods in the past. But to find a starting point, they have put computers to work fitting land masses together with a minimum of overlap and of water barriers between. South America fits Africa south of its westernmost bulge better than any other combination of continental margins. But the Canadian maritime provinces can be brought just off the Moroccan coast, Newfoundland close to Spain, and the whole North Atlantic narrowed until it is a mere epicontinental sea on broad "Laurasia." The south coast of Australia matches well the east side of Antarctica, with India's triangular mass like a keystone fitted between Australia's northwest coast and Africa north of Madagascar.

These calculations, based only upon the outlines of

continents around their coastal shelves, make no assumptions as to details on the land itself. Yet Pre-Cambrian rock outcroppings on one continent are brought amazingly close to those on another; Paleozoic strata line up. So do many fault lines that geologists have regarded as more than 200 million years old. Until this long ago, it may be that there was no Atlantic Ocean, either North or South, and no Indian Ocean because India was in contact with Africa, and no "Roaring Forties" around Antarctica because Madagascar and Australia blocked the winds and the water they might drive.

A different type of evidence challenges the geophysicists. In volcanic rocks (chiefly basalt) that reached the earth's surface and cooled, crystals of iron oxide (magnetite) are oriented in definite patterns. Presumably they point the directions of the earth's magnetic field at each place for the exact moment when the rock solidified from the liquid state. These patterns could be used to trace the wanderings of the continents if only some way could be found to learn where the magnetic poles were at each period of past time. Unlike the North and South Poles, which seem fixed enough to use as reference points for longitude and latitude, the magnetic poles move even faster than continents can, following no route that scientists have been able to reconstruct or explain. The key to their travels is likely to unlock the secret of where the continents have been since the beginning of Mesozoic times.

The Last 230 Million Years

Until the Mesozoic, which began about 230 million years ago, the supercontinents of Laurasia and Gondwanaland may well have been separated by an equatorial sea called the "Tethys," which opened broadly to the east but narrowed in the west where the North American component of Laurasia had just begun to separate from northwestern Africa, creating the first identifiable part of the North Atlantic Ocean. The corresponding expansion of a sea floor between South America and Africa is dated around 150 million years before the present, in mid-Jurassic. Probably the Indian Ocean had its origin about 110 million years ago, rather early in the Cretaceous period, as Antarctica with Australia and New Zealand attached swung away from Africa in one direction, India and Ceylon in another, leaving Africa itself with only Madagascar and Arabia still in close association. Later in the same period, about 80 million years ago, New Zealand gained its independence from Antarctica.

The expansion of an Atlantic Ocean corridor between

the Old World and the New could take place only by crushing the Pacific Ocean's eastern side, shoving down the sea floor into trenches, compacting the western coastline of the American continents, breaking open the earth's crust and forcing up molten volcanic matter, folding and elevating strata into the great cordilleran mountains from Tierra del Fuego to Alaska. By Paleocene times, around 60 million years in the past, the North Atlantic broadened into a seaway between Norway and Greenland. Apparently it remained shallow enough to offer a land bridge in later periods, and to restrict the flow of Atlantic water into the Arctic Ocean.

Australia could have broken away from Antarctica during the Eocene, around 40 million years ago. Apparently it rotated almost a quarter of a turn as it drifted eastward, compressing the sea floor between it and New Zealand and perhaps causing the rise of mountains in western New Zealand and eastern Australia. Eurasia may have been rotating clockwise, crushing the western portions of the Pacific Ocean's floor, creating conditions that have persisted to the present in the form of volcanic outbursts. The Pacific "ring of fire" still has a large number of active craters.

During all of this time, Europe and North Africa seem to have been separated by a broad remainder of

The lizard-like tuataras of New Zealand are the sole survivors of a reptilian order well represented during the Age of Reptiles. (New Zealand National Studios)

the Tethys Sea, which extended eastward to cover Asia Minor and to connect north of India with waters along the coast of Southeast Asia. One reason for this breadth at the Atlantic end was that the Iberian peninsula had not yet swung south to open up the Bay of Biscay, an event assigned to Eocene times. During these millennia, we can think of India plowing across toward the Asiatic mainland, Ceylon trailing behind. The advancing blunt boundary of this sub-continent crumpled into mountains during the Eocene. Apparently the journey ended in the Oligocene period, closing off the east end of the Tethys Sea and producing the Gulf of Aden—perhaps the first rifting to form the Red Sea too. Geologists invoke something akin to momentum to account for the forceful joining of India to Asia, which created conditions that raised marine sedimentary deposits to spectacular heights in Miocene times with the formation of the lofty Himalayas and the plateau country of Tibet.

33

The Miocene must have been a fearful time over much of the world, with volcanic activity in many places. One outburst apparently developed as a consequence of spreading in the sea floor of the eastern Pacific Ocean, which separated the volcanic area of the Galápagos islands from the South American coast. In regions of North America that now are eastern Washington and Oregon, together with parts of adjacent Idaho and Montana, lava spread over an area of 250,000 square miles. Additional flows of approximately the same age in western Canada and the southwestern United States raise the total to well over 300,000 square miles. From Patagonia to Alaska in the New World, in southern Europe, Ethiopia and other parts of eastern Africa, volcanos multiplied.

Iceland seems to have erupted into existence along the northern mid-Atlantic ridge as a product of the early Pliocene. This period did not end before the northern end of the oceanic ridge in the eastern Pacific produced the Gulf of California. This line of activity in the earth's crust continues today along a fault zone that runs up through California to San Francisco. Known as the "San Andreas Fault," it is responsible for the repeated earth movements and threatening quakes in that area. However, none of these so far can compare with the fantastic faulting of rocks that occurred in the Pliocene and produced the sheer escarpment of the Sierra Nevada in eastern California and the spectacular Grand Tetons of western Wyoming. We can only marvel that these events of less than 13 million years ago caused so little obvious change in the living things of the region, as is shown by the fossil evidence from before and after the faulting.

Centers of Evolution

As we reconsider the living things immortalized in the fossil record and adjust to the new ideas about the grouping and location of continents, we realize how little is known about coastal life as it changed in relation to altered geography as well as to passing time. Geography seems to have been of little consequence until the land plants and animals of Devonian and Carboniferous times had a chance to exploit the dry

The maidenhair tree, or ginkgo, is extinct in the wild, but planted widely in Asia and in North America for its shade and peculiar, fan-shaped deciduous leaves. The oldest species of tree extant, it dates back to the Triassic era. (Lorus and Margery Milne)

34

world. But Carboniferous life does show differences; there was a richer fauna and flora on the southern supercontinent of Gondwanaland as compared with that on Laurasia. Presumably these differences reflect a more varied climate in the south.

During the last 50 million years of the Paleozoic era—the Permian period—colonists began moving across the Tethys Sea. The characteristic seed ferns (*Glossopteris*) spread north into Eurasia, while the giant fern allies *Sigillaria* and *Lepidodendron* extended their range to Brazil and Rhodesia. Somehow the giant horsetails failed to reach Gondwanaland, and soon became extinct on Laurasia. Lungfishes, widespread in Gondwanaland, were ancestors of those surviving in Australia, Africa and South America.

The recent discovery of an amphibian jawbone in Antarctica confirms the belief that these animals were about equally represented in fresh waters both north and south of the Tethys Sea. Only one order (the Stegocephalia or Labyrinthodontia) evolved in all the time from the Carboniferous into the Triassic. Reptiles, however, diversified rapidly after their appearance in the Permian into more than a dozen orders. Of these, only the Rhynchocephalia escaped major change or extinction; appropriately, its sole surviving species (*Sphenodon punctatum*—the tuatara of New Zealand) lives on an outlier of ancient Gondwanaland which, in the Permian, had a genus in South Africa and another in Brazil. Far more spectacular "sail lizards" inhabited Laurasia, from which six-foot skeletons with a three-foot crest are often displayed in museums. The supposed ancestors of the dinosaurs, known as thecodonts, had bipedal and crocodilelike members on both supercontinents. But the groups that successfully spread as colonists from Gondwanaland to Laurasia included two less impressive types: one the little mammal-like therapsids, such as *Dicynodon*, and the other the cotylosaurs which so closely resembled the ancient amphibians. One genus of cotylosaur ranged all the way from South Africa to the U.S.S.R.; another is best known from the two-foot specimens of *Seymouria* preserved in Texas.

Ancient Reptiles

We think of the Permian as closing an era; it was the time of the last trilobites and ostracoderms and of the first coniferous trees and beetles. It led naturally into the Age of Reptiles that was to follow, and the era when continental drift most probably occurred. Yet if we compare it with the 73 million years of Cretaceous time with which the Mesozoic era ended, we can ask ourselves how much effect the new isolation of continents had—or to what degree each of them had actually become isolated. By then the reptiles had reached their peak and were about to crash. It was the day of the last dinosaurs and ammonites, the first flowering plants and pollinating insects. Flowering plants, insects and warm-blooded animals were about to inherit the land.

The Cretaceous had representatives of terrestrial and amphibious reptiles in about a dozen orders. The rhynchocephalians were still there in the Southern Hemisphere, as were the medium to large crocodilians that lived as amphibious predators. Since the crocodilians had not yet spread into the Northern Hemisphere, there was a place along northern waterways for amphibious predators of extraordinarily similar form, called champsosaurs. Later, the champsosaurs died out, apparently unable to compete with crocodilian colonists. The continents composing Laurasia had turtles and tortoises too, and these spread to the parts of old Gondwanaland after the Cretaceous ended. Most of them have been sluggish, armored herbivores throughout the 180 million years since these reptiles first appeared.

The Cretaceous reptiles that failed to survive into the Eocene included the flying pterosaurs ("pterodactyls"), such as the virtually tailless *Pteranodon;* they had a wing span of as much as twenty-seven feet, and were exclusively natives of the northern continents. The huge, amphibious sauropods, made famous by the Jurassic *Brontosaurus*, continued their herbivorous ways on all continents except Australia and Antarctica. So did the bipedal predators of upland habitats, the theropods, which reached their most terrifying proportions in *Tyrannosaurus rex*—fourteen feet tall and more than forty feet in length. From the continents of Gondwanaland, additional herbivores spread northward during the Cretaceous: the horned ceratopsids, such as the three-horned monster *Triceratops;* the armor-plated stegosaurs and ankylosaurs; and the bipedal ornithopods, some of which resembled cassowary birds while others were as distinctive as the gigantic spoonbills that ate vegetation in the Upper Cretaceous marshlands of Wyoming.

The ancestors of lizards and snakes appear among the fossils of the Lower Cretaceous and subsequently. Apparently they ate chiefly insects, scorpions, spiders, worms and land snails—resources that did not fail during the revolutionary change in the vegetation of the continents. *Sphenodon* has the same food habits, and so do many turtles and tortoises that prefer animal matter to plants. It is tempting to regard this diet as their salvation while the dinosaurs dwindled into extinction. Neither lizards nor snakes reached New Zealand, which may indicate that they started out in the northern continents and spread south too late.

Many reasons have been offered to account for the disappearance of the dinosaurs at the end of the Cretaceous period. The small mammals, which were becoming more numerous and varied, may have destroyed the reptiles' eggs at night when warm-blooded animals could be active but cold-blooded ones were too sluggish to provide any parental protection. Long-continued erosion of the continents may have lowered the land so far that rivers dwindled, no longer over-flowing their banks each spring with new loads of fertilizing mineral matter on their flood plains and renewing dissolved nutrients along sea coasts also. Chronic malnutrition could have reached both sea-going predatory reptiles, such as the ichthyosaurs, as well as the giants on the land.

Certainly the vegetation would be expected to show the effects of nutritional deficiencies before animal life suffered any large scale disaster on this account. Yet the fossils preserved during Cretaceous times prove that flowering plants (angiosperms) spread out of Greenland and northern continents into the southern ones at a pace that is almost unbelievable. At the beginning of the Cretaceous everywhere, ferns and gymnosperms were clearly dominant, angiosperms virtually unknown. Southern continents had their forests of cycads and, representing the conifers, their podocarps, phylloclads and araucarians. Northern forests were of ginkgoes in considerable variety, and the conifers were pines and spruces, yews, and members of the bald-cypress family such as the then widespread sequoias.

We realize that the angiosperms differ from other kinds of plants in being equipped to use animals to distribute pollen and seeds where the wind cannot serve efficiently. But more than that, they contain an amazing array of alkaloids and other chemical compounds which may penalize animals that eat their buds and leaves, their ripening fruits, and often their seeds as well. It is tempting to think of the dinosaurs as shunning the unfamiliar and perhaps poisonous angiosperms, and starving when the supply of ferns and gymnosperms dwindled too far. In northern continents, at least, the surviving gymnosperms are generally in regions where the weather is unsuited to reptile life. Regardless of this possible consequence among the dinosaurs, we know from the fossils that by the end

of the Cretaceous, trees with flowers comprised about 70 per cent of forests in temperate and tropical regions. Most of the modern families had evolved, giving the northern continents a wealth of magnolias, dogwoods, elms, willows and birches. In the Southern Hemisphere, horsetail trees and screwpines, antarctic beech and members of the fantastic family Proteaceae spread from continent to continent.

The Evolution of Mammals

In the 63 million years since the end of the Mesozoic era, the mammals took full advantage of the ecological niches left vacant by the extinct reptiles. Mammals benefitted further when they could use as food the grasses that spread over the semiarid lands to create the first prairies; they formed a soil and afforded a supply of energy where none had been available to vertebrates before. Of the three orders of mammals that appeared in Jurassic times, the strange pantotheres (order Pantotheria or Trituberculata) had no chance to participate in the new diversification, for all of them vanished before Cretaceous times. The members of order Allotheria (or Multituberculata) continued into the Paleocene period about 60 million years ago, and seemingly left as descendants the egg-laying mammals (order Monotremata) now limited to Australia and New Guinea. All of the other modern mammals apparently came from the ancient stock known as triconodonts (order Triconodonta), which were wide-spread by the end of Mesozoic time. Their diversification, in fact, follows a continental pattern which shows to what extent the continents had already become isolated one from another.

The separation of the pouched mammals from the ancestral stock occurred during the mid-Cretaceous, apparently in South America. But whether these marsupials spread to Australia and New Guinea by way of land connections via Central and North America, through Asia, or across open water from South America by way of Antarctica and Tasmania has not been discovered from the fossil record. In South America the marsupials diversified into opossums of many sizes, shrew-like caenolestids, and big carnivorous borhyaenids, including some that resembled sabre-tooth cats. Only a few opossums survived the influx

Lake Taal fills the vast crater of a volcano on the island of Luzon in the Philippines, but intermittent activity continues on the small island in the lake. (Lorus and Margery Milne)

36

of placental mammals from the north during the Pliocene. In Australia the relatives of these marsupials evolved into mole-like burrowers, banded anteaters, pouched mice, native "cats," wolflike carnivores, bear-like koalas, bandicoots and possums, phalangers, kangaroos and wombats.

Meanwhile, on the continents that eventually became Eurasia and North America, ten different orders of placental mammals survived to the present, while a half-dozen others faltered and became extinct before the end of the Oligocene. The ten orders spread widely, dying off locally, to provide their present distribution. The order Insectivora comprises shrews in North and South America, Eurasia and Asiatic islands as far as the Philippines; moles in North America and Eurasia; hedgehogs in Africa, Eurasia and to the Philippines; tenrecs on Madagascar and solenodons in the West Indies; elephant shrews and golden moles in Africa all the way to the Cape of Good Hope. Bats (order Chiroptera) have flown virtually everywhere. The primates included lemur-like animals by Paleocene times, ancestral to Madagascar's modern lemurs, and monkey-like creatures in the Eurasian and African areas by the Eocene—but no monkeys in South American jungles until the Miocene, apparently by a colonization across the South Atlantic Ocean. Scaly anteaters (order Pholidota) apparently spread to Africa from Asia Minor; they are found as far east as the Philippines. Rabbits and hares (order Lagomorpha) never reached the West Indies or Madagascar, the East Indies or Australia—until man took them to these places. Rodents (order Rodentia) evolved almost explosively after their appearance in Eocene times, yet until the Pliocene less than 13 million years ago, only porcupines represented this big order in South America —perhaps from ancestors that crossed the South Atlantic Ocean from Africa, but more likely from progenitors that ambled south from North America after coming from the Old World. Today the rodents have reached, by their own means of dispersal, only the southernmost islands of the West Indies. However, Madagascar has seven closely related genera of hamster-like cricetids; about 21 genera have colonized New Guinea and Australia—the only nonflying placental mammals to make the trip without human help.

Without a wealth of fossilized bones to record the

The circumboreal trees attain the same form and size among the European windflowers and deciduous birch trees as in North America. (Photographed in Småland, Sweden: Ingmar Holmåsen)

past emigrations of large mammals, it would be impossible to know so surely that horses and other hoofed animals with an odd number of toes (order Perissodactyla) originated in North America, then traveled so far. Today no one is sure whether the first men to reach North America killed off the last horses there. The survivors include the Mongolian wild horse of Asiatic steppes, the domesticated Eurasian horse, and the striped "pyjama horses" or zebras of Africa. The North American rhinoceroses died out too, but not before some of them had spread to tropical Asia and Africa. A similar history accounts for the presence of tapirs in Southeast Asia and the East Indies as well as Central America and parts of South America.

The even-toed hoofed mammals (order Artiodactyla) appeared first during Eocene times on continents in the Northern Hemisphere. But those that evolved into camels left no survivors in their place of origin—North America—which contributed emigrant llamas and vicuñas to the Andean regions of South America and burden-bearing camels to the Near East and North Africa. Hippopotamuses and giraffes seem to have remained in the Old World, but some of the pigs reached tropical America to become peccaries; today Africa supports the greatest variety of modern wild pigs. Cattle, goats, sheep and antelopes (all family Bovidae) are primarily products of Eurasia that are now most abundant on the African plains and in Asiatic mountains. Deer (family Cervidae) probably had a similar origin, but spread across a Siberian-Alaskan land bridge to North America and southward into South America.

Preying upon the meat animals, the members of the mammalian order Carnivora spread similarly. Yet among them distinctions can be recognized: dog-like animals (family Canidae) and raccoons (Procyonidae) originated in North America. Weasels and skunks (family Mustelidae) came from Asia to the New World during the Oligocene period, and bears (family Ursidae) only late in the Pliocene. Civet cats and hyaenas never managed to make the trip. True cats failed to reach South America before the Pliocene; their origin remains a mystery, and they are missing from the West Indies, Madagascar, and the East Indies beyond Bali in the direction of New Guinea.

Somewhere on the northern continents, members of the order Pinnipedia grew progressively better adapted to aquatic life; the seals and similar animals soon ranged widely throughout the oceans. Today they have failed to reach only the more northern part of the Indian Ocean and the coasts of Madagascar.

Two other orders from the Northern Hemisphere have fared less well. After a beginning in North America during Paleocene and Eocene times, the gliding vegetarians known as "flying lemurs" (order Dermoptera) presumably spread to Asia and southward, where two species of *Cynocephalus* still inhabit the tree tops of tropical forests from the Philippines to Sumatra. Aardvarks (order Tubulidentata) have had a similar history, for their remains have been discovered in Eocene strata of North America, Eocene to Pliocene of Europe, the Miocene of Africa, and the Pleistocene of Madagascar. One species, *Orycteropus afer*, remains in Africa south of the great deserts. It is a powerful animal that competes in the northern part of its range with pangolins (order Pholidota). It is the ecological equivalent also of the anteaters (order Edentata) in Latin America and of the spiny anteaters (order Monotremata) in New Guinea and Australia, for it eats principally termites and other insects.

The edentates seem to have arisen from insectivorous ancestors on Central America, which long existed as an island with only temporary connections to other land. Later armadillos spread northward as well as southward, while anteaters and sloths reached only South America. The fossils of giant ground sloths and huge armadillo-like animals in the big northern islands of the West Indies show that water barriers between Central America and the islands were narrower in the past than today, even if no actual land bridge existed at any time.

The remaining orders of existing mammals seem to have originated in Africa and adjacent parts of the Near East—the areas that still are the territory of the puzzling conies and hyraxes (all of order Hyracoidea). These animals combine in the features of their feet and teeth some characteristics that make them resemble hoofed mammals, and others that link them to elephants. Living in loose colonies, they fill much the same ecological niches as marmots and tree squirrels do in other parts of the world. Elephants (order Proboscidea) have left a fossil trail beginning in the Eocene of northeastern Africa and leading in all directions on that continent: across southern Asia to Malaya and the China coast, across northern Asia into North America, through Central America to Argentina, and to shores of the Atlantic at many points in the Americas. Today only the African elephant and Indian elephant remain. Whales (order Cetacea) had their origin in northern Africa, finally forsaking the land to swim out of the remains of the Tethys Sea into the great oceans. The dugongs and their kin (order Sirenia)

proved less venturesome, and now are represented only by tropical and subtropical vegetarians in the Old World and the New.

Scientists who accept the theory of continental drift during Mesozoic times can point to about a dozen orders of reptiles—an average of six apiece for the two big supercontinents of Laurasia and Gondwanaland—of which only three are still represented by more than a single species. They can count the orders of mammals that have largely replaced the reptiles on the land, and see that the three continents of the Northern Hemisphere produced about twelve of existing mammals and another six of extinct ones—an average of six apiece for the continents of the Holarctic realm. South America provided the first marsupials, and six orders of mammals that failed to survive. Africa was the original home to four orders with existing species and of one now extinct. Australia might be credited with the egg-laying mammals, and Central America with the edentates. It is easy to conclude that the fragmentation of supercontinents favored the proliferation of mammals.

The most successful mammals (bats, primates, hares and rabbits, rodents, carnivores and hoofed animals) all came from continents with the greatest opportunities for faunal interchange. Conversely, most of the extinct orders arose on island continents and their members were unable to survive when geological changes in the land areas allowed competitors to arrive from farther north. Concomitant with the joining of South and Central America to North America, and the juxtaposition of Africa and India to Eurasia, the number of relatively isolated land masses shrank from seven to four. The twenty-nine orders of land-based mammals were reduced to about seventeen, a decrease in variety proportionate to the decrease in isolation.

The Seeming Misfits

While the geologists gradually changed from dismissing the idea of continental drift as preposterous to offering evidence that it had taken place, the biologists sought other explanations for the similarities in life that Alfred Wegener had pointed out on the two shores of the South Atlantic Ocean. Drift would have to be quite recent anyway to account for the seeming misfits in biogeography. Other explanations might be better than that of an assumed contact between South America and Africa during the Eocene or since, or a narrow ocean across which natural rafts might float with living vegetation, porcupines, camels or freshwater characin fishes as colonists to the opposite continent.

GEOLOGICAL TIMETABLE

Eras	Periods	Epochs	Millions of Years Ago	Geological Events	Appearance of Early Animals	Plants
CENOZOIC	Quaternary	Recent	0.015	*Glaciation*	Mankind	
		Pleistocene	2.0			
	Tertiary	Pliocene	13	Himalayas	*Age of Mammals*	
		Miocene	25	Alps		
		Oligocene	36	Andes		
		Eocene	56			
		Paleocene	63		Placental mammals	Grasses
MESOZOIC	Cretaceous		135	Rockies	*Age of Reptiles* Birds & Mammals	Flowering plants
	Jurassic		181			
	Triassic		230			
PALEOZOIC	Permian		280	*Glaciation* Appalachians	Reptiles	Conifers
	Carboniferous		345		Insects, Spiders & Centipedes	Mosses, Ferns, Seed Ferns & Cycads
	Devonian		405	Acadian	Amphibians *Age of Fishes*	Psilopsid land plants
	Silurian		425	Caledonian	Fishes with jaws Ammonoids	
	Ordovician		500		Jawless fishes Nautiloids	
	Cambrian		625	Good fossilization *Glaciation*	Trilobites Marine invertebrates	Marine algae

Now that the fossil history of porcupines is better known than in Wegener's day, we could expect to discover at least whether the terrestrial Old-World porcupines antedated the arboreal New-World animals. Instead, the oldest remains are in Oligocene strata of both South America and Europe; none are known from North America until late in the Pliocene. Today Old-World porcupines (family Hystricidae) are widespread from Morocco through most forested parts of Africa, across Asia to southern China and the Philippines, to Ceylon, Java and Borneo; but not to Madagascar. New-World porcupines (family Erethizontidae) include several species in South America and one or two in North. Probably the North American porcupines arrived from farther south fairly recently.

It may be neither group of porcupines is ancestral to the other, that they independently evolved a similar armament of deciduous quills, making the presence of porcupines on both sides of the South Atlantic coincidental. Or, both may have been derived from a northern stock that spread southward, leaving survivors on the southern continents to specialize further on their own, and vanishing in the north. So far, the evidence favors no one possibility.

The fossil history of camels is more complete and conclusive, showing that they originated in North America and crossed land bridges to reach Asia and South America. Probably the freshwater characin fishes, which seem so strangely similar on these two continents of the Southern Hemisphere, originated in

the north too and now seem misfits only because no survivors remain in their continents of origin.

The disappearance of primarily tropical animals from northern latitudes can be blamed on the glaciers during the past two million years. In North America these stretched from ocean to ocean, and southward beyond the fortieth parallel of north latitude. The cool climates of the Pliocene and Pleistocene must have severely limited the kinds of life that could spread across the dry land between Asia and Alaska. This corridor can scarcely be called a bridge for, at times, it had a width of 600 to 700 miles. Through it westward went camels and horses, while in the other direction the bears, bison and elephants reached the New World, along with crows, jays and chickadees. Most of the exchangees were

The nine-banded armadillo, which usually bears identical quadruplets, is the only one from among about 20 kinds of armadillos in tropical America to spread into the United States. (Kurt Severin)

tolerant of cold weather. To this extent the connection between East and West acted like a filter, holding back some potential travelers while letting others pass. The narrow isthmus between North and South America provided more obvious limitations, despite the warm climate there. One kind of bear, and crows and jays got through to South America, while opossums, armadillos and the coatimundi came north. Chickadees, with 42 different races in America compared to about

300 kinds and subkinds still in Eurasia, have not yet reached beyond Mexico.

From Europe to North America, the shortest distance is over the North Pole. But regardless of whether the Arctic Ocean was ever ice-free during the past two million years, the polar route could offer little to living things. Nor was the way from Scandinavia to Labrador by way of Iceland and Greenland an easy route except for creatures able to walk long distances on ice-covered oceans or to be carried passively by gale winds. While critically considering the evidence for interchange, the Norwegian biologist Carl H. Lindroth listed in his fine book *The Faunal Connections between Europe and North America* all of the terrestrial and freshwater animals he knew to occur naturally on both continents: 20 kinds of mammals, 107 of birds, 800 of insects, 141 of spiders, 70 of snails and slugs, 17 of earthworms, but only 3 of fishes, and no amphibians or reptiles at all. Except for the shrubby juniper he could find no woody plant common to both continents. The 908 species in his list are either highly adapted to life in the arctic and subarctic, hence well fitted to spread around polar lands, or else ideally suited for passive dispersal. He suggested that the smaller species, particularly insects, spiders, worms and plants such as fungi, lichens and mosses, might also be older and have had more opportunities to colonize the distant lands.

From Norway to Labrador by way of Siberia is more than 8000 miles. Yet so extensive has been the genetic interchange among the populations of plants and animals under similar climatic conditions east and west that, while the species are mostly different in the Old World and New, the genera and families are generally the same throughout the holarctic realm. Seeming misfits are few, whereas the filtering effect of climate on the Siberian corridor to America becomes more apparent whenever biogeographers pursue comparisons in northern latitudès.

Elsewhere in the world, the actual number of seeming misfits is so small in proportion to the diversity of living things on the two sides of each ocean barrier that continental drift need not be called upon in explanation. Any real proximity in the recent past would have allowed considerable traffic in both directions, and differences in the flora and fauna on the two sides of the Atlantic would have tended to disappear. That this did not happen is quite evident. Only a few colonists made the trip and became successful in their new environment. They match what would be expected for emigrants over a hazardous route, whether by raft or by land corridors thousands of miles long.

For a plant or animal that is intolerant of salt water, as for a freshwater fish that cannot survive in air between one lake or river and the next, there is no gain from crossing 99 per cent of a barrier. Each colonist faces an all-or-nothing trip before it can test its adaptations against the living conditions in a distant place. So small are the chances that any individual will reach a destination before it dies that the distinguished American paleontologist George Gaylord Simpson called this type of spread "sweepstakes dispersal." Variations in the readiness of a species to travel, in the length of time during which opportunities can recur, and in the success of colonists competing in a new area readily account for the small numbers of seeming misfits.

In the long evolution of the world and its life, a number of living things have managed to continue relatively unchanged over immensities of time. Among these "living fossils," whose near relatives have vanished through extinction, animal kinds outnumber the plant kinds and have longer histories. In part this reflects the greater use of lime as hard parts in animals that could make good fossils. The zoologist looks at his modern lancelets (*Branchiostoma*) and peripatuses as little altered from Cambrian times 600 million years ago. The mollusk *Neopilina* and the lamp shells (brachiopods) are from the Ordovician period; the horseshoe crabs date from before the Silurian; lungfishes and cockroaches are from the Carboniferous; the tuatara (*Sphenodon*) is from the Permian; crocodiles from the Triassic; and turtles, tortoises and the *Nautilus* from the Jurassic. Egg-laying mammals, from the Cretaceous, are almost recent—a mere 75 million years old. The botanist has few fossils to show how plants he regards as primitive (the psilopsids, clubmosses and horsetails) fared from the Devonian to the present. He can trace cycads and the maidenhair tree to the Jurassic 200 million years ago, but his fossils from the Cretaceous include a whole forest of familiar types: magnolias, tulip trees, chestnuts, beech, oak, persimmon, sycamore, birch, alder and elm; the cabbage palm and shrubby greenbrier represent the monocot seed plants.

The seeming misfits and the ancient stay-at-homes confer a double benefit upon natural science. They support belief in the comprehensibility of change with time, and the gradualness of most steps in the progression. And they test each interpretation of the geographic distribution of living things, in the search for events in time and space that have produced the living world as we find it today.

3
Life
in the
Old World
Tropics

Almost a century ago, while scientists were arguing heatedly about Charles Darwin's conclusions regarding *The Descent of Man*, a professor of zoology at the University of Jena made a prediction that everyone regarded as rash. Professor Ernst Haeckel pointed out that all of man's nearest relatives—the great apes and the long-armed gibbons—lived in the tropics of the Old World, in Africa and in tropical Asia. It was there that excavations would most likely uncover the remains of man's ancestry. And since Haeckel inclined toward the gibbons as most similar to mankind, he judged the East Indies, where gibbons live, to be best of all places to dig. The fossil to be sought was an ape-man without speech, which should therefore be called *Pithecanthropus alalus*.

Among Haeckel's admirers was a young Dutch physician, who longed to make a scientific reputation for himself while earning a living. He could get an appointment in the Dutch East Indies, and dig for fossil man while practicing medicine. Accordingly, Eugène Dubois, M. D., signed on as junior surgeon in the Netherlands East Indian Army, and sailed for Sumatra in 1889. But the excavations upon which he lavished his spare time yielded nothing more remarkable than the teeth of orangutans.

A fellow officer sent Dubois a human skull from central Java. It was obviously human, although old. Perhaps Java would be a better place to dig. Dubois arranged for a transfer to that island still farther east. He, and the workmen he set to the task, uncovered not only additional skulls of relatively recent people but also, between 1890 and 1892, some fragments of a more ancient primate from Pleistocene times. Two molar teeth from the right upper jaw, a thighbone which Dubois judged to belong to a primitive man that walked upright, and a bowl-like skullcap more than seven inches long, led him to calculate that Java had been home to a human ancestor with a brain volume between 800 and 1000 cubic centimeters. This is about halfway between that of the largest apes (500 for the gorilla) and that of the smallest modern man (1200 for Australian aborigines). Great was Professor Haeckel's joy when in 1894 he received from Dubois a full account of this discovery, written in German and entitled, "*Pithecanthropus erectus*, a Human Transitional Form from Java." He arranged for Dubois to bring the actual specimens to Leiden in 1896, so that scientists from all over the world could examine them during the International Zoological Congress.

Today, Dubois' "Java man" is regarded by many anthropologists as belonging to the same species, *Homo erectus*, as "Peking man" and other fossil remains from the first half of the Ice Ages. Opinion too has changed. Now gorillas and chimpanzees of Africa are regarded as nearer than gibbons to man in body build and molecular details. Since 1920 many fossil primates have been recovered from Africa. They include "southern apes" (*Homo africanus*) from the Transvaal of South Africa and from various sites near Lake Victoria in the equatorial highlands.

Modern biogeographers see plenty of reason to expect the discovery of more fossil men all across Africa south of the great deserts and in Asia south of the Himalayas. This whole region shows an impressive unity, even though it is divided by the Red Sea and Indian Ocean. It is the home territory for all of the world's anthropoid apes, for all but two of the Old-World monkeys (macaques and baboons), and for the shy nocturnal primates known as lorises, pottos and bush babies. It is also the only part of the world that now has leopards, tigers and lions, rhinoceroses and elephants, scaly anteaters (pangolins), bamboo rats (family Rhizomyidae), and diminutive deer-like chevrotains (family Tragulidae). It has all of the known wild peafowl, and most of the earth's hornbills, bulbuls and sunbirds. Its crocodiles and monitor lizards are widespread, as are its pythons and snakes of the cobra

family. Its tree frogs are colorful members of a special type, comprising the family Rhacophoridae.

Fewer of the plants native to this great region, generally called the Old-World tropics, are famous. For thousands of years it has been the source of precious resins from aromatic shrubs and trees, such as frankincense and myrrh. Gardeners all over the world have imported its *Asparagus* species, its *Gerbera* daisies, its *Thunbergia* flowering vines and shrubs, and its snake-plants, which contain fibers known as bowstring hemp. Until poultices went out of favor in civilized lands, it was important as the source of a lowly herb (*Pedalium*

Burchell's zebra and the antelope known as kudu drink together from a waterhole on the Etosha Pan Game Reserve of South-West Africa. (Clem Haagner)

murex) whose winged or barbed fruits contain mucilaginous seeds that are easily made into a paste for keeping human skin wet or holding medicines to it.

Our Nearest Living Kin

Only three completely different kinds of anthropoid ("man-like") apes share the earth with us. These

45

members of family Pongidae are all tailless forest dwellers, but only the orangutan of Sumatra and Borneo spends much time high above the ground. Orangutans travel slowly and carefully, eating foliage and fruits. For the night they build sturdy platforms six to ten feet in diameter of interwoven branches in a tree top, and lie outstretched on it through the dark hours, locked in place by the firm grip of opposable thumbs and of big toes. No African primate builds a nest of this type, or anchors its body so securely for the night. Smallest and most agile are the chimpanzees, males of which sometimes weigh 175 pounds. They range through the tropical forests of western and equatorial Africa, sometimes making a nuisance of themselves by raiding a coffee plantation to eat the half-ripened coffee fruits. Gorillas remain in more secluded places, proceeding at a leisurely pace along the ground, feeding in family groups numbering about a dozen individuals, and building crude nests for themselves each night. The lowland gorillas travel through forests between southeastern Nigeria and the eastern Congo, whereas mountain gorillas had until recently a natural sanctuary among the bamboos of the Congo volcano slopes, between 8000 and 10,000 feet elevation. Adult males may weigh 800 pounds, which is more than three times as big as the largest orangutan.

By comparison, the dozen species of family Lorisidae seem small animals, none of them weighing more than three pounds. The lorises of the Asian tropics, the potto and the bush babies of the African rain forest are all tree-dwellers and strictly nocturnal. In darkness they move slowly among the high branches, picking buds and fruits, but ready to snatch quickly at any insect or lizard they encounter. Small prey is seized and carried to the mouth for a hard, tenacious bite. The lorises are tailless, the potto has a tail of medium length, whereas bush babies almost resemble squirrels because of their bushy, long tails and pointed faces.

All but three of the Old-World monkeys (family Cercopithecidae) have tails, the exceptions being the Barbary ape of Algeria, Morocco and Gibraltar, and two different macaques of Celebes and adjacent islands in the East Indies. Best known are the rhesus monkeys of India, which have long been held sacred by the Hindus and permitted to destroy crops needed for human food even in times of great scarcity. Because rhesus monkeys adjust well to captivity, they have been used extensively in medical research and were the first primates to ride in rockets out into the stratosphere. The Rh blood factor was discovered in these animals before it was recognized in man. In the Malay Penin-

sula, pig-tail macaques are often trained to climb coconut trees and help in harvesting the nuts. Long-tailed macaques roam freely through the famous botanic garden in Singapore, sampling the flowers, fruits and buds from specimen trees and shrubs introduced from many distant lands. We noticed that in each troop of about twenty-five individuals, one adult male is dominant and strides about unhindered, his status shown by a characteristic arching of his tail above his back.

Shortness of nose gives a superficial resemblance among the dozen different vervet monkeys that are so active in the trees and on the ground of Africa by day, and to the commonest of the langurs or leaf-eating monkeys of mountain forests in Southeast Asia. But one of the snub-nosed denizens of tall forest trees in Borneo is known as the proboscis monkey, because old males generally develop a long, soft, pendulous nose while in mature females this part of the face is shorter and generally turns upward.

The longest noses are those of the baboons and mandrills of Africa, which spend most of their lives on the ground. Yet as these animals forage for fruits and grubs over the short-grass plains and among rocky outcroppings, they show an awareness of the bushes and trees that grow at intervals. Generally skirting the tall acacia trees, they often pause to feed in a gall acacia bush. Carefully the baboon crunches the spherical galls, the size of golf balls, on the spiny stems. It avoids the spines but swallows the tiny red ants that make their home inside the gall, as though the baboon delighted in the sour taste of formic acid and was immune to the insects' bites and stings. A similar willingness to eat almost anything edible has long been characteristic of the anubis baboon, which was the sacred monkey of ancient Egypt; its symbol is depicted on monuments, primarily as an attendant to the god Thoth. Only the older males develop the heavy mane on neck and shoulders. Although almost exterminated in Egypt, the anubis baboon continues to find suitable living conditions in Africa from the Sudan to Somalia, and in rocky hill country of the Arabian peninsula.

Big Cats of the Old World

Most widespread of these powerful predators is the leopard, whether black or buff with open rosettes of

Screw-pines, with sturdy prop roots, are among the most characteristic trees in Old-World tropical wetlands. (Emil Javorsky)

black spots. It ranges from West Africa and South Africa across Asia Minor to Java in the East Indies, northward into the Himalayas, and to parts of China and Mongolia. Everywhere it is an elusive animal, concealing itself among dense vegetation wherever possible, but hiding in caves or among rock piles in areas with scanty cover. Leopards clearly prefer the rain forests, in which they can perch on overhanging limbs and pounce on passersby. They often demonstrate almost incredible strength by hauling the carcass into a tree and caching it there, perhaps after making an initial meal from it.

Sometimes a leopard can satisfy its hunger or have a midnight snack without descending to the ground. It creeps upon peafowl that have sought safety on high branches, or upon nocturnal primates that have waited in a treehole until all daylight has faded before venturing out in search of insects and fruits. Every free and wild leopard that we have encountered was spending the day in a tree. Never did one seem at ease, as a lion would be, relaxed and snoozing. In Asia, leopards are often included in a tiger's diet, but in Africa man seems to be their greatest danger.

Tigers remain purely Asiatic animals, ranging from Sakhalin Island and the coast of the Sea of Okhotsk on the east, southward to the islands of Bali, Java and Sumatra, and across southern Russia to the Euphrates River. They are rare in northern parts of this vast expanse, for prey is harder to find here than in southern swamps and forests. There they prey on wild hogs, native deer, domestic cattle and Asiatic peafowl upon the ground. Tigers that become man-eaters generally have grown too old to catch more wary game.

Near tiger territory, but not actually within the tiger's range, lions still survive in small numbers, chiefly on the Gir peninsula of India's west coast. In the Near East and North Africa they were exterminated in Biblical times. Formerly lions ranged far north in Europe. Now they are common only in the high country south of Africa's great deserts, and in the mountains of Ethiopia. Those we have watched near the Equator were all tawny beasts, often climbing into trees to sleep by day where air movement and some shade relieved the heat of the sun. They are adapted to life in the dry thornscrub, on the savannas and open plains, where they can pounce on antelopes, zebras and ostriches—prey that has no counterpart in tiger territory. In South Africa, the lions wear black manes and stride boldly along the ground before sunset and after dawn. While the sun is high, they rest under the trees. At night they enforce a curfew on pedestrian travel along lonely roads, giving good reason for the strict rule that keeps every visitor within the confines of the high fence around the park campground when not protected by the metal and glass of a closed automobile.

The differences between lions and tigers must have had a long history, and have served to keep the two big cats from interbreeding in the wild. While captive in zoos, however, the two species cross successfully. The offspring of a lion and a tigress is known as a "liger," and that from a tiger and lioness as a "tigon." The hybrids show features of both parents, including some of the transverse black striping of the tiger on brow, body, legs and tail—a pattern that is different on the two sides of the same animal, and also from one tiger to another. Only the lion develops a distinctive mane in open country and in captivity. Those inhabiting the thornscrub may lose these long hairs, and are often maneless.

One Nose Horn or Two

In the tropics and adjacent areas of the Old World, five different kinds of rhinoceroses have survived to the present day, each like a monster from the remote past. Those in Africa inhabit the thornscrub and savanna country, whereas those of Asia live in dense forests. Young rhinos occasionally fall prey to the big cats, but the adults have no obvious enemies except man. Largest by far is the one-horned Indian rhino of Nepal and northeastern India, which takes refuge in reed beds and swampy forests. Some of these animals attain a weight of four and a half tons, exceeded among land mammals only by the two kinds of elephants. The smaller Javanese rhino has now been exterminated on the mainland of Southeast Asia and on Sumatra, where it once was widespread; less than fifty survive and those are in a sanctuary on Java. The still smaller Asiatic

Above, far left: Africa's sunbirds, such as this male scarlet-throated species, use their long beaks to probe flowers for nectar so regularly that they are the principal pollinators of many flowering trees. (D. C. H. Plowes) Above left: The red-and-yellow barbet, which has an undulating flight like a woodpecker but feeds on fruits, builds its nest in a termite mound, where the termites will fend off ants and wasps. (J.-F. and M. Terrasse) Left: Asian forests have their venomous green pit-vipers, such as Trimeresurus stejnegeri, *which are well camouflaged by day and able (with their heat-sensitive pit organs) to find sleeping birds and mammals in darkness. (Manuel V. Rubio)*

two-horned rhino is almost as rare. It is hunted illegally, as are other rhinos, over its entire range—from Assam to Malaya, and on Sumatra and Borneo—to obtain its horns, blood and urine, all of which bring a high price in China because of a popular belief that these materials have aphrodisiacal properties.

The larger of Africa's rhinos is the sociable square-lipped "white" species, which grazes on grasses and low shrubs in open country from the eastern Congo to the southern Sudan, south into Zululand of South Africa. Carrying our cameras and accompanied by the warden, Mr. Ian Player, we stalked some of these huge animals on foot in the Umfolosi Reserve of Zululand, where protection has allowed a modest increase in the numbers of this rhino. Suddenly we became aware that square-lips were grazing on three sides of the little clump of scrubby trees in which we crouched. Some were very young, small by comparison with mothers weighing as much as four tons. We do not know whether some vagary of the breeze, which seemed to blow steadily from the rhino's direction to us, let these dull-eyed animals detect a human odor, or whether the unfamiliar whir of the movie camera and the click of a still-camera's shutter were picked up by the rhinos' sharp ears. Gradually the whole group moved away, and some

took up their alarm position, with tails together, facing out like the hands of a clock, as though ready to charge in every direction as soon as any threatening object could be located.

In the nearby Hluhlue Reserve, we had an opportunity at dawn to creep over a tick-infested hilltop into the wind to see a few of the more solitary and dangerous black rhinos, which use their V-shaped, prehensile upper lips to browse on thornscrub. Although weighing at most no more than two tons, they appear to defend their own territories, charging at speeds as great as twenty-eight miles per hour toward any strange sound, scent or sight. These are the animals that attack vehicles, campfires and tents, impaling people on the forward of their two horns, tossing them in the air or trampling them. Our companions—another warden and his assistant—were clearly apprehensive that the morning breeze might shift, informing the rhinos of our presence and necessitating a frantic run to the get-away car parked on the nearest road. This well-earned reputation for aggressiveness may have saved black rhinos from being molested. They are still the most numerous of all rhinos and range across Africa from eastern Nigeria to Kenya, where they sometimes abound in dense mountain forest, and south to Natal in South Africa.

The Vanishing Elephants

Among our greatest surprises in the tropics of the Old World was the discovery that elephants can be inconspicuous. The African kind (*Loxodonta africana*), males of which attain a weight of more than eight tons, are the largest land animals in existence. Yet they can blend with their surroundings among trees or clumps of tree spurge, and walk silently along their narrow dusty trails. African elephants are readily distinguishable from the Asiatic kind (*Elephas maximus*, which weigh no more than five and a half tons), by proportionately larger ears, a fingerlike projection above as well as below the tip of the trunk, and three nails instead of four on each hind foot. The massive skeleton accounts for about 15 per cent of the body weight, just as in a man of slender build; however, the elephant's skeleton supports as much as four tons per leg, and is thus stressed close to the physical limit for bone. To keep

The slow loris sleeps all day in forests from Assam and Burma through Southeast Asia to East Indian islands in the Philippines. At night it moves deliberately in search of insects and plant food. (Lim Boo Liat)

50

Cheetahs, whose long legs let them speed faster than any of their prey on the African savannas, watch for hours before attacking. They select the injured, the diseased and the old among antelopes, zebras and ostriches. (Clem Haagner)

from damaging its skeleton, an African elephant has to move sedately, never jumping or running. The "charge" of these animals is a fast walk on long legs, at about fifteen miles per hour. In savanna country, where these elephants reach their greatest size, we noticed that they rest standing up; this may account for the medieval legend that elephants cannot bend their knees. Often they support their trunk and tusks on a sturdy horizontal limb. The two tusks on an African elephant in the record class weigh in excess of 450 pounds. Even the number of ribs is extraordinary—twenty-one pairs, not nineteen as in the Asiatic elephant, or twelve as in most mammals.

Both kinds of elephants find the grass, tender twigs and leaves, the vines and fruits they eat in a wide variety of situations, from dense forest to grassy plains, from river valleys to the high slopes of equatorial mountains, close to tree line. Elephants near Mount Kilimanjaro in Tanzania and Mount Kenya in Kenya make regular migrations up and down to match seasonal variations in the availability of their favorite foods. When the supply runs low, or thirst goes unsated, or biting insects become too troublesome, the elephants march off in single file to new territories.

Obtaining enough to eat and having the time to eat

the necessary food must be challenging to a big elephant. In the teak forests of the Far East, where Asiatic elephants work eight-hour days for man, they divide their off hours about equally between eating and sleeping. The African elephants are no less tractable or intelligent, but because of their greater bulk they need to eat 500 pounds or more of forage daily, which is about two and a half times as much as the Asiatic animals. Doing so occupies most of their waking time, leaving few hours in which they might work for man. These discoveries, made by Belgians on captive herds in the Congo, make us wonder how Hannibal managed to feed the African elephants he led into Italy over snowy passes during his campaign against the Romans in 218 B.C.

The teeth and skeletal parts of elephants have made lasting fossils, allowing paleontologists to reconstruct the history of these amazing animals. They seem to have originated in Ethiopia and spread from there throughout Africa, across southern Asia into the East Indies, across northern Asia and a land bridge to North America, thence east and south far into South America. These appear to be the same routes that mankind took during the late Pliocene and the Pleistocene. Either because of man's arrival, or through some unknown change that occurred simultaneously, the contemporary elephants—chiefly mammoths and mastodons—became extinct, leaving only the two that survive today in the tropics of the Old World.

Scaly Anteaters and Miniature Deer

Fossils show that the scaly anteaters (pangolins) of the mammalian order Pholidota once had a wider distribution too. But not since historic times have any of these armadillo-like animals lived beyond the tropics and subtropics of Asia and Africa. Three different kinds occur in India and Ceylon, south through Malaya into Sumatra, Java, Borneo and Bali, and across southeastern China to Formosa and the island of Hainan. Four more of the same genus (*Manis*) inhabit Africa south of the great deserts. All of them resemble animated pine cones with a body as much as twenty-eight inches long and an armored tail that may be longer. With their strong front claws, they rip into the nests of termites and ants, capturing the insects by sticking into the opened chambers a slender, sticky tongue nearly ten inches long but no more than one-quarter inch in diameter. Lacking teeth, the anteater swallows the insects whole and relies upon tiny pebbles embedded in the thick muscular walls of its stomach to grind up the hard bodies. When biting and stinging ants or termites attempt to counterattack by getting between the horny scales on a pangolin's body, legs or tail, the insectivorous mammal simply clamps its scales together, crushing and killing the insects. To keep them from reaching sensitive places, the pangolin closes hard lids over its eyes while eating, and constricts the openings to its ears and nostrils. Asiatic pangolins have a ridge to represent the external ears, whereas African kinds lack any such projection on each side of the pointed head.

Pangolins sometimes climb high into trees to reach insect nests among the branches. They cling solidly with their back legs and prehensile tail while clawing into the nests to expose their food. In the Old World tropics this habit seems doubly surprising, first because a pangolin's tail seems too long-tapered, thick and scaly to be twined around a branch, and second because the monkeys in the same part of the world fail to use their long slender tails for anything more than balance.

Among the undergrowth in tropical forests of the Old World live some of the most diminutive, deer-like animals known, chevrotains. Standing no more than twelve inches tall, they are often called mouse deer. The word "chevrotain" comes from the French word *chèvre*, for a goat. The first one we saw we mistook for a short-eared rabbit or an agouti—a big rodent from Central and South American jungles. Then we got a better view and saw that it was chewing its cud. Male chevrotains have enlarged canine teeth in the upper jaw, and on old individuals these project as shining tusks. Yet these animals are so shy that no one seems to know whether the tusks are used in combat at mating time, perhaps taking the place of antlers or horns, which chevrotains lack.

"*Bringing . . . ivory, and apes, and peacocks*"

The Bible mentions that King Solomon regularly imported the most amazing animals he knew to delight his people, as tokens of wealth and signs of splendor appreciated in 900 B.C. The common peafowl had actually been domesticated in India and Ceylon more than a thousand years before his time. The male (peacock) carries long tail-covert feathers as a graceful train above his short true tail, except when he excitedly raises them as a dazzling fan with great eye spots, and turns them toward the plain female (peahen) he is courting. The Javanese peafowl, which ranges from Burma to Java through the jungles, is similar in size and form but has green feathers instead of blue ones on its neck and head, and full webbing on its crest feathers, which come to a point. On a common peafowl these feathers spread apart and have webbing only at the tips.

An African counterpart of the Asiatic peafowl became known in the 1930's through persistent scientific detective work. Dr. James Chapin of the American Museum of Natural History was largely responsible for discovering the shy Congo peafowl, which is bronze and green. The first representative of the true pheasants to be found south of Africa's great deserts, it remains elusive, its habits largely unobserved.

No less than fifteen of the twenty-five different types of crocodiles and near kin belonging to the ancient order of reptiles Crocodilia make their home either in the tropics of the Old World or in the warm temperate regions close by. The two with the widest distribution are also those that most often attack human beings: the saltwater crocodile of the Indian and East Indian region, and the African crocodile, which has settled in most watercourses all across tropical Africa and southward to the Natal province of South Africa. It swims freely in salt water, and is well known and feared on Madagascar. The African crocodile was once numerous in the lower Nile and along Mediterranean coasts to Palestine, but it is now extinct in these places. The Oriental saltwater crocodile travels from island to island, occasionally reaching Fiji in the South Pacific and Cocos-Keeling Island in the Indian Ocean, more than 600 miles from the nearest land. By contrast, the marsh crocodile of fresh water in Pakistan, India and Ceylon is protected as a sacred animal, called the mugger. Some of these animals, kept in a big tank along the road that pilgrims take to Karachi, have become so tame that artists have painted verses from the Koran on their armored heads. In the Ganges, the muggers dispose of the remains from many human bodies that are set free on funeral pyres. Other bodies are devoured, probably in advanced stages of decay, by the most narrow-headed of all crocodilians, the gavial (or gharial), which normally feeds on fishes. Swimming in the Ganges and the Brahmaputra of India, the Indus of Pakistan, and rivers flowing to the coast of Burma, gavials swish their specialized beak-like heads from side to side and snap at their slippery, elusive prey. Although gavials grow to be twenty-one and a half feet long, they seem never to attack living people or domestic animals. Only where funeral procedures send human cadavers floating downstream are gavials occasionally caught with bracelets and other jewelry in their stomachs.

Snakes that Squeeze or Bite

The pythons and cobras of the Old World tropics include several that have earned reputations for attacking people or domestic animals. One, the reticulated python, is the largest of the constrictor snakes; it is widely distributed because it is a great swimmer. A snake of this kind, shorter than the record size of 28 feet and 250 pounds, is recorded to have been the first reptile to reach and colonize the island of Krakatau near Java following the complete destruction of all life there by the great volcanic explosion of 1883. Not even a one-ton saltwater crocodile is likely to attack this python, for it is too formidable in coiling about any other animal and clamping firmly each time its prey exhales, suffocating it by preventing it from getting another breath. Perhaps the deliberate and fearless behavior of the big snake reflects its general freedom to attack. From the coast of Burma down the Malay Peninsula and throughout the East Indies, it commonly

forages around busy waterfronts, hunting at night for cats, rats, dogs, sleeping birds and possibly pigs. Since the python spends the day in a crevice, for example among cargo stacked on a pier, it may travel passively. In distant lands to which marine shipping goes, reticulated pythons ten to fifteen feet long turn up occasionally at wharves and warehouses.

Other Oriental constrictor snakes with similar food requirements tend to avoid competition with reticulated pythons by hunting farther from the sea. The short-tailed python of Malaya and the East Indies to the Philippines, stalks along rivers and streams; because of its brick-red color, it is often called the blood python. The Indian python of India and Ceylon frequents dry woodlands and villages as well as the rain forest, and sometimes conflicts with human interests by catching and swallowing pigs and domestic fowl, along with rats and wild warm-blooded animals of many kinds. Snake charmers often display well-fed Indian pythons that are sluggish but brightly patterned and obviously powerful.

The commonest African python ranges over most of Africa south of the great deserts, in open plains and savannas as well as in forested country. It is more slender than the Indian python, and often twenty feet or more in length. If surprised in the wild, it usually tries to escape. If seized carelessly, it may strike with its mouth wide open and inflict a terrible wound. Yet men of the Bavenda tribe in the Transvaal of South Africa go out each year to capture unharmed a few of these giant snakes for the marriageable girls to carry in a circular snake dance. Afterward, the snakes are set free again.

The venomous front fangs of cobras and other members of the family Elapidae make us fear them far more than any python. We think of the 30,000 deaths from cobra bite alone in India each year. Yet snake charmers all over the Orient make a specialty of the Indian cobra, and exhibit healthy-looking specimens five and a half to six feet long. While watching these famous reptiles rear up out of a wicker basket, spread their distinctive hoods and flicker their tongues while the man squats right in front of them, swaying and tootling the horn (which no snake can hear) we never know what to conclude. The snake's venom fangs may have been removed and its mouth sewn shut at the sides. It may merely be unlikely to strike with mouth agape by day. We would be far more alarmed to meet a big king cobra on a forest path and have it rear up more than four feet of its twelve-foot length, poised to strike in our direction. The king cobra is a snake-eater, found

from Burma to the Philippines. Yet it is feared less than some of the kraits no more than five feet long. The common krait, which ranges from India to South China, has an extraordinarily potent venom; normally it hunts and strikes only at night and, when disturbed by day, it often curls up into a ball with its head at the center and can be killed with a stick.

We know that cobras can be found all the way from Australia's east coast to the Atlantic shores of Africa, and that the Australian regions have a larger proportion of these poisonous snakes than the Asian or African tropics. Yet it is the Oriental and African cobras that receive most of the bad publicity. The mambas of equatorial Africa often attack without obvious provocation, and with no possibility of eating the large-bodied victims of their virulent poison. Some of these snakes are eight feet long. All of them have much-enlarged teeth at the front of the lower jaw, opposite the poison fangs, which help them hold on to their prey until it dies of the venom—usually in just a few minutes. In South Africa, the spitting cobras manage to eject their venom for eight or ten feet with astonishing accuracy, into the face and eyes of approaching animals; the poison causes immediate, intense pain and, if not washed out promptly with water, permanent injury to the eyes. By contrast, the Egyptian asp gives an almost painless bite and was recommended in ancient times as a suitable means for putting to death a political prisoner; it is believed to be the snake with which Cleopatra committed suicide, and is represented with its hood slightly spread in the headdress of familiar sculptures from Egyptian tombs.

Subdividing the Old-World Tropics

Attempts to find orderly patterns in the geographic distribution of life are relatively recent, because farther back in history no one person had traveled widely enough or had access to a large enough collection of the world's plants and animals to recognize these generalities. Among the first to offer a convincing analysis was the British ornithologist P. L. Sclater. In 1857, two years before the appearance of the *Origin of Species*, Sclater presented before the respected Linnaean Society of London a remarkable account, "On the general geographical distribution of the members of

The square-lipped "white" rhinoceros grazes in open country, but now exists mostly in South Africa, where it is given protection. Like the three rhinos of Asia, it is in danger of extermination. (Wolf Suschitzky)

55

the Class Aves." Because the six separate regions he recognized did not correspond to cultural or political areas, he gave them new names: the Ethiopian, for Africa south of the great deserts, plus the Arabian peninsula; the Indian, to include India and Southeast Asia; the Australian, to include also New Guinea and islands of the South Pacific; the Palearctic, for the rest of the Old World; the Nearctic, for America north of Mexico including Greenland; and the Neotropical, for the rest of the New World.

Eighteen years later, the widely-traveled English zoologist Alfred Russel Wallace brought out his *The Geographical Distribution of Animals*—meaning land animals—and incorporating many of his own observations on mammals and insects, particularly in the tropics of the Old World. Wallace agreed with Sclater that the Ethiopian region was separated from tropical Asia by the deserts in the Near East, the Red Sea, the Persian Gulf and the broad Arabian Sea. But he felt dissatisfied over the vague boundary between Sclater's Indian region and the Australian. Wallace noticed a gradual transition in the East Indies, between islands with land animals characteristic of Australia and islands with a more Oriental character. He renamed Sclater's Indian region, calling it the Oriental, and commenting that the Asiatic tiger, for example, ranged southward across the 100-mile Strait of Malacca to Sumatra, and along the island chain as far as Bali. It did not, however, cross the deep Lombok Strait to the east, although this channel is less than seventy miles wide. Marsupial mammals lived on islands westward from New Guinea to Celebes, but not to Borneo. Accordingly, Wallace drew his line between Bali and Lombok, between Borneo and Celebes. Later biogeographers extended Wallace's line northeastward, to include the Philippines in the Oriental region.

The Oriental Region

Heavy rainfall causes a great portion of the Oriental region to be covered by dense forests, whereas the Ethiopian region contains enormous deserts and some of the most extensive grasslands and savannas in the world. Nowhere is a western biologist likely to feel farther from home than in the rain forests of the Oriental region. Yet explorers have found there a great wealth of useful and decorative plants, many of which we recognize instantly, at least by name. The trees include teak, one of the most durable woods known, in forests of Southeast Asia, where trained Indian elephants instead of machinery are used to harvest the

logs. Fragrance and fine grain are combined in white sandalwood from India, Malaya and the East Indies; red sandalwood from the islands; and the East Indian satinwood of Indonesia. Sal yields a timber second only to teak in value, as well as a natural dammar resin, and foliage that in parts of India is used as food for lac scale insects, which are the source of shellac and ingredients for varnishes. Camphor comes from trees of Malaya and Borneo. Henna dye, both orange and red for personal adornment, is obtained from leaves of a bush ranging from East Africa to Australia today, but probably of Indian origin. Indigo dye, from an East Indian plant, supported a great industry until cheaper synthetic pigments supplanted it.

Foods and flavorings from the Oriental region have influenced eating habits all over the world. It is the source of lemons, limes, sweet and bitter oranges, tangerines, citrons, and pomelos (an ancestor of grapefruit), all members of the big genus *Citrus* from Southeast Asia. Eggplant, a fruit that is also a vegetable, is from Southeast Asia and southern China. Dasheen and its cultivated form (taro), whose huge leaves resemble elephant ears but grow from thick underground stems, is a major source of edible starch in much of the Far East. Breadfruit, like jackfruit, grows on a tree; based upon Captain Cook's observation that people in the East Indies ate these fruits in quantity, an expedition ship that became famous as H. M. S. *Bounty* was sent to the South Pacific islands to get plants for introduction into the West Indies. Less exportable are the plants that yield the ill-smelling, delectable-tasting durians and sweet mangosteens, which rank among the most highly prized of edible fruits in the Old-World tropics; or bilimbi fruits and carambolas, which have a delightfully acetic flavor and grow on short stalks from the bark of the trunk and branches of low trees. Litchee nuts for the cocktail table are the dried fruits of a tree that is native to the Oriental region from the Philippines to southern China.

Black pepper is still among the most popular condiments in the world, after a long history as a prime reason for international commerce in the Orient. Betel pepper leaves are chewed with betel nuts from an Indonesian palm, and preferred to other "chewing gums" by more millions of people than is generally realized. Ginger comes from tropical Asian herbs of swampy places, cardamom as a spice from seeds, and cinnamon from the inner bark of a Ceylonese tree. Tea is brewed from leaves of low-growing trees native to highlands from India to southern China.

For decades chaulmoogra oil from a shrubby tree

of Burma and adjacent India remained the only known medicine with which to help lepers. Gutta percha from Indonesia and India rubber from the latex of the India-rubber tree lost their importance only recently through new progress in rubber technology.

The *Coleus* plants that housewives cherish all over the world because of their beautifully colored foliage and ease of propagation from cuttings placed in water, came from the Oriental region. So did the kudzu-vine, which may provide a wall of greenery for the front porch or hold the soil of road embankments from eroding in parts of the world where frost is mild or rare. In these same states and countries, we find the tree known variously as chinaberry or pride-of-India producing flowers and then fruits that dry to become white spherical beads for necklaces, whence the additional name of beadtree. Denser shade in moist subtropical and tropical countries is provided by introduced

Elephants, pelicans and African open-bill storks associate peaceably around a waterhole in Tsavo National Park of southeastern Kenya. (Alan Root)

banyan trees, which were named in their native India for the markets, or banyans, held under the shelter of their many out-thrust leafy branches; these trees are supported by stout, adventitious roots. In Thailand, we were grateful for shade from a related tree called the bo tree or peepul, which grows naturally over much of Southeast Asia. It is held sacred to the memory of Lord Buddha, in the belief that he meditated extensively beneath a bo tree, protected by it from the hot sun.

The world's tallest grasses, the giant Indomalaysian bamboos, are as high as 120 feet and grow as rapidly as 15 inches daily. These and the largest of flowers—the Sumatran *Rafflesia arnoldi*, with 20-pound blossoms

57

37 inches in diameter—are oddities to be hunted out in this region. *Rafflesia* is named for Sir Thomas Stamford Raffles, who served as lieutenant-governor of Sumatra and founded the city of Singapore in 1819. The flowers rest on the ground and the plant has no foliage, for it is a parasite on the roots of rain-forest vines and lianas.

Some of the handsomest of flowering trees are native to the Oriental region. One from Burma, *Amherstia nobilis*, has been voted the "queen" among decorative vegetation of this kind, and is cultivated throughout the tropics for its pendant sprays of orchid-like blossoms. The orchid trees (*Bauhinia* species) are related members of the pea family, now widely introduced from their home range in India and Southeast Asia. The ornamental golden-shower tree is of Indian origin. So are the great *Koompassia* trees that in Malaya tower as much as 330 feet high above the rest of the rain forest.

The palm family is well represented in the Oriental region; these easily recognizable trees have an almost endless variety of local uses. For example, the talipot palm of Ceylon and Southeast Asia provides leaves that are emergency umbrellas; unopened leaves yield a fiber called buntal, which is woven into hats, and also an Indian substitute for paper, known as olla, upon which messages can be written. Sago palms have a pith rich in edible starch. Even the palms that take the form of climbing lianas in the rain forest produce marketable items: malacca canes, from stout stems of the Malaysian *Calamus rotang*, and the photoengraver's varnish known as "dragon's blood" from the Southeast Asian climber *Daemonorops draco*. The gomuti palm of this same region yields most of the world's palm sugar, or jaggery. For the Palmyra palm, Hindu literature lists 801 uses, including timber, thatch, brushes, mats, baskets, paper, fruit, sugar and wine. Yet the Chinese fan palm, actually from Indomalaysia but grown in tubs in hotel lobbies of many countries, is almost the only one of these Oriental palms encountered by most people in other parts of the world.

Strictly Oriental Mammals

Limited to the Oriental region are two families of primates (the tarsiers and the tree shrews) and the graceful gibbons, all suggesting important steps in the evolutionary sequence that led to monkeys, apes and man. One whole order of mammals (the Dermoptera) is restricted to the region.

The three kinds of tarsiers (family Tarsiidae) are strictly nocturnal. They have enormous forward-facing eyes and prominent membranous ears, as well as enlarged pads on the tips of their fingers and toes, with which they can cling to almost any surface. Sometimes they stalk a lizard or other small animal as prey, walking slowly toward it on all fours. At other times they leap for surprising distances, and pounce on the unsuspecting victim. Tarsiers are well known to people who live where the rain forest is regenerating after being cleared, on islands from the Philippines to Sumatra, including Borneo but not Java. Apparently the ancestral tarsiers branched off from the hereditary line that led to monkeys.

Tree shrews (family Tupaiidae) are more widespread, from the forests of India to southwestern China, south through Malaya to Borneo and to the Philippines. Although active by day and squirrel-like in appearance, they are omnivorous primates, using their paws like hands to bring food to their mouths. Zoologists regard them as the least changed and simplest of all primates, showing what style of animal split away from ancient members of the order Insectivora. Their nearest non-primate kin may well be the present eighteen kinds of elephant shrews (family Macroscelididae) native to Africa south of the great deserts.

Far more spectacular primates in the Oriental region are the long-armed gibbons. These seven different kinds of slender, tailless, anthropoid apes measure no more than three feet in length. They live in small family groups, defending home territories in forests from Assam to Burma, to Sumatra, Java and Borneo. Warning intruders with a loud *hoo-hoo-hoo* call, they walk along high limbs with their arms upraised to balance themselves, or swing through the trees with unmatched dexterity. Gibbons feed chiefly on plant materials, such as buds, leaves and fruits, but take an occasional young bird, egg or lizard they chance to find. Sometimes they carry uneaten food clutched in their feet while traveling, supporting themselves alternately from one arm, then the other. Their fingers may take no hold at all on a branch, the gibbon swinging with its wrist bent to form a hook that keeps it from falling.

The colugos or "gliding lemurs" of order Dermoptera explore for vegetable foods in forests from Southeast Asia to Sumatra, Java, Borneo and the Philippines. Colugos are better equipped than any other mammal for riding on a cushion of air. They extend their long legs and tail, stretching taut a membrane that links tail tip to toes and fingers on each side, then glide from tree to tree. Mostly nocturnal, they usually hide by day in the darkness of large tree holes, with several animals side by side, each hanging head up, its hind legs, tail and gliding membranes wrapped around it. The generic name, *Cynocephalus*, connotes "dog-headed," but no common kind of dog has such short ears, prominent eyes and conical snout. To us, colugos appear more squirrel-like, although their teeth are distinctively different. A fully-grown colugo may be twenty-six inches long, ten inches of this being tail.

Forest Birds

We think of the Orient as the home of the jungle fowl that thousands of years ago became "chickens" under domestication. It is the native territory too for the magnificent Argus pheasants, named for the multiple eyespots on the tail-covert feathers, which give the birds a length (including the "train") of six to seven feet. Yet only one family of birds (Chloropseidae) is exclusively Oriental. Known as leafbirds, fairy blue-birds and ioras, these brilliantly-colored songsters, five to ten inches long, live in the tree crowns of great forests from northern India to Southeast Asia, in the highlands of Sumatra and the Philippines. Leafbirds are particularly fond of the nectar and fruit of tropical mistletoe, and serve both to pollinate its flowers and to distribute its seeds where the parasitic plant can reach new host trees. Fairy bluebirds also prefer fruits, especially wild figs, but ioras are primarily insect-eaters and frequent the forest edges and canopies where insects are most abundant.

The gentle little bleeding-heart pigeon of the Philippines is just one of an extraordinarily large number of different birds of this family native to the Oriental region. It is difficult to realize their beauty or the variety of types that exist until one sees such a collection as that at Yale University. They range widely across the Old-World tropics, mostly as fruit-eaters that rarely come to the ground. Green pigeons, some of them Asiatic and the rest African, have partially feathered feet and surprisingly wide mouths. Imperial pigeons, some twenty-two inches long, can disengage the elastic sockets of their jaws (much as many snakes can) to swallow whole fruits that are larger than their heads. These birds frequent the forests from Malaya through the East Indies to Australia, and on eastward to some of the larger islands of the South Pacific. Although they do not migrate, they manage to ignore Wallace's line and the water barriers in the Old-World tropics.

Cold-blooded Animals

Despite the wealth of animal life found in rain forests anywhere, the Oriental region has just one exclusive family of reptiles—the big-headed turtles—and none of amphibians or of fishes. The special turtles live in rocky mountain streams of Southeast Asia, from Burma to southern China and the island of Hainan. They grow to be only six or seven inches long in the shell, with a tail that is scarcely shorter.

The Oriental tree frogs of the family Rhacophoridae include some members that have counterparts nowhere else in the world. The Malayan flying frog is actually a glider, not much smaller than a flying squirrel. It can plane diagonally downward from one tree trunk to another, for distances as great as fifty feet, by spreading the extremely large webs between its long fingers and toes, and by holding its under surface slightly concave to gain extra lift. By tilting its gliding vanes the frog can control to some extent the direction and length of its trips. A non-gliding relative, *R. dennysi*, which is brightly colored, is revered in some parts of Malaya. On religious holidays one of these frogs is given a prominent place in a procession down the village streets, carried in its own sacred chair. Most of these rhacophorid frogs lay their eggs in masses of foam, instead of in ponds. But their toes end in enlarged adhesive pads with which they can climb even on vertical wet glass, just as tree frogs of family Hylidae have in the Americas, Europe and Australia.

The Ethiopian Region

So vast is the African continent that its climates and types of life seem almost endless. The Equator cuts across it almost midway, and the tropics cover the bulk of the land—nearly 5000 miles east to west, and 3300 north to south. Beyond this for another 750 miles or so, the Temperate Zones extend to the Mediterranean coast and to the Cape of Good Hope. But they are warm temperate, almost subtropical, both along the shores from Morocco to Egypt and around the blunt tip of South Africa. All of Africa, in fact, lies closer to the Equator than does any part of Europe, or than the northern boundary of the southernmost states in the U.S.A.

Nearly half of Africa is deserts lying under the Tropic of Cancer and the Tropic of Capricorn—not under the Equator. Yet it is almost impossible to prove today that the bare deserts are due to the climate during the last 5000 years as much as to the repeated visits of nomadic people, who were often accompanied by goats and camels, which will eat almost any plant that grows. Leafless stems and recognizable bits of root become fuel for a nomad's fire. Almost anywhere in

Beyond the narrow coast of the Mediterranean, where the flora and fauna are more like those of European coasts than like the rest of Africa, the stony and sandy deserts spread from the Atlantic side to the Red Sea. Yet wherever the underlying strata bring soil water to within sixty feet of the surface, wells have been dug and date-palm oases developed. The precious liquid may be brackish or bitter, and have come by slow subterranean flow from the Atlas Mountains in Morocco, or the Hoggar massif in Algeria, or the Tibesti massif in Chad, or even high country still farther south. Yet it will satisfy a camel's thirst and nourish a surprising assortment of useful plants in the shade of date palms. The amount of water that can be hauled up and used locally to irrigate the desert during the trees' most active growing season determines the population of the oasis, for the products of one date palm (*Phoenix dactylifera*) will support one person on a perpetual-yield basis at the standard of living that desert people have refined to match their isolation. It was a wise choice, for under cultivation, other date palms produce less fruit and sugar, thatch and other building materials. They require more water too, as is shown by the wild palm (*P. reclinata*) which grows along stream banks all the way from the tributaries of the upper Nile near the Equator to the eastern Cape Province of South Africa.

Across the Sahara and on similar Asian deserts between the Red Sea and the Persian Gulf, the sparse native vegetation is highly adapted to prolonged drought and salty soils. The desert cauliflower (*Anabasis* species) forms dense tussocks, like low gray cushions, while a mignonette and a Saharan lavender remain straggly and fragile-looking. Nor do stiff stems and sharp thorns keep hungry camels from browsing on the Sahara broom or the camelthorn (*Alhagi* species) that takes its place in Arabian deserts. The human inhabitants are more attracted by shrubby trees, both for firewood and for the fragrant resins that can be gathered from those of the balsam family (Balsaminaceae)—myrrh and Mecca balsam and bdellium from various kinds of *Commiphora*, and frankincense from *Boswellia carteri*. Toward the southern fringe of the African arid lands the deeply rooted screw-bean acacias grow thirty feet tall, in the shape of an inverted cone. Signs of subterranean moisture can be recognized in the occasional good-sized terebinth tree, or a thorn-wood such as the one called a desert date or a doum palm with a uniquely branching trunk.

The red-hot poker (Kniphofia) *is native to eastern and southern Africa and Madagascar. (Emil Javorsky)*

these barren lands, an area that is securely fenced and policed for ten years or more becomes a grassy patch or a copse of thorn bushes. Migratory birds and wandering wild animals of many kinds congregate in it as though it were an oasis.

Fully a third of Africa is semiarid plateau land, where trees might grow if given a chance. But under the continued influence of people, fire and wild animals, this high country has less true forest than grassy plains and open savannas with scattered trees, and large areas of thornscrub. Although the Dutch word *veld* (or *veldt*) refers to a grassland, the South African terms "high-velt" and "bushvelt" usually distinguish places that are mostly grass on the one hand and mostly shrubby trees on the other.

The remaining sixth of the great continent includes the beautiful coast at the southern tip, the river valleys and the rifts, the rain forests (particularly in the Congo), and the slopes of spectacular mountains.

This is the land of animals that can burrow underground to escape the heat of day, which is the habit of the jerboas and many insects, as well as of the desert cat, the fennec fox and the puff adder that feed on the jumping rodents. The lanner falcon and the eagle owls get their moisture from their prey. Addaxes with spiral horns and camels that have been temporarily freed to fend for themselves are extraordinarily able to tolerate the heat and to get the water they need from vegetation. The wild ass is a desert dweller too, but needs to drink more often. So do the several kinds of gazelles that enter the deserts only after enough of the annual ten inches of rain has fallen in quick succession.

Southern Deserts and Semi-Deserts

Arid areas of impressive size spread eastward from the "Skeleton Coast" of southern Angola and South West Africa, across northern South Africa and through most of Botswana. They constitute the dreaded Namib and Kalahari waste lands, and give South Africa unproductive territory called the Big Karoo and the Little Karoo. Some of the strangest plants survive there despite the drought. One is a unique gymnosperm with a huge carrot-shaped underground stem. During its long life, it produces just two leaves, each a broad straplike extension eight to ten inches wide, which grows a little more each year and at the same time erodes at the tip from the battering of desert winds. A member of the yam family called elephant's foot or Hottentot bread has a similarly enormous stem barely projecting from the ground. It is covered with waterproof cork, and gives rise each year to a slender, climbing stem with large leaves and a spike of small regular flowers. The wit-gat tree, with a branching trunk that appears to have been whitewashed, stands bare of leaves most of the year, as does the ghostman which rises as a tapering, slightly bent column as much as twelve feet tall, and a foot in diameter at the base. Woody aloes, four feet in diameter and twenty feet high, bear stiff crowns of sword-shaped leaves at the top. The spurge family (Euphorbiaceae) is represented by cactus-like types, such as the candelabrum-style *Euphorbia tetragona* which is fifteen to twenty feet tall.

After one of the rare rains, and sometimes in the spring of the Southern Hemisphere, these desert plants produce conspicuous flowers. The Namaqualand daisy may carpet many acres, opening its two-inch blossoms by 10 A.M. and closing them again by 5 P.M. when the chance of a visit from pollinators decreases. The lily-like *Nerine bowdeni*, a close relative of South Africa's so-called Guernsey lily, follows a similar schedule.

Between rains the more obvious vegetation is composed of shrubby, spiny acacias and the narras melon, which grows on sand, reaching water with a root forty feet long and extending thorny tendrils over the bare desert. Bushmen gather the gourds of the narras for food, for the water they contain, and as disposable containers.

The desert plants support small populations of antelopes that are well adapted to life where fresh water is virtually nonexistent. The magnificent gemsbok, with horns as much as four feet long, and the springbok, which hurl their gazelle-like bodies in great high leaps when excited or frightened, wander as nomads in this dry territory, browsing on whatever they can find. Insects, too, feed on the vegetation; then spiders, scorpions and burrowing reptiles attack the insects. Seed-eating weaver-birds seek out the big acacia trees and sometimes the tree aloes as the support for their massive apartment houses, constructed from dead plant materials like haystacks well above the ground.

The Karoo has its own acacia known as sweet thorn from the scent of the yellow rounded flower heads that may appear several times a year, after each rainy period. But the truly distinctive plants are "stone plants" (*Lithops* and other genera of the carpetweed family, Aizoaceae), which resemble large pebbles on the dry ground. Other succulents (especially *Mesembryanthemum* species, and various kinds of *Crassula*) are spaced out by carrion-flowers of the milkweed family, which resemble cacti but have five-parted leathery flowers with an odor attractive only to the flies that serve in pollination.

Highvelt and Bushvelt

This is Africa's big game country. People and domestic livestock are still excluded from much of it by diseases to which the native animals are almost immune. To them the biting flies that transmit the diseases are mainly a nuisance: the tsetse flies that carry the blood parasites (trypanosomes), causing sleeping sickness in people and nagana in cattle; and the blackflies that transfer infections, bringing blindness to both man and livestock.

It is a land in uneasy balance. If the animals are few, tree seedlings sprout, survive, and soon grow taller than most browsers can reach. If many elephants stop in the shade and rub their gargantuan itches against the bark, the trees die. Just prior to the rainy season, when the tops of the dormant grasses are parched and most flammable, a wild fire can sweep quickly across the plain, igniting bushes and scorching trees to death right through their bark. But if domestic cattle are introduced and herded for safety from predators, they selectively remove the kinds of plants they like, leaving thorn bushes to transform the scene into a bushvelt.

Bushvelt and highvelt cross Africa today south of the great deserts from the Atlantic coast near Dakar, to touch it again in Dahomey and Togo, and once more far south, from the mouth of the Congo River to the Namib Desert in Angola. This open country, with a reasonable amount of rain at least once each year, skirts the rain forests of West Africa and of the Congo basin, extends all the way to the coast of the Indian Ocean, and continues narrowly beside it almost to Port Elizabeth in South Africa. Ecologically it cuts Africa in two and surrounds the continent's great equatorial Lake Victoria—second largest of all fresh-water lakes. It deteriorates to bushvelt and semiarid badlands along the north from Mauritania past fabled Timbuktu and Lake Chad to Somaliland, and along the south from Angola through the valleys of the Zambesi and Limpopo Rivers almost to the eastern coast.

The native grazers and browsers minimize the inevitable competition from living on the same land by choosing distinctive diets, combinations that utilize most efficiently the assorted plants available in each season without putting undue pressure on just a few kinds. The stately giraffes use their long tongues and longer necks to reach over the flat tops of the acacia trees, or to browse from the sides twelve feet or more above the ground. The long-necked gerenuks stand on their back legs to reach branches at the ten-foot level. Hartebeests and topi are among the big antelopes that graze among tall grasses, such as elephant grass and Kikuyu grass. The brindled gnu, often found in great herds mingled with zebra and even ostriches, prefers the short-grass plains and savannas. Perhaps there the excellent vision of the zebras and ostriches, and the keen sense of smell of the gnus, can detect a pride of lions in hunting formation before it is too late.

Men with a special interest in range grasses disregard the tall *Pennisetum* species, despite the fact that people harvest pearl millet in the Sudan and an early millet in Nigeria and parts of West Africa. They disdain too the common kinds of *Andropogon* as sour grasses, and look for the sweet grasses cattle prefer, such as the widespread African redgrass or an *Eragrostis* of some kind. In Ethiopia, tef is a favorite cereal for human use.

Baboons show a similar awareness of the bushes and trees that grow at intervals among the short-grass plains, rarely stopping their foraging for fruits and grubs to climb a tall acacia tree, but often pausing to feed in a gall acacia.

Over much of this vast area, the conspicuous animals are the hoofed cud-chewers. In eastern Africa alone, our friend C. A. W. Guggisberg, in his *Game Animals of Eastern Africa*, describes and illustrates no less than sixty-five different kinds that the visitor is likely to meet. There are wild sheep, Cape buffalo, an incredible assortment of antelopes with distinctive silhouettes and horns: gnus, hartebeests, duikers, klipspringers, steinboks, dikdiks, waterbucks, lechwes, kobs, reedbucks, impala, gazelles, oryxes, elands, bushbucks and kudus. Add to this the African elephants, hyraxes, rhinoceroses, zebras, wild hogs of various kinds, hippopotamuses, lions, leopards, cheetahs and smaller cats, hyenas, jackals and other wild dogs, and the outdoorsman is thrilled beyond measure. After an experience with the wildlife of the African highvelt and bushvelt, other continents offer only anticlimaxes.

The best still photographs and detailed dioramas communicate few of the distinctive features of the various animals. Probably these clues to kind are more conspicuous outdoors because the animals themselves rely upon them, and flaunt them as communication. The twitching black tail on a Thomson's gazelle becomes a signal quite unlike the swishing white tail of a Grant's gazelle, which, if young, may be no bigger than the common "tommies." As these graceful creatures turn warily to watch for danger, they let us notice the nearly parallel horns of the Thomson's and the diverging horns of the Grant's. The differences are slight, yet consistent, and visible despite the shimmer of heated air over the grassy plains at midday.

In this region, where large grazing animals are today more numerous than elsewhere on earth, predators are also more in evidence. Yet they cull the breeding stock much as an animal husbandryman might, removing as prey those maimed by accident, or sick with a disease, or loaded with parasites; they satisfy their own needs without endangering the prey species. In open highvelt, from northern Nigeria to the Sudan and Ethiopia, south to Zululand in South Africa, and eastward in Asia across suitable country as far as western India, cheetahs run down their victims in solitary chases at speeds up to sixty miles per hour. Over distances of half a mile or less, these predators are probably the fastest animals on earth. But they watch the herd and seem to pick out a particular individual—the one least likely to get away. Similarly, a healthy antelope or zebra or ostrich can usually escape from a pride of lions despite the stealth with which the lionesses space themselves out on one side of a herd while the males walk into the herd to drive them into the ambush.

62

When bluffing does not settle the relative dominance of two gemsbok in their prime, they battle for a while —until one dashes away, pursued only a short distance. Others watch the encounter. (Clem Haagner)

In the drier bushvelt, packs of hunting dogs seem less selective of victims showing signs of weakness. They run to exhaustion the prey they take after, then tear it to pieces. Generally so many hunting dogs share the same carcass that none of them is satisfied for more than a few hours. Wardens assured us that these dogs are very destructive to game. Yet neither type of animal is new to the bushvelt, and a balance of some sort must exist. If the dogs did not drag down so many antelopes, the greater populations would almost certainly overeat their thorny range and starve. The bushvelt itself would deteriorate, perhaps letting the desert extend a little farther—to the benefit of few kinds of life.

The helpless young antelopes and the aging individuals seem the obvious victims. Yet the one type of prey is seasonal, and the supply of the old unwary animals is erratic. Predators must spend most of their time finding small mammals, birds, reptiles and insects to stay alive between feasts. Otherwise, Africa could

scarcely have so many smaller members of the cat family: the caracal, with long tufted ears and short tail, hunts by day, avoiding only the deserts and the rain forest; the serval, with large untufted ears, long legs and short tail, patrols the bushvelt, chasing short distances after some prey, leaping upward or climbing swiftly after others; the golden cat of forests from Sierra Leone to Kenya is twice as big and powerful as a domestic cat, but hunts stealthily by night; and the blackfooted cat of southern Africa, which has a narrow pointed face but larger ears and body than a domestic cat. These and other cats that rarely come south of the great deserts may continue lines of inheritance that

63

early man combined into the ancestry of the domestic animal, which was venerated in Egypt as far back as 1300 B.C.

The small carnivores in Africa include more kinds of mongooses than in all the rest of the world put together, plus civets and various members of the weasel family, especially the ill-smelling polecats and the remarkable honey badger. Unlike beasts of prey on other continents, the yellow mongooses live in colonies like prairie dogs, generally with a sociable near-relative, the slender-tailed meerkat. Many members of the colony wander off to hunt by day; others stay home, basking in the sun, improving the burrow system, or socializing.

Honey badgers may well be more powerful for their size than any other mammal. Native to the whole area south of the great deserts, and at home even in the Kalahari, they use their short legs and strong claws to dig for burrowing animals, to roll stones aside, to rip open rotting wood, or to climb into trees. So thick and tough is the honey badger's skin that insect stings and snake fangs seem unable to penetrate.

The honey badger's freedom in raiding the nests of bees makes it the natural ally of one of Africa's strangest birds—the honey guide. Named for its habit of leading a honey badger (or a human being) to a bees' nest by a fairly direct route, the bird waits quietly until the raider has ripped the honeycomb to shreds, eaten most of the sticky honey, and left the bees nothing to defend. Then the honey guide flies down to eat the beeswax, for the digestion of which it has the necessary rare enzyme. While a honey guide is young, these habits are hidden in its heredity and it has not yet begun to produce the necessary digestive agent. Inner guidance, not imitation of birds around it, induces the honey guide to seek the aid of a mammal toward enjoying a feast of beeswax for, like European cuckoos and American cowbirds, parent honey guides lay their eggs in the nests of other birds and let foster parents rear their young.

The same highvelt and bushvelt hide habits of growth that produce some of the world's most gigantic types of life. Its elephants are the largest mammals on land, its ostriches the largest birds. Some of its trees seem comparably gross and grotesque, particularly the baobabs, with a trunk as great as thirty feet in diameter and huge spherical fruits known as monkey bread. These monsters tend for a while to distract attention from less bizarre kinds of life, some with special beauty. The "flame trees of Thika," with large reddish orange flowers set off against dark green foliage, are among the handsomest trees to be found anywhere. The common mopane, which forms great open forests, has twinned leaflets like the wings of butterflies where other trees have ordinary leaves. The akee produces pinkish-orange fruits the shape of a green pepper, each containing three marble-like brown seeds with edible soft parts attached; at the right stage of ripeness, they make a delicious vegetable with a nutty flavor; at other times they are a deadly poison.

The Coast of the Cape

From Port Elizabeth to Clanwilliam around the southern tip of the continent, Africa has the greatest wealth of plants for any area of this size in the world. Formerly its forests produced quantities of stinkwood, an immensely hard wood useful for cabinetwork but one which impatient man is unwilling to let regenerate at its own slow speed. On the coast of the Cape, the silvertrees suggest poplars with narrow leaves silvered by a coating of white hair. The wild flowers are now the garden and greenhouse favorites of the world: *Agapanthus*, *Amaryllis*, asparagus-fern, bird-of-paradise flower, calla lily, Cape marigold, *Freesia*, common geraniums, *Gerbera* daisies, *Gladiolus*, ice plants, *Lobelia erinus*, montbretias (actually a hybrid between two kinds of *Tritonia*, of the iris family), *Nemesia*, *Plumbago*, red-hot-poker, and tuftybells. The diversity of wild heaths and of spectacular flower patterns among members of *Protea* and related genera make this coastal area additionally outstanding.

The striped tiger of Asiatic swamps and forests is a powerful animal whose habits and geographic range distinguish it from the closely-related lion. (E. Hanumantha Rao)

Overleaf: Above left: In the rain forests of the Congo and coastal West Africa, the flame lily grows vine-like through the shadows or more erect in partial sunlight. (J.-F. and M. Terrasse) Center left: The 10-inch armadillo lizard sometimes is not fast enough to escape from a snake or other predator. It then flails about with its heavy, armored tail and often is released after being half swallowed. (Manuel V. Rubio) Below left: Emerging from its underground hiding place, a scorpion of the bushvelt in the South African Transvaal finds a long-horned green grasshopper full of nourishing juices. (Anthony Bannister) Right: The lioness does most of the hunting for the lion pride. Formerly native from central Europe to northern India as well as Africa, these animals are now common only in Africa. (Clem Haagner)

The rolling plains and hills of the Cape still have some less appreciated mammals, such as burrowing mole-rats, which attack garden crops, and Cape gerbils, which multiply in grain fields. But the South African zebra was exterminated in the wild more than a century ago. Equatorial zebras have not roamed free there since 1910. The black wildebeests used to accompany these zebras on the coastal plains, but like the white-stockinged blesboks of the forests, they are now represented by only a few dozen that are held captive on provincial game farms, fenced in for their own good.

Rivers and Rifts

Well watered as the Cape coast is, none of Africa's great rivers empty there. Instead, the major drainage pattern of the continent is strange. Only the Congo flows for its whole length through rain forest. The Niger rises in relatively dry areas and gains volume as it goes to the sea. The Nile and the Orange Rivers begin among misty mountains and later dawdle through deserts, losing half of their water before it can reach the river mouth. The Zambesi and the Limpopo come through highvelt and bushvelt on their way to the Indian Ocean.

Africa's deepest lakes and some that are shallow and salty owe their existence in the eastern third of the continent to seismic events about 70 million years ago. A change of almost global proportions double-faulted the continent along a twisting line from somewhere south of Madagascar to the foothills of the Taurus Mountains in southern Turkey. A strip of rock 10 to 100 miles wide sagged abruptly, while the sides of the fault lines rose, producing a deep rift valley. It remains as a channel that could not have been formed by erosion, and that erosion has not yet obliterated.

At its southern end, the rift lies deep under the Indian Ocean. As it approaches the African southeast coast, it follows the bottom of the Mozambique Channel, with a side spur to the northeast that may indicate the way and the time that Madagascar separated from the

Above left: The black spot on each fetlock and a readiness to leap across a road in a single arc identify the impala, one of the commonest of African antelopes from the Congo and Kenya southward. Left: Secretary birds, standing 4 feet tall, can walk faster than a man ordinarily runs. In pairs, or family groups, they stalk about in search of snakes and other reptilian prey in most of Africa south of the Sudan. (Both by Clem Haagner)

mainland of the continent. Halfway up the Mozambique coast, the valley is obliterated briefly. But it appears inland as a narrow basin holding Lake Nyasa. Elevated once more, yet with signs of side branches (one to the north being particularly significant), it deepens to hold Lake Tanganyika, the world's second deepest lake (4730 feet) with a bottom 2300 feet below sea level.

Mountainous upheavals in the eastern Congo produced a cluster of volcanos, the Virunga group, of which the craters called Nyamlagira (10,028 feet) and Nyiragongo (11,385 feet) have been active since 1947. Seemingly these movements blocked the rift and stopped the northward drainage of Lake Tanganyika through Lake Kivu into the White Nile. Presently the current flows in the opposite direction, and Lake Tanganyika has no river outlet—only a high-level overflow region where excess water can spill westward into the Congo basin.

North of these volcanos and in the morning shadow of the jagged Ruwenzori peaks, one of them 16,763 feet high, the White Nile begins. It crosses the Equator to Lake Albert, still in the rift valley. It receives the overflow from Lake Victoria, second-largest freshwater lake in the world, and cascades through the narrow rocky cleft at Murchison Falls before continuing northward in a channel of great antiquity. It must penetrate a maze of Sudanese marshes, called the Sudd, before joining at Khartoum with the Blue Nile from remote high peaks and ridges of Ethiopia. Thereafter the great river flows onward through the desert, past Aswan and Thebes and Cairo, to empty quietly at Egypt's Mediterranean coast.

An eastern branch of the rift, hinted at in Mozambique, straggles past the old volcanic cones of Mt. Kilimanjaro (19,340 feet) and Mt. Kenya (17,058 feet), both evidences of disruption along the line of rock faulting. This part of the rift system crosses by way of salty Lake Rudolf in northern Kenya into Ethiopia. It is believed to account for much of the depression filled by the Red Sea. Beyond Africa, in the Near East, it holds the Gulf of Aqaba, disappears briefly only to reappear as the basin of the Dead Sea and the valley of the Jordan River, including Lake Tiberias—the Biblical "Sea of Galilee." The present level of the Dead Sea is 1286 feet lower than the waters of the Mediterranean, and the bottom as much as 2366 feet below sea level. No other continent in the world has been scarred so extensively by a double faulting of its rocky underpinnings. Yet the plants and the animals on the two rims of the rifts are virtually identical at each latitude, showing how minor a barrier it has been to their dispersal.

The marshes and shallows along the rivers and the lakes in the rift system, like the borders of Lake Victoria, offer special opportunities for life of many kinds. The shorelines and channels often seem in danger of being choked by dense stands of tall papyrus, with a scattering of ferns and clambering vines of morning-glory. In other places, the banks are lined by reeds and bulrushes, and the water surface clogged with floating Nile cabbage. These abundant plants seem more suitable as places to hide, rather than as food for the animals we can see. The hippopotamuses that soak themselves eye-deep by day all leave the water at night to graze and browse as much as a mile distant, on terrestrial vegetation that attracts them more.

The fishes of these African waters include some that have become famous. Four of the world's six kinds of lungfishes are ready to gulp air into their lungs if the water becomes foul, or to wall themselves up in a cell of mud to fast and barely breathe while waiting out a drought. The Congo has both *Protopterus dollei* and the widespread *P. aethiopicus*, found also in the Nile and the large lakes; the West African *P. annectans* and the East African *P. amphibius* show the same habits and take about the same kinds of food—mostly small fishes of other kinds. The electric catfish of the Nile valley and equatorial Africa is a voracious carnivore too, able to hunt in muddy water and stun its prey with a jolt of electricity. Often the victims of these predators are the bass-like *Tilapia* fishes which scavenge in the shallows, much as carp might do on other continents. *Tilapia* are mouth-breeding cichlids, which provide parental protection by holding their eggs in their mouths until the young hatch out. Nile perch leave their spawn unattended and, although fewer young may survive, some grow old and to the fabulous weight of 200 pounds on an omnivorous diet.

Long-legged wading birds and flying scavengers take advantage of the fishes, but often make meals of snails, aquatic insects and the many small rodents that feed among the marsh vegetation. The giant goliath heron may stride along the bank, right through a resting flock of pink-backed pelicans and yellow-billed storks. Carrion has greater appeal for the black-headed sacred ibis and for marabou storks, both of which lack feathers on their necks. The handsome fish eagles are often conspicuous, perching on bare branches overhanging open water. Their white heads and black wings show clearly as they fly at a modest height, watching for dead fishes or unwary live ones at the surface.

The Rain Forests

In the basin of the Congo, and along the Atlantic coast of West Africa from Ghana to Sierra Leone, high rainfall and low elevation combine to sustain some of the densest forests in the world. As in other rain forests, trees of the same kind rarely grow near one another. Yet plant explorers have found valuable species among the mixture. This is the source of cola nuts, taken from trees of modest size. They are recognizable by their simple glossy leaves, small clusters of red flowers among the foliage, or star-shaped bunches of egg-shaped fruits, each containing eight reddish seeds. Lowland or Liberian coffee and the important oil palm grow naturally in these wet forests. So do the African mahoganies and the pepper bush whose spicy seeds are the source of melegueta pepper. Where the forest is regenerating or the steepness of the land prevents excessive shading by tall trees, the undergrowth includes African violet and the gloriosa lily. Clockvine clambers up the pioneer trees, quite unlike the woody lianas (such as *Hugonia*) which grow downward from epiphytic plants high in the forest and serve as adventitious roots to get more nutrients from the soil.

The stratification of the rain forest, from ground level to topmost canopy, allows many kinds of birds and monkeys, reptiles, amphibians and insects to remain hidden among the topmost tiers, out of sight to anyone walking below. Many of their sounds come down, but rarely can they be identified. Occasionally the sound-makers descend part way and turn out to be *Colobus* monkeys, or forest hornbills, or a great blue touraco. The commonest monkeys are vervets, gray little animals that travel in troops of from thirty to fifty individuals.

At night the sounds that reach the ground may be fewer, but they seem louder and more mysterious. A crescendo of little chirps ending in a scream is usually the call of a tree hyrax. Wailings may signal the activities of nocturnal primates, either bush babies or pottos. All of these animals clamber and leap among the branches, feeding on buds and fruits primarily, but taking such insects and cold-blooded vertebrates as they encounter. Occasionally they must be unfortunate enough to meet a tree cobra, for these poisonous snakes are not rare and are well equipped for getting a warm meal high above the ground.

The Barbary ape of northwestern Africa reproduces most successfully along the Mediterranean coast. (Fred Bruemmer)

The Indian open-bill stork (Anastomus oscitans) *has relatives in Africa as well as in tropical Asia. With its thick beak it feeds on mussels and freshwater snails. (M. Krishnan)*

Animals on the forest floor are generally quiet, slipping amid whatever greenery can grow where so many layers of foliage higher up have already absorbed much of the light. Columns of driver ants forage there, with no fixed abode. Nomadic in the forest, they bivouac wherever night overtakes them, and push on soon after dawn, carrying their larvae and pupae with them. Hunting columns of these ants attack and overpower almost any kind of animal in their path; they shred the flesh and either devour it or feed it to the larvae on the march. Only at regular intervals of many days is there a pause, during which the queen lays new eggs and the eggs hatch, and new adults emerge from

the pupae, hardening quickly to be able to travel on their own.

No one is quite sure how many kinds of animals remain to be discovered in the African rain forests. Dwarf buffalos and bongo antelopes browse in the shadows, the one a shy relative of the formidable Cape buffalo, the other an elusive animal with white stripes narrowly marking a chestnut-red coat and, like the kudus and elands of the highvelt, wearing horns in both sexes. The cow-sized okapi, which is the nearest living relative of giraffes, became known to science only in 1900. The giant forest hog, which eats fallen fruits and digs for roots in the soil between the trees, turned up in 1904. A pygmy hippopotamus lives in West African rain forests, but little has been discovered about its habits in the wild.

Learning more about these creatures will depend upon scientists who are brave and tactful, as well as competent in their subject. They must rely upon many kinds of assistance from native people—who can seldom be expected to understand a scientist's interest in live animals that serve for food, or to be unsuspicious of the motives of visitors. Nor can they easily comprehend an observer going out at night, when most of the wild life is active and a human being is in constant danger from poisonous snakes, several kinds of pythons, leopards and other hungry predators. Biting insects and scorpions are abroad in force. Nor is the giant porcupine to be ignored if it waddles over, rattling its long quills while searching for plant food. It is always ready to chew on equipment with the salty flavor of human sweat. Animal life in many shapes and guises is every bit as effective in protecting the privacy of Africa's forest dwellers as the continuous heat and high humidity that almost melt the will to work.

The Mountain Slopes

The humidity is no less, but the temperature is lower on the sides of mountains that rise above the high plateaus of central Africa. Arab traders found one of the most important beverage plants in the world growing somewhere between 4000 and 6000 feet above sea level in the Ethiopian highlands. It became known as "Arabian coffee" and found a market so vast that East African plantations could not match it. New plantations were begun in Java, Colombia, Brazil, and many other mountainous tropical countries of the world.

At slightly higher elevations, the nights are often too cool for profitable agriculture. Yet primitive farmers continue to clear the forest higher on the mountains.

In the eastern Congo, among the Virunga volcanos, they are encroaching on the belts of bamboo that grow between 8000 and 10,000 feet. Until now, this has been the principal sanctuary for mountain gorillas, and higher than the chimpanzees will venture.

Above the bamboo level on the high mountains is a misty wilderness where few animals make themselves at home. It becomes a spectacular world on the rare clear days, when the "Mountains of the Moon" become distinguishable peaks—the Ruwenzoris—and when Kilimanjaro and Kenya temporarily lose their cloud cover, letting the sun glint from wet foliage and tree branches hung with the lichen known as old-man's beard. Here are forests of a tree (*Hagenia abyssinica*) belonging to the rose family and rarely met at lower elevations. Generally it grows alongside a cedar (*Juniperus procera*), and the two together form thickets through which travel is exceedingly difficult and slow.

Between 10,500 and 11,000 feet, the Hagenias and cedar generally give way to more open slopes dotted with giants representing genera that elsewhere in the world are lowly herbs. A giant heath (*Erica arborea*) and a close relative (*Philippia excelsa*) of the same family rise ten to twenty feet tall, each plant in seeming isolation. Its nearest neighbor is perhaps a giant groundsel, resembling an enormous cabbage on a palm trunk, or a tree lobelia of comparable height.

Still higher on the slopes, where the weather is colder and wet with still more frequent rain or mist, the mountainside becomes irregular, with crags clad only in lichens and soggy grasslands with peat moss, lady's-mantle, sage, and everlastings of many species, often in separate tussocks. Among the rock piles, small rodents and the rock hyrax may be surprisingly common. This may explain the occasional visit of a leopard to these heights. At least one has been found frozen in the snows on Mt. Kilimanjaro. Cape buffalo and elephants sometimes crash through the forest vegetation lower down and die on the mountain meadows. Game wardens suspect these animals of being blind. Their bones provide nutritious marrow for the bearded vultures (or lammergeiers) that soar among the peaks on rare clear days.

Bradfield's hornbill finds a mixed diet of fruits, insects and small reptiles on the ground in Rhodesian savannas. Other hornbills are native to most parts of the Old-World tropics. (Clem Haagner)

4
Life
in the
New-World
Tropics

The first time we set foot on South America was from the gangplank of a bauxite ore boat at the mining town of Moengo in Surinam (Dutch Guiana). Within hours we were afloat again in that paragon of primitive luxury, a dugout canoe powered by an outboard motor. Slowly we chugged upstream on the Cottica River beyond any sign of civilization.

On both sides the dense rain forest came right to the river's edge, presenting a vertical wall of vegetation to the height of truly tall trees. Monkeys and macaws screamed at us from high vantage points. Over the water, brilliant iridescent *Morpho* butterflies five inches across flitted erratically, only to rise over the jungle canopy and disappear from view.

At intervals we came to narrow clearings along the river bank, where African-style houses had been built close to the waterway by small groups of Bush Negroes —the Djuka people—whose ancestors had escaped after being brought to the New World as slaves. Silently the black people paddled their own dugouts from one community to the next, or fished for food from the water that served them in so many ways.

This same rain forest continues southward beyond the Guianas and across much of Brazil for perhaps 1400 miles. Most of it has soggy soil, tall trees and few people. Through it flow the tributaries of the largest

river on earth, the mighty Amazon. This one broad basin occupies nearly a third of South America, collecting the water from 2,050,000 square miles. Rainfall averages eighty inches or more annually, causing regular and prolonged floods in the eight tributary rivers, and sending to the ocean the equivalent of fourteen Mississippi Rivers, many Niles, or several Congoes. Following the heaviest downpours in January and February, the Amazon spreads out to a width of fifty to sixty miles, rising thirty to forty feet above its level in August or September. Precipitation on similar land farther north causes a corresponding rise in the Orinoco River, which empties close to Trinidad.

The Amazon and the Andes Mountains are generally the first features of South America to reach out of a geography book and clutch at the imagination of anyone reading about this continent. Yet both are consequences of events during the last 65 million years, affecting a land of great antiquity. The stupendous folding of the rocks that raised the cordillera of the Andes began in the last days of the dinosaurs. It divided the continent like the peak of a roof, into a narrow Pacific and a broad Atlantic slope.

Under the Equator the tilt of the land still takes the moisture from westbound trade winds, producing the Amazon River. The air itself continues over the lofty Andes more than 2000 miles west of the river's mouth, then down the western slopes, getting warmer and absorbing whatever water is available. If the trades diminish, air moves ashore from over the Pacific Ocean, but brings no rain to the lowlands because the coastal waters are so cold all the way north to the Equator. Together these influences create along the western seaboard of South America some of the world's driest deserts, most of them within sight of the ocean.

The great cordillera, the remains of ancient plateaus, the shape of South America, and the exact location of the continent in relation to winds and ocean currents all combine to establish the pattern of rainfall. The rains, in turn, determine the living conditions that animals and plants must match to survive. Between the rain forests and the desert, the product is a unique mosaic of mountain country, highland savannas, scrublands and areas that alternate between annual flooding and real aridity, great grasslands (the famous

Low clouds often fill the valleys above the rain forest on the eastern slopes of the Andes Moutains in Ecuador. The relative humidity is close to 100 per cent all the way to the ground. (Rolf Blomberg: Full Hand)

pampas of Argentina), and a small region of south-temperate forest in Patagonia.

The shape and location of the continent have much to do with the ocean currents, warm and cold, that flow along the coasts. To the east, this triangular land mass projects into the Atlantic Ocean just south of the Equator. Against this rounded eastern point of Brazil, from Natal to Recife, the trade winds blow all year. With them they drag westward the surface waters, which South America divides as though it were the prow of a giant ship. Part of the steady current is diverted northwest past the Guianas into the Caribbean Sea, only to be deflected by Yucatán, resisted by the waters of the Gulf of Mexico, and poured out between Cuba and Florida as the mighty Gulf Stream. The other part of the current turns southwest toward Rio de Janeiro as part of the South Atlantic eddy. But before it reaches the latitude of Buenos Aires, it too is deflected, in this case by a stream of cold water from off Antarctic shores. The cold flow there is a consequence of the great Palmer peninsula on Antarctica, which turns eastbound currents northward to be split by the south corner of the continent—Cape Horn and Tierra del Fuego. The major part of this Antarctic water forms the Humboldt Current, streaming northward along the west coast of South America all the way to the Equator. The minor current along the east coast merely helps to keep Patagonia cool all year, before turning to join in the South Atlantic eddy.

Lands Adjacent to South America

Only about 700 miles of open water separate Cape Horn from the Palmer peninsula of Antarctica. No other continent reaches so far toward the South Pole. Geologists suspect that in the past the gap may have been much narrower, even if no actual land bridge existed. While Antarctica was warmer, living things in considerable variety may have reached South America from the south, where only penguins are likely to make the trip today.

Now a causeway for land life is open from South America through Panama and Central America, Mexico to the expanse of the North American continent. Across this connection, which is less than 100 miles wide in Panama, plants and animals continue to spread in both directions. The land bridge was probably broader during the Ice Ages, because sea level fell then all over the world while so much of the water from the oceans was frozen atop the continents and the northern oceans. At that time, living things would have found travel easier than now between the mainland of Central America and the northern islands of the West Indies (including Cuba and Jamaica), and between South America and Trinidad.

Far fewer plants and animals spread into or out of North America, and far more within the area of the New World south of the Tropic of Cancer. There, amid a wondrous variety of opportunities, they evolved some of the most extreme adaptations known anywhere. Consequently the area from Cape Horn to the West Indies and to lowland Mexico (but not its plateau country, which is definitely North America) is justifiably regarded as a single biogeographic realm—the neotropical.

The Multistory Forest

For a tree to grow on the flood plains of South America's great rivers, it must have a sturdy trunk to resist the current as well as an extraordinary tolerance for having much of its trunk submerged for weeks or months each year. Yet under these conditions, literally thousands of kinds of trees hold their place, representing many families of plants. To any visitor from the temperate zones walking over the muddy floor of the forest in dry season, the most astonishing feature is the scattered distribution of each species. No two alike may grow on the same acre of ground. Just to identify the various kinds of trees becomes a major task, since only their bark-clad trunks and buttresses are generally within reach. The foliage and flowers may be above an understory of other branches and completely hidden from view. Unusual clues to identity take on new importance, such as the odor of the bruised bark and the nature of the sap or gum that will ooze from a small wound.

Trees important in commerce come from these rain forests, although usually with the help of local Indians who have learned to distinguish the valuable kinds before the tall tree is felled. It is the source of the most desired among mahoganies; of the Brazilian rubber tree, which has been cultivated so extensively elsewhere for its white latex; of the Brazil-nut tree, which yields much valuable timber as well as Brazil nuts; of the silk-cotton tree, which yields kapok as a stuffing material; and of the cinchona trees, whose bark the Incas learned to extract for its bitter quinine as a

The ocelot is a forest cat that hunts both on the ground and in trees. It ranges over suitable parts of Latin America and into the southwestern United States. (Willis Peterson)

treatment for malaria. Brazil got its name from a timber tree whose wood, called *braza* by the Indians, found immediate favor among cabinetmakers of Europe.

Swimmers in the Forest

During the wet season, no one walks through the forests on the flood plains, for the longest legs would not reach the ground. Instead, fishes and other animals swim freely among the tree trunks. Skates and rays come far upstream, and share in the food exposed on inundated land. The native freshwater fishes spread out too.

Some of these rain-forest fishes provide unusual hazards. Among the twelve whole families of catfishes, one species is especially dreaded. It is the candiru, a very slender creature about two and a half inches long that lives as a parasite on the gills of larger fishes. It shows an unexplained readiness to swim to and squeeze itself into the urethra of a man or woman bathing in the water, causing excruciating pain and necessitating expert surgery for removal.

Four kinds of voracious fish known as piranhas attack wading animals and quickly strip them of flesh. An electric eel, whose six-foot body is largely specialized for production of electric pulses as powerful as 600 volts, uses this means to immobilize its prey or to defend itself. Related fishes in these rivers use similar pulses of lower voltage as a means of finding food in the muddy water.

Crocodiles and caimans lurk at the surface of the rivers, with only their eyes and nostrils exposed while waiting for prey to come within snatching distance. One of the giant constrictor snakes, the anaconda, drapes its heavy body, as much as twenty-five feet long, over sturdy tree limbs that arch out over the water. Its victim may be a giant otter of the rain forest, measuring seven feet in length from nose to tip of the long flat tail. These otters prey on catfishes and other fishes, or whatever aquatic birds and mammals they can catch. Only the manatees, which are as much as thirteen feet long and weigh to 400 pounds, seem too big to be disturbed as they scull around in shallow water, feeding on submerged vegetation or reaching out to browse on low overhanging branches.

Perching Plants and Clinging Animals

Most of the plants and animals in the rain forest live far above the water and, in dry season, above the ground. Every horizontal branch is loaded with perching plants, many of which produce showy flowers. Orchids are prominent among them. Some have enlarged storage organs for water and starch, letting them tolerate a dry season and then grow and bloom when the humidity increases. Orchid fanciers learn to distinguish favorite genera among the neotropical kinds: the giant-flowered *Cattleya* species; the smaller members of *Epidendrum*, which Latin Americans call *pajaritos* ("little birds"); the various kinds of *Laelia*, *Masdevallia* and *Oncidium*.

Flower clusters often almost as brilliant as orchids grow on the perching bromeliads (family Bromeliaceae), which are close relatives of pineapples. Many of these perching plants cling to bark or high branches, holding great quantities of rain water in the whorled bases of their stiff leaves. Known as "tank epiphytes," they provide places for aquatic animals high above the ground. (Only one of these epiphytic bromeliads grows well outdoors beyond the neotropical realm. It is the Spanish moss, whose soft branching stems and silvery color are familiar in the southeastern United States as far north as coastal Virginia. Spanish moss will thrive when supported only by a telephone wire to which it has been blown by a gust of wind.)

The water held by the tank epiphytes slakes the thirst of animals among the tree tops. Frogs and toads use these miniature ponds as sites for eggs and tadpoles. Mosquitoes and other aquatic insects complete their life histories by accepting the limitations of the water in each tank. On rainy days and nights, small crustaceans often climb the trees and make themselves at home in the tank epiphytes.

High in the forest, the orchids and perching bromeliads vie for space on the limbs with ferns, mosses, leafy liverworts and luxuriant lichens. In combination, the plants perching on a single sturdy branch may weigh a ton. They hasten the day when the whole branch will break off and, with its load of life, crash to the ground. Often the event is spectacular, and the evidence

Above left: Few people see the bright pink face and forehead of the rare ukari monkey (Cacajao rubicundus) *that lives high in the rain forest of the Amazon basin near the border between Brazil and Peru. (Kurt Severin) Far left: Tree frogs of the genus* Atelopus *wear warning coloration matching their poisonous skin. This seems to save them from attack by reptiles, birds and mammals. (Manuel V. Rubio) Left: Golden conures and other gaudy members of the parrot family produce unmusical squawks that echo through the rain forest, however the birds themselves often blend inconspicuously into their tree-top environment. (Harald Schultz)*

clear enough to reconstruct the steps that led to the sudden downfall. Stains and channels in the fractured wood show where fungi, wood-boring beetles and termites weakened the structure. During a rainstorm, while the outstretched limb and its perching plants were drenched with an added load of water, the wood tore apart. Sometimes a giant tree that has declined in health due to fungus infections collapses all at once. Generally, as it topples, it knocks down several smaller trees in its path.

In dry season, these calamities in the rain forest benefit the wild pigs known as peccaries. For a few nights, they have an abundance of food among the broken branches and perching plants from the fallen trees. No longer must they search widely for fallen fruits, or for those the monkeys in the forest canopy have discarded half-eaten. But soon the forest floor is clean again, just bare slippery mud, perhaps shielded by a prickly tangle of wild pineapple. This bareness is sometimes attributed to the dimness of the light that filters down through the many tiers of foliage overhead. It is due too to the speed with which organic material, such as fallen leaves and wood, decay where the humidity and temperature are uniformly high. Moreover, the almost-daily rains (even in dry season) leach from the soil the soluble products of decay, leaving it a deficient mud, unsuited to the growth of most seedlings. Where flooding is frequent, low growth tends to be swept away by the current, and conditions do not favor its quick replacement.

Animals that climb well or fly find plenty of food far above the ground, both from the foliage, flowers and fruits of the trees themselves, and from the clambering vines that tie together the forest canopy. Some of these vines get extra mineral nutrients by extending long woody roots 150 feet or more to the ground; these are lianas. Few of the mammals, birds, reptiles, amphibians or insects of the tree tops have any reason to descend far. The monkeys and marmosets are well adapted to this high level, and represent families (Cebidae and Callithricidae) with no members elsewhere in the world. Five families of bats are uniquely neotropical. Nearly half of the birds belong to families encountered nowhere else. Many of the insects are equally distinctive.

Only in these rain forests do mammals of so many kinds have prehensile tails, used like an extra hand while climbing among the tree branches. The conspicuous monkeys, such as the howlers with their special sound-resonators, the black-capped capuchins, and the spider monkeys, are all experts at hanging by their tails or steadying themselves by wrapping this appendage around a support. The tree porcupines, which are rodents; the kinkajous, which are carnivores; the tree-dwelling anteaters, which are edentates; and most of the opossums, which are marsupials, are equipped in a corresponding way.

The little marmosets use their furry tails only in balancing as they run and jump. The big weasel-like tayras and smaller raccoon-like coatimundis climb agilely, switching their long tails to maintain equilibrium while foraging for birds' eggs, lizards, insects, fruits and other miscellaneous food. On the other hand, the slow-moving sloths have virtually no tail; they suspend themselves upside down below the branches while reaching out with utmost deliberation for edible leafy twigs. So well do a sloth's hooked claws support its body that it may hang in place for days or weeks after it dies.

Even some of the snakes seem especially equipped for clinging among the tree tops. The brilliant-green emerald boa suspends itself by its prehensile tail from tree branches and sways to and fro, ready to snatch an unwary bird in flight or a big lizard moving through the forest; the boa has unusually large teeth, as though to help it seize its prey; without much change in position, it can constrict the life out if its prey and swallow it whole. The larger and more famous boa constrictor, which often grows to be ten feet long but rarely longer, is tan and dark brown, making it blend with the shadows lower in the forest where it lies in wait for mammals. Constrictor snakes find their food by day, and are unique among snakes in having two lungs instead of only one. Unlike the pythons of the Old World, which lay eggs, boas bring forth active young.

Fliers of the Rain Forest

Neither the agility of the monkeys nor the camouflage of the lethargic sloths, which often resemble greenish brown masses of hay because of green algae growing in the grooves of their long, shaggy hairs, keep these mammals safe. The large, crested harpy eagle catches monkeys and sloths as its favorite prey. Rarely does the bird fly *over* the rain forest. Instead, it darts among the branches, dodging expertly and seizing what it wants to eat.

Probably most birds in the neotropical realm are equally specialists, since there are so many different kinds. Of nesting hummingbirds, it has 307 species, as compared with 12 in the United States and Canada, and with none elsewhere in the world. It has 365 species of tyrant flycatchers, whereas only 30 are known regularly north of the Rio Grande. It has all of the antbirds on earth—222 different kinds of them—and about an equal number of tanagers, while only four live in the North Temperate Zone. It has 215 kinds of ovenbirds in a family of their own, quite different from the ovenbird of North America (which is a warbler). Many of the smaller families are unique and striking as, for example, the 30 different kinds of toucans and toucanets with their enormous, colorful but lightweight beaks, used both for picking fruit and for signalling to one another. In structure and habit, each kind of bird fits a special level in the forest.

In darkness, more kinds of bats become active in the neotropical rain forest than in any other equal area of the world. Some find caves in which to hide by day, but a larger number choose sleeping sites in the trees. They cluster on the trunks between the great buttresses with which the forest giants keep themselves from toppling. The tent-building bat often uses its sharp teeth to cut a broad or fan-shaped leaf until the leaf folds over and provides a shelter where none would otherwise exist.

Most of the bats in the rain forest feed on fruit and nectar. Many are important to the plants, disseminating indigestible seeds or pollinating flowers. A few of the bats show extraordinary adaptations, such as the fishing bats which skim over the surface of streams and rivers, seizing fishes as much as three inches long without stopping, and the vampire bats which lap blood from warm-blooded animals whose skin they have slit open by a painless bite.

Moths that fly by day and butterflies that are active at night all have a place in the tropical rain forest. Owl butterflies (*Caligo*) may frighten away birds and lizards by exposing the huge brown eyespots on their underwings. Two different kinds of owlet moths, with wingspans of eleven to twelve inches, hold this record for size in modern insects. Some of the beetles, such as the Hercules beetle and related giant scarabs, are runners-up to Africa's goliath beetle as the heaviest insects known. The delicate and dainty are present too, in moths and butterflies with wings so transparent as to be almost invisible, and in the great family of heliconiid butterflies with myriad colors, seen at all levels as they fly on outstretched quivering wings or as they settle sociably in groups to sleep at night.

The View from the Forest Floor

Often, as we have walked and skidded on the floor of a great rain forest, we have tried to imagine what birds and other animals were making the sounds we could hear. Sometimes we could recognize the call of a toucan, the raucous voice of a macaw, the sweet whistle of a tinamou, the whir of a hummingbird's wings, or the whine of a big cicada. Yet our level of the rain forest seemed strangely empty, the animals all hidden by the dense foliage above. Actually, watchful eyes were everywhere, their owners motionless, whether predator or potential prey. We worried lest we step on one of the camouflaged pit-vipers, perhaps a fer-de-lance or the largest of them, the bushmaster, which is as much as twelve feet long. Rarely are these nocturnal snakes seen before they have a chance to use their deadly venom. Smaller animals remain equally inactive by day, relying upon protective resemblance to foliage, twigs, bark, bird droppings or animals that do have some way to defend themselves.

In these rain forests of the neotropical realm about a century ago, the English naturalist Henry Walter Bates discovered that many kinds of insects moved around in full view without being attacked. Certain of these were wasps, bees, ants armed with a stinger, stink bugs with a malodorous secretion, or kinds known to be poisonous. Many unarmed insects resembled the armed ones, and were equally shunned by monkeys, birds and lizards. In each case, Bates saw that the model the mimic matched so well was more numerous than the mimic. Insectivorous animals always were more likely to get the wrong one if they chose to sample the look-alikes.

In the same general part of the world a few years later, the naturalist Fritz Mueller sought to explain why the bad-tasting, poisonous and armed insects show more similarities in color, form and behavior than should be expected where so many different kinds live side by side. He concluded that this phenomenon was a mutual form of mimicry. So long as thousands of different species of bees and wasps bear similar yellow body hair banded with black, a predator has no trouble learning to leave untouched all insects with these markings. If each species had a distinctive pattern, the predator would need an almost endless series of lessons, and probably learn nothing at all. "Muellerian mimicry," as it is now called, benefits the unrelated look-alikes that are similarly armed. The only requirement is that members of each species retain some means for recognizing potential mates in the community, without becoming confused by the obvious similarities.

A few kinds of ants nest underground. Among these are the leaf-cutters, which use their sharp jaws to scissor off bits of foliage in trees, then carry the pieces to the ground and along well-worn trails three or four inches wide. The trails lead to subterranean passageways and chambers as big as a bushel basket. In these the foliage decomposes, becoming a compost. Special fungi grow on this material and produce nutritious nodules, which are almost the sole food of the leafcutter ants. Among the inherited features of these insects is not only an inherited program for cultivating the fungus, but also a habit shown by the virgin females before they go out on their mating flights, to stuff into little cavities like cheek pouches a generous sample of the living fungus with which to inoculate the compost in a new nest.

Other ants have gone into apparent partnership with trees, hiding and raising their young in small chambers reached through holes where the leaves are attached. If anything jostles a branch of the fast-growing, abundant trumpet-tree or an ant-tree, the ants pour out by the hundred, ready to bite, sting, and drive away the intruder. Still other ants, and many kinds of termites, build bulky nests high in the trees, and repair them after an anteater has ripped in to get a meal.

The legionary, or army, ants make no nest of any kind, and carry their young with them wherever they go. At night, and when their queen is laying tiny eggs, they bivouac on the forest floor. Otherwise they travel by day in columns a foot or more across, with scouts spreading out in all directions, foraging for any animal small enough to be overcome and stripped of meat. The food is shared along the marching column, and fed to the young that are being carried by the workers. Often the air overhead and any bushes nearby are full of neotropical birds known as ant wrens, ant thrushes and ant pittas, which are taking advantage of the ants' activity. Each insect that flies up from the ground is food for the birds, some of which seem to get meals in no other way.

The Gallery Forests

In much of South America, the rivers and streams are still the principal highway for travel at ground level. And from a boat a traveler may gain a false impression of the country through which he goes. Just as a modern parkway for automobiles is landscaped with shrubs and

On San Martín Island off the Peruvian coast, millions of the cormorants known as guanay birds pant, with beaks ajar, in the hot sun not far from a seated human visitor to their nesting grounds. (Rolf Blomberg: Full Hand)

82

trees that hide the towns and farms on either side, so too the rain forest extends along both sides of the neotropical rivers where they flow through savannas and grasslands. Through the soil the water spreads out enough to support a narrow belt of trees called a "gallery forest," which may be only ten to fifty yards wide. This is the realm of the cacao tree, on whose trunk and major branches inconspicuous flowers are followed by huge fruits; these contain seeds known as cocoa "beans"—the source of chocolate. Trees of lower growth also include the sapodilla, which yields chicle —the latex base for chewing gum—as well as edible fruits, and the calabash tree, whose large spherical fruits grow on short stalks from major branches. The fruits harden externally to become calabashes which, when dried and hollowed out, serve as containers for water and other materials.

The Andean Slopes

In most of South America, west is the direction to the great cordillera of the Andes. The continent is unique in having a continuous mountain range more than 4600 miles long. It extends from close to the Caribbean coast in Colombia and Venezuela all the way south to the tip of Tierra del Fuego. Volcanic activity in Ecuador has disrupted the pattern somewhat, producing the highest active volcano in the world, 19,487-foot Mount Cotopaxi. Pounding seas have breached the mountain ridge at the narrow Straits of Magellan. Yet for more than 50 million years, plants and animals have been able to spread north and south along the slopes as on no other continent. They could find tolerable climatic conditions by simply going higher or lower while progressing to other latitudes. Quite literally the Andes have offered a highroad for living things.

The effects of the mountains are simplest in Ecuador, and in Patagonia far to the south. There their crests form a single line, dividing the drainage of the continent into a western and an eastern slope. The direction of the prevailing winds can then determine which slopes receive the rain, and which get little or none. In the tropics, the trade winds come from the east, whereas south of the Tropic of Capricorn (which crosses close

The tree porcupine of tropical America uses its prehensile tail to hold itself in place, and can descend a small tree head first. Related porcupines in North America have short, quill-studded tails and must back down trees. (Lorus and Margery Milne)

to Rio de Janeiro), the westerlies blow from over the Pacific Ocean. But in tropical Colombia, the cordillera forks into three—a western, a central, and an eastern lofty ridge, with broad high valleys between. In Bolivia and western Peru it is double, bounding a great *altiplano* that extends as a semiarid grassland for 1500 miles southward before tapering into a jumble of peaks and narrow valleys along the western border of Argentina for another 1000 miles. Altitude, latitude and winds combine among these multiple crests to provide an amazing variety in climate. The living things seem to have matched or evolved their adaptations to each and every grade of weather.

Along the whole length of the Andean chain, the condor hunts for living prey. In this habit it differs from most of the related vultures. With a wing span exceeded only by that of the wandering albatross and a few birds in the Old World, this giant soars at equatorial latitudes between 7000 and 16,000 feet above sea level. In Patagonia it roosts on cliffs along the coast. Everywhere it is now menaced because of its fondness for fawns of deer, such as the brocket that people would like to hunt, and its acquired taste for the lambs of introduced sheep on ranches.

The largest of all flamingoes is the Andean, which congregates particularly in northern Chile, where the tiny crustaceans and other aquatic life they eat abounds in shallow alkaline lakes high in the mountains. This is a much larger species than the common Chilean flamingo or the James's flamingo of the Bolivian Andes.

Tiny hummingbirds venture on warm days as high as 16,000 feet. Many of them seem able to take shelter among boulders and in small open caves if the weather turns cold. They wait in semitorpor, conserving their reserves of fat as fuel, until the sun beats down again. Among these birds that brave the heights in search of nectar and miniature insects in flowers is the largest of all hummingbirds, well named the giant since it is fully eight inches long. Unlike most hummers, the giant beats its wings so slowly that they can be seen while it is hovering.

Between 7000 and 9000 feet elevation is the preferred territory for a much smaller hummer, the racket-tailed, which remained one of the world's rarities until its nesting territory was discovered to be the single small valley of the Rio Utembamba in Peru. This is the altitude, too, at which the spectacled bear wanders. It prefers the dense cloud forests, and is the most herbivorous of all bears, as well as the only kind in South America.

We have only to look at Ecuador and Colombia to see what feathered wings make possible, letting birds find whatever elevation suits them best between ground level and the tree tops and between sea level and the mountain peaks. Ecuador is larger than Arizona but smaller than New Mexico, whereas Colombia's size is between that of Alaska and Texas. Yet these two countries in their great range of habitats have more than half of the kinds of birds that nest in South America—more than 1350 kinds in Ecuador, and 1550 in Colombia. By contrast, the continental United States and Canada south of the Arctic Circle have only 691 kinds of nesting birds, which is about average for world areas of this size.

The Altiplano

Between the Andean chains near the boundary between Peru and Bolivia, the sloping semiarid grasslands of the altiplano drain toward Lake Titicaca, the highest of all lakes in the world. Its fresh waters extend 110 miles long and 35 miles wide; it is 12,506 feet above sea level and provides the greatest attraction for Indians in the high country. Around its shores grows a bulrush commonly referred to as the totara "reed," which the Indians use as material from which to weave their little fishing boats, as well as the sails to catch the wind.

Peculiar frogs of the genus *Telmatobius* swim in the cold waters of the lake. For them neither the ability to flip out the tongue to catch insects nor to breathe the thin air have any importance; as they develop from the tadpole stage, they gain neither a tongue nor lungs. They respire, instead, through a loosely baggy skin while swimming languidly or reaching with extraordinarily long fingers into the bottom ooze for particles of food. All members of this little genus live exclusively in the high Andes, where their extreme adaptations make them helpless on land. Yet their ancestors must have hopped up the mountain slopes, for *Telmatobius* is just one of the strange frogs in the large family Leptodactylidae, represented only in tropical America and Australia.

No one knows how fishes got into Lake Titicaca. Most of them have been there long enough to evolve into distinctive types, the most peculiar being about twenty different members of the unique tribe Orestiinae in the family of egg-laying topminnows (Cyprinodontidae). Having to compete only with each other and with a species of lake catfish, these topminnows diversified in their food requirements and reproductive habits until each kind held a niche of its own. Some are herbivorous, others carnivorous. Some mature at small size, others grow to be ten inches long. All are now threatened because lake trout from North America were introduced into Lake Titicaca about 1940. The success of the newcomers, which are predators, was at the expense of the orestiines, which seemed doomed. But as the number of native fishes diminished, the trout too declined. The Indians were deprived of an important food resource, one they had relied upon for untold centuries.

The Indians cultivate the land around the lake, which is the most fertile part of the altiplano. Its native plants are mostly grasses, such as ichu grass and various reed-bentgrasses and fescues. These and less obvious kinds of vegetation provide seeds that sustain many of the birds. The Peruvian song sparrow and the Andean miner are often the most conspicuous. The miner is a neotropical ovenbird that acts like a miner (or a bank swallow) in digging a horizontal tunnel into a hillside as the entrance passage to its nest chamber. The song sparrow generally chooses a more exposed site in a queñoa tree, a member of the rose family and the only woody growth on the altiplano. The little flightless grebes of Lake Titicaca, found nowhere else, build floating nests among the totara bulrushes. These habits keep the birds fairly safe from predators, such as the long-legged Andean foxes, which sometimes climb this high in the mountains.

A casual inspection of the bleak landscape in the altiplano reveals nothing that seems worth introducing to the rest of the world. Yet this is the native home of the "Irish" potato and of two other crops that the Indians cultivate: quinoa (of the goosefoot family) for its edible seeds, and oca (of the oxalis family) for its small starchy tubers. Neither of these exotic foods should be overlooked in a world that is hungry for nutritious plant products. In the Indians' gardens the potatoes are still marble-sized, and give no hint of the development that has been possible with this vegetable at lower elevations.

The Paramos and the Punas

At 12,000 feet elevation and above, there are subtle differences between the cool damp grasslands in Ecuador, Colombia and Venezuela, and the drier steppes with scattered clumps of grass in southern Peru, Bolivia, Chile and western Argentina. Something of the distinction can be seen in the nest-building habits of the two kinds of coots in these mountains. The giant coot of paramos in northern Peru finds plenty of vegetation around the little lakes where it congregates, and builds a nest of floating weeds. The horned coot of the puna country gathers pebbles and constructs a volcano-shaped mound in shallow water. Later it lines the crater top with dry pieces of flowering plants, which it air lifts to the site from considerable distance.

At these elevations, we envy the rabbit-like mountain vizcachas along the little permanent streams, and the Andean members of the camel family—the wild guanacos and the fleet vicuñas. All of these herbivorous animals have special adaptations for getting oxygen into their blood despite the low atmospheric pressure—half of what we are used to at sea level. They can engage in strenuous activity, such as running, which quickly leaves us breathless. We need no encouragement to move slowly while examining the living things, both animal and plant, in the neotropical highlands.

The Outer Slopes of the Cordillera

Above 12,000 feet, many of the mountainsides support no woody growth at all. Perhaps it is lack of trees in which to excavate a nest hole that induces the three kinds of Andean rock flickers on the outer slopes to nest underground; that they feed on the ground, eating chiefly ants and other insects, makes them obviously flickers. But that they find insects to eat is due largely to the fact that more rain falls on the outer slopes, facing the Pacific Ocean on the one side and the distant Atlantic on the other, than in the altiplano. At slightly lower elevations, streams are numerous. In the pools of deep water between one cascade and the next, the Andes harbor six different races of torrent ducks whose habits are scarcely known.

On some of the bare slopes, a tall member of the pineapple family attains a height of about thirty feet. It is the strange puya, whose flower stalk is clad in thorny leaves and the only conspicuous part. Bare-faced doves roost sociably in puya columns, and the black-winged spinetail builds its basket-like nest among the lower leaves. But this peculiar plant is also deadly to small birds as they fly close in the gusty air. If they

Above right: The kinkajou, or honey bear, although classed among the Carnivora, feeds chiefly on fruits, insects and other small animals in trees of the rain forest from southern Mexico to Brazil. (Willis Peterson) Far right: Cattleya orchids bear large handsome blooms in the rain forests of Central America. (Emil Javorsky) Right: A diminutive Jamaican tody, known locally as "Robin Redbreast," awaits a flying insect to dart after. (J. A. L. Cooke)

are tossed against its leaves, they are likely to be caught by the sharp spines, which are hooked and only work deeper as the victim struggles to get away.

At lower elevations on the mountain slopes, the Indians find various plants and animals of value. Long ago, in prehistoric times, they domesticated the short-tailed, plump-bodied little rodents that became known elsewhere as guinea pigs. These animals adjusted to living in primitive houses, eating vegetable scraps, keeping out from underfoot, never biting the children, and reproducing so prolifically that the housewife could usually find a one and a half-pound adult to skin and cook. These Indians discovered the food value in seeds of the native kidney bean and of the lima bean. They still relish products from two other native plants that have been developed horticulturally elsewhere: the tomato for its fruits, and the tobacco plant for the smoke obtained from burning its dried, cured leaves. To numb their hunger and to tolerate their fatigue on long trips down to market and home again, the highland people chew the leaves of the coca tree, which contains the narcotic cocaine. They shield themselves from the intense sun with "Panama" hats made from narrow strips cut and dried from the young leaves of the jipijapa, a palm-like vine. And they wash their hands with the foaming sap of the soapbark tree.

Lower still on the Andean west slopes, plants tolerant of a temperate climate include many that have been appreciated for decorative use on other continents. Botanists sometimes compare these slopes to Mediterranean coasts, and speak of Chile's "Mediterranean flora." It includes native nasturtiums, heliotrope, climbing Chilebells (Chile's national flower), the verbena-oil shrub, apple-of-Peru, painted-tongue, butterfly flower, several varieties of fuchsias, slipperwort and carpet plant. The balsa tree is a member of the forest community, yielding its extremely lightweight wood to many human enterprises, including the floating of Thor Heyerdahl's raft *Kon-Tiki*. In thickets, called

Above left: Damp sand along a jungle stream, such as a sandbar exposed for a day or less, attracts hundreds of butterflies. These (Phoebis *species) are among the most migratory, and were reported flying in clouds off the South American coast by Charles Darwin aboard* H.M.S. *Beagle. (Karl Weidmann) Left: Green and motionless one minute and easily overlooked, a tropical praying mantis may suddenly display its colorful wings to startle an approaching predator or to attract a mate. (Harald Schultz)*

carrascos, grows a climbing vine whose aqueous extract is the crude arrow-poison known as curare.

Variations in topography affect the climate too much to predict for any latitude the elevation on the mountain slope at which particular kinds of life will thrive. The most wide-ranging kinds are the tolerant ones. Anywhere between 700 and 10,500 feet above sea level may be suitable for the terrestrial snake *Leimadophis bimaculatus*, which reacts to any threat by spreading its body until its red skin shows between its scales. The false chameleon, which has a prehensile tail as well as sticky pads on its toes, and a sirenid salamander are both found as high as 11,400 feet. One of the gaudy frogs, whose relatives live mostly in the tropical rain forest, manages to reproduce in the Andes as high as 15,600 feet. These are all cold-blooded animals, unable to maintain their body temperature despite chilly weather. This makes us doubt that the Andean tapir follows its unusually long flexible snout to heights as great as 15,000 feet because it is the woolliest tapir on earth. More likely it tolerates the climate and, high up, is less subject to attack by pumas.

The Desert Coast

Compressed between the cordillera of the Andes and the Pacific Ocean, one of the longest, narrowest, driest deserts on earth extends from about halfway along the Chilean coast nearly to the Equator in Ecuador, a distance of about 1600 miles. In the tropical portion, this land is in the rain shadow of the mountains. South of the Tropic of Capricorn, where the prevailing winds come from the west, it is dry because the waters of the Humboldt Current are so cold; they limit evaporation into the air above. Yet the spray of breaking waves does produce a salty mist that drifts inland and settles. Without rain to dissolve the salt, the deserts are salty as well as dry. Often they show no plant growth except a few encrusting lichens year after year.

In June and July, the pattern of winds below the Equator changes enough to bring winter fogs known as *guaras* in Peru and *camachancas* in Chile. Their moisture sustains new growth on cacti, such as the one called torch thistle, which puts out extra branches and flowers from its single upright stem until it is about thirteen feet tall. Thorny shrubs, mostly *Acacia* and *Prosopis*, open finely-divided green leaves that can be dropped when the supply of moisture becomes critical. That day is put off a little as these members of the pea family (Leguminosae) extend their roots to greater depths toward whatever subterranean water may be waiting. Both of these genera have members in similar deserts on other continents, and employ the same adaptations wherever they live.

During their brief winter growing season, the plants are fed upon by insects, and these in turn by little lizards and a surprising variety of birds that migrate to these low elevations on a regular schedule. The Peruvian song sparrows and the miner birds arrive from the altiplano, the long-tailed mockingbird and some of the brightly colored small flycatchers from more moderate altitudes.

Along the shore line, birds that rely on fishing or scavenging congregate. Most famous of them are the guanays (or Peruvian cormorants), which nest on adjacent islands and leave vast quantities of droppings rich in nitrates. The guano that is now dug and exported all over the world as fertilizer is also contributed by nesting boobies, including the variegated and blue-footed, and by brown pelicans. All of these birds dive expertly, and consume unimaginable quantities of small fish, particularly anchovies. Fishermen are not allowed to compete vigorously with the birds, because the Peruvian and Chilean governments can see more gain in foreign exchange from sale of the guano to people in rich countries than from any market for fish.

At intervals that recently average seven years, both the fishing and production of guano collapse because of an unexplained change in the pattern of ocean currents in the Pacific. The cold Humboldt Current weakens, and warm waters from the latitude of Panama push down the coast of South America. Usually this happens around Christmas time, and coastal residents have come to call the disaster *El Niño* after the Christ child. Not only do the fish descend to cooler depths, out of reach of diving birds and fishing nets, but the weather takes on an unaccustomed form. Onshore winds from over the warm water drop torrential rains on the foothills just beyond the deserts, causing floods and tremendous erosion. The sediments carried to sea change the chemical nature of the coastal waters and make them so turbid that light cannot penetrate to normal depths. The drifting plants upon which the fish are feeding die without the light. Decay bacteria use up the oxygen, suffocating fishes that are already starving to death. Microscopic organisms called dinoflagellates, which are normally present and harmless in small numbers, abruptly multiply, turning the waters blood red as a "red tide." In this abundance, the dinoflagellates are poisonous to most kinds of marine life, including sea turtles.

The fishermen who are idled by El Niño have learned to expect such a calamity from time to time. Yet they know now that the change will not be permanent. The cormorants, the boobies and the pelicans that fly away will come back as soon as the Humboldt Current returns to normal. The Peruvian penguins that swim far from the coast to find cooler water with abundant fish will swim shoreward again.

Meanwhile scavenging birds come to clean up the rotting carcasses of fishes and turtles that the waves throw upon the shore. Black vultures squabble with turkey vultures and gray gulls. Franklin's gulls with black heads arrive, the only North American gulls that winter south of the Equator. Even the Andean condors may descend to the shoreline to feast when El Niño upsets the customary balance.

The Scrublands and the Savannas

Living things find more winning combinations of latitude, altitude and direction of prevailing winds than any gambler can hope for on the three dials of a slot machine. This trio of physical features controls the temperature variations and the precipitation month by month, which are the climatic realities that plants and animals must match or perish. Some species are tolerant enough to be present over a considerable range. But those whose special adaptations fit them best to a particular situation are the ones we tend to notice first. Often they are dominant in the local community of life.

In widely-separated areas, the neotropical realm has warm semiarid lands where cacti, thorny shrubs and armored animals that burrow seem to be the principal types of life. Largest is the scrubland just east of the Andes in northwestern Argentina and southern Paraguay, called the *Gran Chaco* from the Indians who live there. Others are in eastern Brazil and northwestern Venezuela, clad in thickets known as *caatinga* from two Indian words signifying white forests—the appearance of the trees while leafless. These scrublands receive too little rain from the erratic winds to support a more productive forest. Yet the moisture comes all through the year, not during one or two rainy seasons, and grasses that can go dormant have no advantage. Woody or succulent plants that resist desiccation, store water, repel herbivorous animals, and benefit quickly from each major shower by putting out fresh, expendable leaves and root hairs are most likely to succeed.

The Gran Chaco

Lying just south of the Tropic of Capricorn, the Gran Chaco is warm enough for cacti and many kinds of

woody vegetation to excel. Thorny trees are fewer, and rarely grow taller than twenty feet. Like the shrubby, spiny undergrowth, they stand leafless most of the time.

Animals that dig for food have an advantage. Among them are the smallest armadillos, each about six inches long when mature and weighing less than a pound. Chaco Indians who speak Spanish call these little animals the "pichiciego menor" (*Chlamyphorus truncatus*) and the "pichiciego mayor" (*Burmeisteria retusa*) —the "minor" and the "major"—and claim that both eat mostly ants. These midget armadillos are unusual in having their hind quarters protected by an especially large plate of armor, which becomes the only part exposed to the outside world if the animals are disturbed. They quickly dig a vertical burrow and remain in it, head downward, effectively plugging the opening they have made. The short tail of *Chlamyphorus* fits a notch at the lower edge of this large armor plate, and cannot be raised while the animal is walking. Instead, the tail drags along the ground and scores a continuous trail.

The largest of armadillos, the tatu, is particularly numerous in the Gran Chaco, although its range extends from northern Argentina to southern Brazil. This giant, which often weighs 130 pounds, seemingly subsists exclusively on animal matter, taking insects, spiders, worms, snakes and carrion.

Into the Gran Chaco from narrow savannas on the mountain slopes along its western fringe, predators come frequently into the scrublands to hunt. For some reason the savanna fox and the fox-like maned wolf are active primarily at night. We marvel that they can

The silky anteater has a prehensile tail slightly longer than its six-inch body and head. Sleeping by day, raking into termite nests for edible insects at night, this pygmy of anteaters is seldom seen. (Lorus and Margery Milne)

pounce on their prey, such as rodents and small sleeping birds, without tearing themselves on the formidable tangles of spiny bromeliads, the thorny shrubs, and the many cacti. It is not as though they had long previous experience with prickly hazards, as they might if they came from savannas on the far western side of the Andes; in comparable parts of Chile; there the trees are called monkey-puzzles (*Araucaria araucana*) because almost every surface of trunk and branches is covered by short, stiff, spinetipped leaves. On the eastern slopes the dominant trees of the savannas are the related Brazilian pine (*A. angustifolia*), named for its narrow flat leaves. It too is an evergreen conifer, but its trunk and branches are smooth after the soft leaves fall off. It also grows quickly, and then has all of its branches well above the ground, spreading out like the ribs of a parasol.

The Caatinga

Geologists recognize the scrublands clad in caatinga as eroded parts of an ancient plateau that once extended from southeastern Brazil to central Venezuela. Less eroded parts of the plateau are now grassy savannas, both in the Brazilian highlands and more isolated areas to the north. The broad valley of the Amazon

A termite nest built among the spreading adventitious roots of a fallen screw pine offers an attractive nest site to tegu lizards, which claw out a cavity and let the termites guard the eggs. (Emil Javorsky)

only one—the three-banded—can roll itself into an almost perfect sphere, with the flat top of its head and the flat top of its tail side by side, closing the gap where its fore and hind plates come together. The six-banded armadillo has a less flexible body, a longer tail, and a preference for running from danger. The nine-banded armadillo is the same kind as has spread into the United States since 1900. All of these animals dig burrows in which to hide by day, using their strong clawed feet, which are also well adapted for excavating roots and insects as food. Matching the gritty diet, the small peg-like teeth of armadillos have persistent roots and continue to grow as the exposed surface wears away.

The Pantanal

Adjacent to the Gran Chaco, right under the Tropic of Capricorn, lies a vast marshy territory in northern Paraguay, southeastern Bolivia and a part of Brazil. In this area of almost 98,000 square miles, the rains come in one continuous rainy season, and then cease. But the land is so nearly level in this broad upper basin of the Rio Paraguay that the water accumulates instead of draining off. In a week or less after the rains begin each year, the soil is saturated and the area becomes one great continuous wetland. The only way for men to get around is in small boats. Yet, when the rainy season ends, the water evaporates and the sun bakes the exposed land, killing most kinds of plants that cannot go dormant. Animals migrate to other regions, or dig underground for a period of estivation.

This schedule suits the program in life for local aquatic insects and for many kinds of amphibians. It attracts a wonderful assortment of waterfowl. Boat-billed herons arrive from their usual habitat in mangrove swamps along the sea coast. Whistling tree ducks come from the Amazon basin. Roseate spoonbills nest side by side with egrets, herons and bitterns. The wood ibis, which ranges northward to the Florida Everglades, is the most abundant marsh bird, sharing territory as well as one scientific name with the related jabiru stork. The American sun-grebe represents a little family of birds known otherwise only from the tropics of the Old World.

Giant rodents known as capybaras, weighing 110 pounds or more, swim and dive to reach favorite water vegetation, or hide among the floating plants with only their nostrils exposed. Sometimes they get stepped on inadvertently by spur-winged jacanas, which scamper over the lily pads on very long, widespread toes. The Pantanal matches perfectly the adaptations of the widespread lungfish, which waits out the annual

now separates these parts of South America and, while the flat plateaus still get enough rain for grasses and a few trees, the eroded country captures little moisture from the erratic winds.

Many of the cacti in the caatinga are tree-like, or resemble inverted candelabras in the style of *Cereus jamacaru* and *C. squamosus*. One, the xique-xique, grows vertically for a few years, then sprawls and branches in tangled masses which cattle will eat if the sharp spines are burned off. Cowhands, who set the fires to benefit their herds, regard themselves as *vaqueros* rather than as *gauchos*—the Argentinian's preferred name. They wear heavy leather clothing to protect themselves from the spiny plants.

Armored animals are much in evidence. Three different kinds of armadillos frequent the *caatinga*, but

drought in a mud cell, letting its urea accumulate in its blood. When the rains soak the soil, the lungfishes emerge, get rid of their urea, feed and breathe fresh air, then find mates and raise families.

The Pampas and Patagonia

South of the latitude of Buenos Aires and Montevideo, between the Gran Chaco and the Atlantic Ocean, the *pampas* spreads broadly as one of the world's most important temperate grasslands. Most of its grasses belong to the big genus *Paspalum*, and often grow in tall tufts. Curiously, the name "pampas grass" has been applied to two less valuable native grasses: one, *Gynereum argenteum*, forms high dense hedges along the borders of narrow winding watercourses, and the other, *Cortaderia selloana*, which ranges far into Brazil, waves great banner-like clusters of flowers and seeds ten to fifteen feet in the air over clumps of tangled leaves.

The English naturalist W. H. Hudson, who spent his boyhood and youth observing the wildlife on the pampas, made this region famous. It is the distant goal of countless birds of North American uplands and shores, including the golden plovers. They arrive in drab plumage, spend the months of northern winter, and depart again early, without revealing to the South American people the brighter colors and distinctive mating behaviors they will show on their northern breeding grounds.

The pampas plants serve birds from farther south as well. Migrants from Patagonia fly north after the travelers to the Northern Hemisphere have gone. There are always birds available to the gray foxes of the pampas and the weasel-like huróns. These predators vary their diet too with mouse-sized tucotucos and the occasional fawn of the pampas deer.

Over considerable areas of the pampas, the cattle ranchers have succeeded in replacing most of the native life with beef animals. Once there were far larger flocks of the plains rhea, five feet tall and weighing to fifty pounds, which is the lowland equivalent of the ostriches in Africa and the emus of Australia. Dwindling populations of plains vizcachas live colonially, burrowing underground but emerging at night to feed on the plants. The amount eaten by these sturdy rodents must be tremendous, for adult animals are eighteen to twenty-six inches long, with a six to eight-inch upcurled tail, and weigh as much as fifteen pounds apiece. Some of their burrow systems, known as vizcacheras, have apparently been in continual use for centuries. Burrow-

Fruits of the cacao tree in all stages of ripeness hang from the trunk and branches. Each fruit, when mature, provides up to two dozen seeds called cocoa beans. (J. Allan Cash)

ing owls, lizards and snakes often take shelter in the same interconnected tunnels, without seeming to unduly disturb the vizcachas.

In marshy places, a still larger rodent—the nutria (or coypu)—burrows into the banks or builds nests of vegetation on the shore or in shallow water. Like giant muskrats, weighing fifteen to twenty pounds, they vary their vegetable diet with frequent meals of freshwater mussels.

South Temperate Forests

Matching an increase in the rainfall during the drier parts of the year, the pampas grade southward imperceptibly into the forests of Patagonia. The largest of the evergreen trees are the Chilean arbor-vitae and an antarctic beech known locally as the coihue. Both belong to genera that are widespread in the temperate portions of the Southern Hemisphere.

Still farther south, the evergreens give way to de-

93

ciduous kinds of antarctic beech: the rauli and the nirre, whose broad leaves resemble more those of birch and oak. These trees have a greater tolerance for strong winds, for cold rain and sleet. Seldom, however, are they weighted down by snow. Where snow is frequent, the trees diminish, both southward and toward the Andes from which the cold night wind blows regularly all through the year. As the trees shrink, the grasses (chiefly *Agrostis* and *Poa*) take over between the rocks. At the Strait of Magellan the transition is almost complete, and on the clustered islands of Tierra del Fuego trees of any kind are scarce.

It seems strange to find among the antarctic beech forests one of the world's largest woodpeckers, the Magellanic, which is a close relative of the imperial woodpecker of Mexico, the ivory-billed of the southern United States, and the smaller (but still huge) pileated woodpecker of the United States and Canada, coast to coast. The Magellanic is just as particular about the trees in which it excavates its nest, rarely choosing a site less than fifty feet above the ground. To us it seems even stranger to discover hummingbirds and members of the parrot family in bleak Tierra del Fuego. Yet the Chilean fire-crown hummer ranges to this southern tip of the habitable world, as well as to altitudes as great as 6000 feet in the Andes. The Tierra del Fuego parakeet is the most southern representative of its family, flying in great flocks whenever the sun breaks through the clouds.

To preserve a sample of this remote part of the world, the Argentine government has set aside a national park bordering Nahuel Huapí Lake, near the village of San Carlos and both the Chilean frontier and 11,660-foot Mount Tronador. The forest there contains many magnificent specimens of the native timber tree called alerce, the Spanish word for larch, as well as of Chilean cedars. Dissatisfied that the park had only the Patagonian deer known as the huemul and the smallest of all deer, the pudu, the manager introduced both the European fallow deer and the red deer. The foreign animals are showy and doing well, but may crowd out the local kinds. Already this part of the world has lost the biggest of all armadillos, whose fossilized armor indicates a body length of at least fifteen feet.

The Bleak Coasts

At the same latitude as Nahuel Huapí, Chile has a natural sanctuary in big Chiloé Island, 110 miles long, where some of the trees (*Fitzroya patagonica*) attain a diameter of fifteen feet, a height of 180 feet, and an age estimated at about 2000 years. Some of the island is high enough to bear thickets of the same bamboos that grow at intermediate heights on the Andean slopes.

Nearly thirty different kinds of North American waterfowl spend the northern winter months on Chiloé. It is the southern limit for the Chilean slender-billed parakeet, but the northernmost part of the range for the Patagonian kelp geese. The kelp geese, males of which are white and females dark brown with white only on wings and tail, feed in the intertidal zone when the tide is out. They hop and skid with their webbed feet over the slippery exposed seaweeds, eating principally a leafy red alga known as laver in the Northern Hemisphere, where it grows in correspondingly cold waters.

Closely-related geese, all members of the little tribe known as shelducks and sheldgeese, live inland on Patagonia and the islands of Tierra del Fuego. There their grazing habits bring them into competition with introduced sheep. The upland (or lesser Magellan) goose, which is found on both slopes of the Andes, is like the kelp geese in having different coloration on males and females; the males, in fact, are white-breasted as well as white headed on the higher meadows, but white barred with black at lower elevations. Females are darker, and so similar in markings to both sexes of the smaller ruddy-headed geese that we have difficulty distinguishing them at any distance. The ruddy-headeds travel to Tierra del Fuego for the antarctic summer, and then north to a still-undiscovered destination as the weather worsens. The ashy-headed geese, both sexes of which are gray instead of reddish brown on head and neck, migrate similarly, but in the southern summer occupy a far larger range in both Chile and Argentina.

Near-shore waters around these coasts seethe with the flexible blades of coarse brown algae, whole sub-oceanic forests of kelps that are tethered by root-like holdfasts to rocks on the bottom. The upright, flexible stalks are often four inches or more in diameter, like lianas in the sea reaching up through as much as sixty feet of water to the flat leaf-like and strap-like parts near the surface. Floats filled with gas buoy up the kelps, the giant of which is the same one found in cold waters from California to Alaska and down the Asiatic coast of the Pacific Ocean. A slightly smaller kelp is limited to the Southern Hemisphere.

Among the kelp beds and farther off shore, crustaceans and small fishes and squids are so abundant that they attract great numbers of sea birds, particularly cormorants and penguins. Handsomest of the cormorants is the one called the king shag, which has an

orange tuft of feathers on its brow as well as bright red feet. The red-footed shag lacks the brow patch. These birds differ slightly in their requirements for nest sites and their methods of catching food, just enough to minimize competition with related neighbors such as the rock shag, the blue-eyed shag and the bigua.

Largest of the penguins that feed and nest along the coasts of South America's tip are the king penguins, more than three feet tall, which take advantage of most shores this close to Antarctica. Like the still larger emperor penguins of the polar continent, the kings make no nest, incubating their single eggs atop their warm feet, under a feathered flap of belly skin. By comparison the rockhopper penguin and the Magellanic penguin, only twenty-five inches tall, seem little birds. Both normally lay two eggs in a crude nest and lie, belly down, to incubate or brood their young.

Shellfish in shallow coastal waters attract the attention of the flightless Magellanic steamer duck, whose only near relatives are an equally flightless bird of the isolated Falkland Islands and a flying steamer duck that ranges northward and inland on rivers and lakes. All of these heavy-bodied ducks paddle with their wings as well as their feet when they rush over the water surface, calling to mind the old-style paddle-wheel steamships.

The stormy weather for which the tip of South America is famous seems logical when its latitude (55 degrees) is matched with places an equal distance north of the Equator, such as the Aleutian Islands off Alaska and Hopedale, Labrador. We see no reason for surprise when southern fur seals and South sea lions seek out Tierra del Fuego's deserted stony beaches and rocky points in season as pupping and breeding grounds. Sea lion bulls may weigh more than 1200 pounds when first they haul out in early spring, the bull seals only about half as much. Just as in the Arctic, the big males battle for territory and then wait for the females to come ashore to give birth and join the harems. Bachelor males wait around the periphery, and today are harvested in large numbers under careful supervision. Many of these animals are bought by the government of Uruguay as the source of an oil rich in vitamins, which is distributed to hospitals. The pelts are used in the fur trade, although they bring a lower price than those of the northern fur seals. A century ago, before furriers became so critical of quality, the demand for pelts from seals and sea lions of the Southern Hemisphere almost led to the extinction of these animals.

The Islands of the West Indies

Often we wonder what the European explorers found when first they came ashore on the various islands, large and small, in the West Indies. These were their early landfalls because they had sailed southwest from Spain and Portugal to the tropics so as to have the trade winds behind them in their search for a new route to the Orient. The islands and the people they found fell far short of expectations.

Their journals help us recognize one tree they learned to dread and shun: the manchineel, which provides inviting shade along and near the shore. Some of the men died from eating its poisonous fleshy fruits, which resemble green apples. Others developed great itching blisters where sap from the bruised leaves and broken stems dripped on their skin.

The Spanish arrived just as the Carib Indians from the Venezuelan coast were conquering the peaceful, agricultural Arawak Indians on the large northern islands of the West Indies. Already the Caribs had eliminated the Arawaks from the smaller, southern islands by taking the women prisoner and killing —generally eating—the men and children. *Carib* is the Arawak word for cannibal. Caribbean became the name for the whole area the Caribs raided in their dugout war canoes.

Today, a fuller knowledge of the plants and animals on the islands and on the north coast of South America allows fresh interpretations of the behavior of Arawak and Carib Indians. Linguistic studies show that the Arawaks were among the most widespread types of Indians in South America, with about ninety tribal groups over the area from the coast of Colombia and the mouth of the Amazon to Paraguay and Bolivia. Throughout this area the native Indians still rely upon manioc, (known also as yuca and bitter cassava) as their principal starchy food. They cultivate the plants, dig up the thick underground stems, and prepare them elaborately to get rid of poisonous substances. What could be more natural than for the Arawaks to take

live cuttings of manioc with them when they colonized the island chain all the way from Trinidad to Cuba and across to the tip of Florida? Apparently they introduced also from Brazil the peanut and the cashew tree, both of which have seeds ("nuts") which need to be roasted to reveal the delicious flavors. Today these plants are grown so widely that their neotropical origin is often forgotten.

With a knowledge of wild plants and how to use them, the Arawaks had plenty to eat wherever they went. But the Caribs who followed wanted meat. Once they progressed beyond Trinidad, which has most of the game animals and birds native to northern South America, the Caribs found almost no mammals to eat, too few birds and turtles, and an abundance of Arawaks. When the supply of edible Arawaks gave out, most of the Caribs disappeared.

Trinidad and the Adjacent Coast

In many ways the island of Trinidad is still a part of South America. It has the same high hills and narrow valleys, dry uplands and permanent swamps, and patches of rain forest much like those on the north coast of the continent. Yet its present separation from the mainland has lasted long enough for some kinds of life to disappear. Trinidad's only native cat now is the ocelot, whereas the coastal forest in Venezuela only fourteen miles away has little margay cats, the jagua-rundi, as well as the giant jaguar, which often weighs close to 300 pounds.

On the other hand, Trinidad has fishes found on the mainland but not elsewhere in the West Indies. Some of the small native fishes manage to survive in the island's famous asphalt lake, living in cracks that contain warm rain water. Trinidad's birds include members of fourteen neotropical families with no representatives on the other islands. Its thirty-four kinds of bats outnumber all of its other mammals; many of them fly no farther from South America.

In the great Caroni Swamp of Trinidad, as in remote backwaters of the Orinoco River in Venezuela, the mangroves with their arching adventitious roots provide nesting and roosting sites for the most spectacular of small herons, the scarlet ibis whose uniformly red plumage has led to the name "flame bird." We cherish the memory of one late afternoon when, with special permission, we accompanied the warden of the sanctuary area in the Caroni Swamp to witness the glorious homecoming of scarlet ibis at sundown. The first few arrived in the light of the setting sun, and seemed ill at ease. Then the great flocks came like clouds from the offshore sandbars where they had been feeding all day. Silhouetted blackly against the brightest part of the sky to begin with, they wheeled and the sun brightened their feathers. Blood-red, the birds circled between our vantage point and the hazy distant mountains beyond Port-of-Spain. Suddenly the fliers settled, clamoring noisily, shining like feathered rubies among the dark green foliage of the mangrove tops.

Trinidad shelters another kind of bird well worth going far to meet: the guacheros, or oil birds, which were discovered first nesting in the Caripe Caves of Venezuela—now a wildlife reserve set aside for the preservation of this amazing creature. Slightly larger than a sparrow hawk and with a stout hooked beak, the oil bird is a fruit eater that feeds at night and hides from daylight. Limestone caverns on forested slopes are their favorite hideaways, but in Trinidad they accept deep, dark clefts in the rock. Like bats they venture forth in twilight and return before dawn, able to find their way in complete darkness by echo-locating their position with high-pitched clicking sounds.

Young oil birds become extremely fat on the regurgitated palm fruits and other foods the parents bring back to the shelf-like nests. Local people have learned to raid the nests and to render the young birds for the oily fat they contain. Ironically, the oil from these birds of darkness is used in place of kerosene to burn in lamps to light the homes of the poor.

Missing from Trinidad but still numerous along river margins in northern South America are some of the strangest birds in the world: the pheasant-like stink-birds or hoatzins. Like no other living birds, nor any known since the days of the dinosaurs, they hatch with two functional finger-like claws on each wing. For the first few weeks, until the claws disappear and the naked young birds grow feathers, the hatchlings crawl about actively like lizards in the branches near the nest. If they drop from the overhanging shrubbery into the water, they swim to the nearest bush and clamber up it, clinging with claws and toes as though four-handed. Yet by the time the hoatzin is mature, its wings have lost all indication that once they served for climbing. The stinkbird flies to reach its favorite food, which is the leaves of plants in the arum family (Araceae). These have roots spread out in the soggy mud along the river margins, while their pointed foliage rises into air and within easy reach of the birds.

Surinam Toads, Paradox Frogs and Million Fish

Efforts to drain or fill in the swamps and marshes of Trinidad and adjacent South America may endanger

A young wood ibis, not yet fully feathered out, teeters on its heels instead of standing erect. These gregarious American storks range from southern United States to most of South America and a few northern islands of the West Indies. (Harald Schultz)

native creatures of particular renown. They include some of the world's most aquatic amphibians, toads and frogs that rarely venture onto land and that feed on worms or insects among the submerged vegetation. One is the tongueless Surinam toad, about which nearly everyone hears because the female waits patiently while her mate spreads her fertilized eggs over her back and keeps them there until her skin crinkles up and holds the eggs securely. Within the jelly of each egg, the embryo goes through its tadpole stages and transforms into a miniature toad before escaping from the protection of its mother. Meanwhile she moves about, like a living perambulator.

We searched out some of the equally amazing paradox frogs which, after growing to be enormous tadpoles

97

fully ten inches long, transform and mature into exceedingly slippery frogs while shrinking surprisingly in size. Never again will they be longer than about two inches. Local people cannot believe that any animal would grow smaller to this extent. They prefer to accept as truth the fallacy, or paradox, that the little frog turns into the big tadpole, instead of the other way around.

In many of the same marshy places, swamps and streams on Trinidad and along the coast of northern South America, beautiful little topminnows propel themselves by waving their long streamer-like tails. Known as mosquito fish because they eat so many insects at the water surface, they caught the eye of the Reverend John Lechmere Guppy in 1866 when he was visiting Trinidad. He took home a few live ones in a jar to share with friends who had a liking for jewel-bright tropical fishes in heated aquarium tanks. Soon the two-inch mosquito fish became popular as the "guppy" or the "million fish" because, according to the publicity, one pair can be the ancestors of a million within a year. A female guppy, after mating once, can bring forth several broods of active young in succession. Each brood of twenty to fifty young follows the preceding after just a few weeks. The young mature quickly and soon have families of their own. The population of guppies does increase at an amazing pace unless predators or diseases or lack of food puts a limit on the number of survivors.

Guppies swim easily in waters that are almost clogged by floating masses of the native water hyacinth, which has gas bladders in the stalk portion of each leaf. Its roots dangle freely, with no need to reach into soil. The giant water lily, by contrast, has its log-like stem embedded in the bottom mud, its circular leaves floating in the water surface. A strong meshwork of veins supports each leaf, which may be six feet across, and sharp prickles below protect it from attack. A thin, raised rim keeps its upper surface dry. So buoyant and firm is a big leaf of this water lily that it will support a small child safely.

The Antilles—the Lizard Islands

North of Trinidad on the islands of the West Indies, the influence of South America is much less obvious. Native mammals are either few or missing entirely. The number of kinds of nesting birds is small in comparison with the variety of migrants that spend the winter months on the islands or that pass twice a year between breeding grounds in North America and places farther south. Only one little family of birds, that of the todies,

is restricted to the islands, with a unique kind in Cuba, another in Jamaica, two in Hispaniola (one a mountain dweller, the other nearer sea level), and still another in Puerto Rico. Although related most closely to king-fishers, todies are flycatchers. They dart out from bare branches to pursue their flying prey, and suddenly flash brilliantly in the sunlight. Not much bigger than wrens, they have vivid-green upper parts, a bright-red throat, and various patterns of yellow or pink or blue on chest and flanks.

The difference between Trinidad and the islands to the north is not in space alone, for several of the islands are bigger than Trinidad, which has only about 2000 square miles. Cuba, at the opposite end of the archipelago, is around 40,000 square miles, Hispaniola (with Haiti and the Dominican Republic) about 30,000, Jamaica 4500, and Puerto Rico 3400. These four constitute the "Greater Antilles." All of the more southerly islands, the "Lesser Antilles," are much smaller. But on the smaller islands and the big northern ones too, snakes are rare and lizards common. On the four big islands, for example, there are about three kinds of lizard for each species of snake, whereas on Trinidad (and the South American mainland and Central America) the snake species outnumber lizards two to one.

While exploring Jamaica, we came to realize that virtually every square foot of ground, every tree limb, every bush is part of some lizard's territory. The lizard may hide from a human being, a domestic animal or a bird, but it is ready at any hour of the day to bluff or battle with any lizard in its size range within the boundaries it patrols. There only the lizard, and perhaps a mate, may freely feed on the insects, worms and other food.

The whole community of living things on these islands is far simpler than on the mainland, and more vulnerable to change. In fact, the smaller the island the fewer are the opportunities for animals and plants. Dr. E. R. Dunn, who studied the reptiles and amphibians on the islands north of Trinidad, suggested that for every tenfold increase in the area of an island, the number of kinds of these animals it can support approximately doubles.

Warm-blooded Animals with Nowhere to Go

Many of the individual islands in the Antilles have distinctive kinds of warm-blooded animals, or local non-migratory subspecies of birds found elsewhere. Not counting the subspecies, a tally shows twenty-eight unique kinds of birds on Jamaica, twenty-three on

Cuba, eighteen on Hispaniola, nine on Puerto Rico, but no more than four on any island in the Lesser Antilles. Those restricted to Jamaica include the tody, three members of the parrot family, three different fly-catchers, two of thrushes, two of the pigeon family, a brown owl, and two distinctive hummingbirds, one of which is the wonderful streamertail known locally as the "doctor bird." The male of this hummer has two long tail feathers that whir as he flies, adding an extra sound to that of his wings.

For various reasons, a good many island birds have become rarities. Unless given special encouragement, a number more seem likely to slip into extinction. Fortunately, the smallest of all hummingbirds—the bee hummer of Cuba—is not yet on the danger list. Cuba's red macaw vanished about 1885; the threatened bird life there now includes the Cuban hook-billed kite, Gundlach's hawk, Fernandina's flicker, and both the rail and the wren that are limited to the Zapata Swamp across the island from Havana.

Jamaica today seems more hospitable to birds than in the past, for none of its distinctive kinds is in obvious jeopardy. The island lost its green-and-yellow macaw early in the nineteenth century, its last pauraque in 1859, its diablotin (a petrel that nested in the mountains) in 1880, and its wood rail in 1890. The Hispaniolan diablotin (another mountain petrel) is now threatened, but the big island has lost only a little parrot called a conure, which it shared with Puerto Rico, in 1892. Puerto Rico's bigger parrot is now in danger, as is the island's unique whippoorwill.

The largest parrot in the West Indies is now the imperial, a twenty-inch bird of Dominica. Today it is a rarity, one that we sought in vain to meet. We sailed southward to St. Lucia, and there succeeded in making friends with a handsome eighteen-inch St. Lucian parrot which watched us with orange eyes while we admired the violet-blue feathers on its head, its red chest patch and maroon undersides, its green back and blue tail. On that particular day, we could see from the St. Lucian mountain top the French island of Martinique immediately to the north, and the British island of St. Vincent in the opposite direction. We wondered why the St. Lucian parrot made no visits north or south, why the St. Vincent parrots stayed home, and why none of any kind lived on either Martinique or Guadaloupe, just north of the island of Dominica. Later we discovered a satisfactory answer: Guadaloupe's parrot and a conure and a red macaw were exterminated early in the eighteenth century, its violet macaw back in 1640, its distinctive burrowing owl in 1890, and its wren in 1914. Martinique had a parrot too until early in the eighteenth century; it lost its macaw in 1640, and a house wren in 1886.

Ever since man arrived, the Antilles have been dangerous places for animals that had no way to escape. During the past two centuries, more kinds of warm-blood animals have become extinct in the West Indies than on all of the continents of the world in the last 2000 years. Jamaica once had monkeys. It still has little rodents, called hutias, the size of a cottontail rabbit, equally vegetarian and extremely elusive. They have living relatives on Cuba and Hispaniola. Cuba has a rare little insectivore, *Solenodon cubanus*, with coarse fur and a long snout; its only surviving kin, *S. paradoxus* of Hispaniola, is equally scarce and slow moving, nocturnal in habits, and threatened by introduced beasts of prey, including dogs, cats and mongooses. Hispaniola and Puerto Rico once had ground sloths, for their skeletal remains have been found in caves, associated with fragments of pottery, pig bones and pieces of human skeleton. One (*Parocnus*) on Hispaniola may have weighed close to 200 pounds, as the largest native animal known from the Antilles.

The Green Antilles

The general lack of fruit-eating mammals other than bats on the islands has apparently influenced the types of plants that are native in the Antilles. Almost none of them produce fruits that need to be dispersed by four-footed animals. The trees and shrubs are of kinds, instead, that ripen small fruits attractive to ants or birds, or buoyant enough to float, or light enough to be distributed by wind. By contrast, on Trinidad and the mainland of the neotropical realm, there are plenty of monkeys, pacas, agoutis, native rats and squirrels, and plants with fruits that can be carried away.

Despite these limitations, the West Indies gave the world such favorite flowers as fragrant night jasmine with tubular yellow blossoms, the showy lantana with its flat clusters of small flowers that change from pink to pale yellow as their day of opening continues, and Spanish bayonet. The unripe fruits of the pimenta tree can be cured and dried to make allspice. Only the perishability of custard-apple and soursop (or guanabana) keeps them from being enjoyed more widely. Hogplum (or yellow mombin) and the fruit of the climber known as Barbados-gooseberry are delicious when fresh, whereas anchovy-pears are more commonly preserved to serve as pickles. Some of the trees yield wood of spectacular hardness. Jamaicans call one breakax (or ironwood). Others are lignum-vitae, the wood of which sinks in water and is used in construction of wear-resistant bearings. A resin from these trees is gum guiaicum, long used by detectives as a test for stains that might be blood, since the powdered gum changes color when oxidized by hemoglobin.

Central America and Mexico

Any good map of the lands that connect South America and North lets us recall experiences we have had there while camping, hiking, riding on horseback or based in our own expedition car. Yet seeing reality shrunk to a single page charted in colored inks lets us gain a different perspective. We can think of the seven countries from Panama to Mexico as parts of a single exciting bridge between two larger, different worlds. It is exciting because it has its own personality, for it is no mere corridor for traffic in two directions. It reminds us of some medieval bridges in the Old World, such as the Ponte Vecchio in Florence and street bridges in Kashmir, where people live to take care of their own little shops right above the river in the middle of a span. The land bridge between the two big Americas has many native residents for whom the continents at the two ends have no modern significance, since their ancestors came from one or the other so long ago.

The only warm-blooded animals we met along the land bridge, and which live free nowhere else, are a bat and a trogon. The one is the sheath-tailed bat, known generally as the El Salvador bat because it does not fly beyond the isthmian parts of Latin America. The bird is the glorious quetzal, for which we hunted diligently, determined to glimpse its magnificent feathers somewhere in its range between western Panama and Mexico's eastern state of Chiapas. It is a creature of misty cloud forests on mountain peaks of medium height, where there are tree holes in which to nest as well as enough insects and fruits to eat.

It was the male quetzal that earned this trogon a sacred place among the ancient Aztecs and Mayan Indians. He is brighter green than his mate, wine-red below, and wears enormously long peacock-green feathers just above his tail; they stream out behind him and shimmer as he flies, and hang down, slightly twisted, when he perches, making him nearly four feet long from beak to the tips of these wonderful ornaments. By contrast, his mate seems to have hardly any tail at all, and her under parts show blackish where he is red. This was the feathered part of the old bird-serpent god of peace, Quetzalcoatl. It is still an emblem

of Guatemala and its name is the unit of currency. But Guatemalans have cleared all of the places where quetzals used to nest, and now have none. Salvadoreans may have lost these birds too. Now the quetzals is an endangered bird, restricted to a few parts of Mexico, Honduras, Nicaragua, Costa Rica, and Panama.

While we searched and, succeeding in our quest, watched motionless as pairs put on nuptial displays of acrobatic flying, we kept an eye out for the other half of the bird-serpent god—the *coatl*, which is a short, squat pit-viper as deadly as a rattlesnake. It too lives in the cloud forests, but on the ground. Fortunately, like other pit-vipers, it is abroad mostly at night when it can find frogs and warm-blooded prey asleep, distinct from their surroundings because of their difference in temperature. Its pit organs, one on each side of the head between nostril and eye, let the snake get into perfect position for a successful strike without alarming its victim or giving the slightest warning.

Friends had asked us to find for them some of the cold-blooded animals that are equally restricted to the land bridge between the Americas. In rivers we were to seek the river turtle, a large species that is eaten in Guatemala. Two kinds of little three-keeled mud turtles with lengthwise vanes atop a flat eight-inch shell proved more elusive, for nothing is known of their food habits. Three different kinds of local lizards were on the want list, for few have been seen alive by scientists; they have two sizes of scales, the commoner small and granular, embedded in the skin, the others larger and flatter, arranged in a chevron pattern along the back. Rarest of all were the eight-inch skink-like, blind, earless, and virtually legless lizards, which our friends hoped we could locate; these lizards have been found only three or four times, usually under a rotting log and by someone who was hunting for quite other quarry.

Our explorations turned up a burrowing toad in southeastern Mexico. It is one of the few amphibians restricted to this region and adjacent Guatemala, gaining special attention because its tadpole has pointed barbels around its mouth, making it resemble a two-inch catfish. The most spectacular tree frog we ever met ranges from central Mexico to Panama; unless disturbed, it hides its bright red eyes by day under lower eyelids that, while green like the frog's body, have about a dozen little transparent windows through which the frog can watch for danger. The underside of this frog is brilliant orange, and its flanks—hidden while it crouches on a tree limb—blue with yellow stripes.

The slow rivers, where turtles might lurk, often show multiple V-shaped wakes where four-eyed fish cruise

Jamaica   Hispaniola   Puerto Rico

Highland Savannas

Scrublands

Rain Forest

Andes Mountains

Desert

Pampas

Savanna

Temperate Forest

Altiplano

Lake Titicaca

Gran Chaco Pantanal

Falkland Islands

South Georgia

along. They travel with the water line crossing their prominent eyes, while they watch events in air through one pupil and whatever happens in the water through the other pupil of each eye. When we maneuvered these fish into shallow water, they readily flipped themselves across muddy shores and sand bars to elude us. In deeper water they dive to safety. These four-eyed swimmers bear active young, instead of laying eggs, but cannot mate at random. Males are slightly twisted,

right or left, and so are females. A right-twisted male can mate with a left-twisted female, or a left-twisted male with a right-twisted female. Other combinations are unsuccessful. Fortunately the population of each sex is about equally divided in its asymmetry, and four-eyed fish are in no danger of becoming extinct.

When we investigated Lake Nicaragua, which is 96 miles long and 39 miles wide, we stayed strictly on land and took no chances with the man-eating sharks

101

Short-horned grasshoppers flying from savanna land sometimes alight on flowering vines that lace together the top branches of trees in the rain forest. (Rolf Blomberg: Full Hand)

that live there. These voracious fish feed mainly on the bottom and grow to a length of eight to ten feet. Ichthyologists have not yet decided whether the differences between the Lake Nicaragua shark and the cub shark of the Atlantic Ocean, the Caribbean Sea and the Gulf of Mexico are due to thousands of years of being landlocked in fresh water, or merely to a few consecutive months in this habitat. It may be that the supply of man-eaters in the lake is replenished almost yearly when the San Carlos River, which drains the lake, is in flood. At other seasons the rapids and falls along the seventy miles of river would seem impassable barriers to a sea-going shark, even to the cub shark which seems tolerant of low salinity.

Native Plants of Middle America

So many of the trees and lower kinds of vegetation on the land bridge between the Americas range into the Amazon rain forests or the deserts farther north that the most limited species must be sought out to be appreciated. They are numerous enough, mostly because geological changes in the past have repeatedly inundated the isthmuses of Panama and Tehuantepec,

isolating Central America ("Middle America") as an island. Dry climate of long duration may have isolated parts of Mexico too, for some of the plants north of Tehuantepec are native nowhere else. Today, with so many additional species that have spread into Middle America from still farther north and south, this part of the neotropical realm shows striking richness—one of the most varied floras to be found in the whole world.

A surprising number of outdoor and indoor garden plants originated along the land bridge. Zinnias and "French" marigolds, cosmos and dahlias are all members of the daisy family that came from there. The area gave the world poinsettias and shrubby fuchsias, wandering jew, the climbing ceriman (or hurricane palm) with its big, slotted and perforated leaves and edible fruit, the tigridias of the iris family, scarlet-flowered salvia, and frangipani (or temple tree), with its perfumed flowers. The nearly leafless coral plant with its red, tubular flowers, and coral vine, which seems too showy to be a member of the buckwheat family, are native to this part of the neotropical realm. So are avocados (or alligator-pears), scarlet runner beans, two kinds of sarsaparilla, and a broad-leaved climbing orchid whose slender fruits yield true vanilla flavoring.

Although all of these plants are now cultivated far from tropical America, vanilla fruits (known as "beans") are still collected by Indians from wild plants in Mexico. Near Veracruz, the bean hunters bring their trophies for payment according to quality. Then the beans are cured, graded and packed for shipment. Those of the highest value sparkle in the sunlight with tiny crystals of vanillin. The fragrance in the curing sheds gave us a heady sensation, which led to questions and our personal discovery of *creme de vainilla* liqueur.

Other native orchids are appreciated most for their beauty. Guatemalans have chosen *Cattleya skinneri*, sometimes called the white nun, as their national flower, and Costa Ricans the Turrialba orchid. Panama honors similarly the dove plant (or Holy Ghost orchid), in which the central white flower parts resemble a little dove with wings upraised, poised behind a cathedral pulpit.

Much of the world has benefited because this part of the New World tropics had maize to share. Of all the trophies the Spanish conquistadores took back to Europe, none had more lasting value. Yet long before the discovery of America, the Indians of Latin America knew maize and developed it according to their liking. Today no wild ancestor of the plant can be found, unless it is the annual grass known as teosinte (or centeotl), which is now harvested as fodder for cattle introduced from the Old World.

Mingling along the Land Bridge

The two-way traffic along the land connection between South America and North is in spread of species rather than travel of individual plants and animals. But progress is evident in the distances reached by pioneers that have settled far from the mainland population. From South America, two of the mammals—the "Virginia" opossum and the nine-banded armadillo—have extended their range thousands of miles in recent times, the one to Canada and the other to states from Kansas to Florida. One kind of tapir and one kind of howler monkey are now found as far as Mexico's state of Veracruz; the spider monkeys have spread to the state of Chiapas, the night monkey to Nicaragua, the capuchins only to Costa Rica, while the squeaky-voiced little marmosets have gone no farther than western Panama. One of the two-toed sloths has crept laboriously from tree to tree all the way to Nicaragua, the slower three-toed sloths (Bradypus) to Honduras. The silky-furred pygmy anteater now rips into termite nests high in forests of southern Mexico, as does the larger collared anteater. The terrestrial giant anteater, which sometimes weighs more than fifty pounds, waves its big bushy tail in open country and swamps even in British Honduras.

Where the land bridge is narrowest and now cut by the Panama Canal, a permanent biological observatory is maintained on Barro Colorado Island, a forested hilltop full of native plants and animals, situated in Gatun Lake, the high-level part of the canal system. The hilltop was isolated by the rising waters of Gatun Lake when the canal was built, and has been administered as a sanctuary for all living things almost ever since by the Smithsonian Institution of Washington, D.C.

From visiting Barro Colorado Island for three months we know that its fourteen square miles support several tribes of howler monkeys, many capuchins, a few tapirs, some sloths and anteaters, a bewildering assortment of birds, the fer-de-lance pit-viper, some caimans, tree iguanas, insects huge and small, chiggers in abundance, and a natural botanic garden of neotropical vegetation. To our surprise, we found the "Virginia" white-tailed deer there, and learned that it

has spread to northern South America, running along the forest edges and swimming narrow bodies of water to reach new territory.

Many northern animals have used the land bridge, lured on by mangrove swamps at sea level, rain forests, mountain meadows and semideserts. Placental mammals took this route many millions of years ago, and almost replaced the marsupials that previously had had the South American continent to themselves. A few shrews have colonized Guatemala and Honduras, but the western mule deer and various kinds of moles have reached only northern Mexico and still have no part in the fauna of neotropical countries. Crows, by contrast, have conquered South America to its very tip, apparently in the last few thousand years, whereas chickadees stop at the isthmus of Tehuantepec. We are astonished that these two kinds of birds differ so much in their ability to travel south, for both of them entered North America from Asia during the Ice Ages, probably while a land bridge and food were available between Siberia and Alaska.

For birds to emigrate from North America to South —not merely to migrate there while winter affects the Northern Hemisphere—strikes the biogeographer as somewhat ridiculous. South America is the bird continent, with more families and species than can be found on any equal area elsewhere. Yet hardy pioneers can make a place for themselves, taking food and nest sites that neotropical birds cannot adequately defend. In this dynamic way, through emigration and resettling, the living things of the whole world are constantly readjusting their distribution.

Many lasting changes have already come to the neotropical realm. New challenges are sure to come in the future, just as in other parts of the world. Yet only on the fringing islands are the living things so few in kinds and so vulnerable through special adaptations as to be seriously threatened. Inner inheritance and open water may prevent them from finding elsewhere the conditions they require for survival. On the mainland of Latin America, from the deserts to the mountain peaks and into the lushness of the tropical rain forests, the opportunity for progress is greater. By changing their addresses to take advantage of the different combinations of climate within reach, the plants and animals gain the time they need to develop new tolerances and maintain a hold on the future.

103

5

Life
on All Sides
of the
North Pole

Prehistoric human colonists came across from the Old World to the New at a point where the two are closest. From the Chukchi peninsula of Siberia to the Seward peninsula of Alaska, the Bering Strait is only seventy miles wide. It may have been narrower when people of Mongoloid stock crossed during the Ice Ages, perhaps 40,000 years ago. They were adaptable pioneers, used to the conditions in Siberia. They spread and changed their ways to match the country they found. Some stayed along Arctic coasts as Eskimos. Others traveled over North America, or through it and to the farthest reaches of Tierra del Fuego, or the West Indies, establishing tribal territories.

How soon these first citizens to settle America managed to establish the Beothuc tribe in Newfoundland or the Acadians in the Maritime Provinces of Canada may never be established. The Viking explorers who visited briefly between 900 and 1000 A.D. saw no resident people and no signs of their presence. Nor did the community of trees and wildlife along the shores seem different enough from that in Scandinavia to be worth establishing a colony there. Yet the route of the Vikings, touching the tip of Greenland on the way, is actually the shortest between Europe and eastern America. The first transatlantic passenger planes flew from Shannon airport in Ireland to Gander airport in Newfoundland for that reason.

The trips Christopher Columbus made were twice as long. On each of his four expeditions in search of a route to the Far East, he set out from the Canary Islands and crossed about 4000 miles of open water to reach the American tropics. He could head westward no more directly because he needed propulsion by trade winds behind him. Better ships were necessary before men could fight the northwesterly winds of higher latitudes and found colonies from Virginia to Quebec.

The naturalists who came on these expeditions in the 1600's and 1700's noticed a wealth of plants and animals that were familiar from temperate parts of Europe. They marveled to see rich resources of kinds of life that had been so largely destroyed in their homelands: abundant fishes, such as cod; fur-bearing animals, including beaver; edible products from the land; and an apparently endless forest full of magnificent pines and oaks. Only a few types of animals, such as muskrats and horseshoe crabs, were completely new to Europeans.

As the frontiers were extended northward in the New World, the naturalists encountered still greater similarities with the flora and fauna of northern Europe. A great forest of spruce and fir, interspersed with dark peat bogs, is limited on the north by a tree line (or timberline) beyond which the weather is too severe to permit upright growth of woody vegetation. In America from Labrador to Alaska this boundary crosses the continent, just as it does Eurasia from Siberia to Scandinavia. Beyond it to the shores of the Arctic Ocean is a great barrens—a circumpolar tundra. The spruce-fir forest, which Russians call the taiga, is referred to as circumboreal, a name honoring Boreas, the classical god of the north wind.

The distribution of tundra and taiga around the North Pole follows a pattern that has no counterpart in the Southern Hemisphere. To some extent this reflects the fact that the northern continents extend far into the polar region, whereas the southern ones (except Antarctica) stop short of the Antarctic circle. The northern coasts of Greenland and of Ellesmere Island come within 350 miles of the pole itself. Yet, for a few weeks each summer, the ice along these shores usually breaks up, exposing open water. Seaweeds and other marine plants grow rapidly in the continuous daylight, nourishing a host of animals.

The low sun accomplishes this annual greening of

A bull moose wades to reach plants in shallow water near spruce forests of the Far North. (Willis Peterson)

Left: The receding tide on the southeast coast of Sweden allowed large rocks to push up through the translucent ice. (Ingmar Holmåsen) Above: The polar bear, an expert swimmer, hunts for seals and fish it can out-maneuver in the clear water of Hudson Bay. (Fred Bruemmer)

the Far North with the help of an ocean current—the Gulf Stream. This current brings warmth from the tropics into the Arctic Ocean between Greenland and Scandinavia at a rate that gives the north polar regions a touch of summer for which the Antarctic has no equivalent, even though antarctic coasts extend in several places into the temperate zone.

From the tip of Florida, where the Gulf Stream begins, to the nearest point of the European continent —appropriately named Cape Finisterre ("end of land") in northwestern Spain—is over 4100 miles. The west coast of Scotland is farther still. Yet at 58 degrees north latitude, where the Highlands come close to the Irish Sea, the Gulf Stream keeps the weather mild enough

for introduced palms and other plants of the tropics to grow outdoors. Ireland is the "Emerald Isle" for this reason.

That any coast can be virtually frost-free at a latitude so far north of the Equator shows how extremely irregular is the correspondence between climate and distance from the tropics. It impresses us all the more because, north of the fortieth parallel, any level land annually loses more heat by radiation to outer space than it can gain from the sun. Only the vagaries of water currents and winds save places north of Madrid, Peking, Denver and Philadelphia from becoming progressively colder until no life at all can be active. The same redistribution of heat energy prevents places nearer to the Equator than the fortieth parallel from getting progressively hotter until they too become intolerable to plants and animals.

The Arctic Tundras

One feature suffices to identify these barrens near the North Pole: summer warmth is never enough to thaw

107

the frozen soil more than a few inches below the surface. Under this is permanently frozen ground, the permafrost. Drainage streams rarely cut into it, but remain instead shallow and inefficient. Nor can the polar winds evaporate the water from melting snow and ice. Consequently, in summer the tundra becomes a patchwork of marshy pools, a haven and vast nesting area for migratory waterfowl and shorebirds.

By the time the birds arrive from their winter quarters to the south, the days are lengthening, the thin covering of snow melting, the pools spreading. In them tiny plants and water animals grow quickly. Seeds germinate, and perennial plants send out new greenery. Where all seemed lifeless, there is food in abundance.

Northern Waterfowl

No less than fourteen types of ducks, four of geese, and two of swans find conditions so comparable in the far north of Europe, Asia and America that the same species are present in all three. Individual birds do not shift from one of these continents to another, and the populations show small differences that are recognized scientifically by designating them as separate sub-species. Most travel for the winter months to more southern lakes or marshes or sea coasts, where bird-watchers get to know them by local names.

The goosander of Europeans is the common merganser of North Americans, but the red-breasted merganser is known by the same name in both regions. The mallard, pintail, gadwall, shoveler and green-winged teal dabble and tip for food in shallow water and nest on all sides of the North Pole. Eiders, the greater scaup, black scoter and white-fronted scoter, harlequin duck, old squaw and common goldeneye dive for food, often in brackish or salt water. The snow goose, white-fronted goose and emperor are equally familiar big birds in both Eurasia and North America; so is the smaller brent goose, which is known as a brant in the New World. The Bewick's swan of northern Russia is equivalent to the whistling swan of northern Canada. The whooper of Iceland and arctic coasts of Eurasia has its counterpart in the once-widespread trumpeter swan of North America.

By comparison, the far north in the Old World has only five common kinds of ducks unrepresented in America—the smew, the European widgeon and three special species of eiders. It has the greylag goose, from which the common barnyard geese were developed,

The snowshoe hare takes advantage of whatever shelter it can find. (U.S. Forest Service)

108

and the swan-goose, which is the wild stock from which Chinese geese were domesticated long ago. It has bean geese, pink-footed geese, a lesser white-fronted goose, the barnacle goose and the red-breasted goose. As a counterpart to these five, the American arctic has Ross' goose, which winters in California, and Canada geese from coast to coast. It had a distinctive Labrador duck until 1875, and still has hooded mergansers, buffleheads and Barrow's goldeneyes that are recognized only as vagrants in the British Isles and in Eurasia.

Shorebirds of the Tundras

The great order of shorebirds (Charadriiformes) includes tundra nesters that follow some of the longest migration

When winter comes, the caribou must migrate into the taiga, where the evergreen trees give them protection from the wind-driven snow. Often they get caught by a storm while traveling. (Widerøe's Flyveselskap A/s)

routes known. The champion is the circumpolar arctic tern. Often the members of this species arrive so early in the short polar summer that they must hollow out a place for their eggs through drifted snow. Before the days grow short again, they and their young are off. Those nearer Bering Strait than to Greenland fly to the west coast of Alaska and follow the Pacific side of the Americas southward to Cape Horn. Arctic terns

109

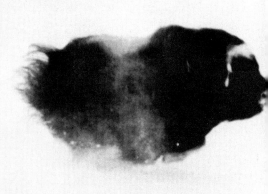

Three muskoxen galloping through deep snow on Cornwallis Island in arctic Canada. (Fred Bruemmer)

nearer to Greenland rendezvous off the Atlantic coast of Europe, and then travel southward to the westernmost bulge of Africa. There the flocks sort themselves out, some birds crossing the South Atlantic and continuing to Cape Horn, others staying close to Africa until they reach the Cape of Good Hope. Thereafter, for the duration of northern winter and southern summer, these terns flit and feed over antarctic waters until it is time for them to wing their way north again. The minimum for a round trip of this kind would be 25,000 miles. By the erratic course the arctic terns

follow and in their shuttlings to bring small fishes to their young, they must fly twice or three times this far each year.

Ornithologists are still seeking explanations for the migratory patterns of birds that nest on the tundras of the Arctic, then cross the Equator to far-south destinations, there detecting somehow the week when the northbound trip should begin. The sandpiper-like phalaropes spend the winter at sea, using their lobed toes to swim while feeding. The red phalarope goes all the way to the South Atlantic, South Pacific and southern Indian Ocean—almost to Antarctica; whereas the slightly smaller northern phalarope winters far from shore in the North Atlantic and North Pacific. Both kinds are circumpolar at nesting time, spinning in

circles on tundra ponds to stir up the food they want.

During the summer of the Southern Hemisphere, turnstones and knots often frequent the same shores as sanderlings in Tierra del Fuego, in South Africa, Australia and New Zealand. Six months later they are all raising young on the arctic tundras. The golden plovers and black-bellied plover travel from American tundras to southern and central South American grasslands, while those from Eurasian tundras are seeking similar places in warm parts of the Old World—even in distant Tasmania and New Zealand. Until 1948, in fact, the nesting sites of the bristle-thighed curlew remained unknown, although the bird was familiar on islands of the South Pacific during the northern winter. It breeds in central Alaska on mountain tundras. The closely-related whimbrel is a circumpolar nester and winters even more widely: along the Pacific coast of tropical America, the Atlantic coast of the Guianas and Brazil, the shores of the Mediterranean, the Indian Ocean and the China Sea.

Hazards of the Far North

On the tundras, the nesting waterfowl and shorebirds are attacked by hordes of mosquitoes whose wrigglers mature in the marshy pools. But they meet few predators other than the skuas and jaegers that fly in where eggs, nestlings and pirated fish can be had so easily. The arctic foxes cannot become numerous, despite this incredible abundance of food, because they take more than one season to mature and only a few can find enough to eat in winter. Those that are alive to see the great flocks of migratory birds arrive have no need for stealth. There is no gain from much activity in raising a large family, for soon the birds will fly south, leaving behind nothing of use to predators. The foxes and skuas stand around, half asleep, digesting their easy meals. The birds on their nests are equally unwary, for only a minute percentage of them will be killed. Adaptations relate to avoiding the winter or surviving its harshness, not to being inconspicuous. For this reason the tundra in summer is obviously alive with animals, completely unlike the tropical rain forest.

Greenery above the Permafrost

The plants are no less colorful and attractive. Peat moss of many kinds and colors, from orange-red to tan and bright green, forms a soggy, spongy shore and bottom to the shallow water. On slightly higher ground, arctic buttercups spread as low fleshy mats, raising small, pale-yellow flowers to attract insects. Yellow or white cross-shaped blooms appear on similar mats of wintercress and of rock-cress, showing that they are members of the mustard family. Flies and bumblebees (which are the only bees in these latitudes) come as pollinators. Wind carries the pollen for sedges whose triangular stems are distinctive; they grow a few inches tall, their leaves forming spiky tufts. Northern bentgrass sometimes grows as tall as two feet as it ripens its grains of the year. Then it collapses completely under the weight of the first wet snow.

The permafrost sets a limit in depth for growth of plant roots, and for burrows of small animals. The winter winds prune away any twig that grows upward. Consequently the polar willow produces its tough branches horizontally, each clad in brownish-red bark and bearing glossy, green leaves on branchlets that are

surprisingly brittle. A plant many years old may be ten feet across and no more than two inches tall, its main stem thick and twisted—hardly suggesting a tree. Arctic willow manages to thrust its roots a little deeper into the thawed ground, and to attain still greater thickness in its prostrate trunk. Its branchlets are flexible, its leaves paler green, with no gloss at all. Both of these willows drop their foliage as the summer comes to its early end, and are bare many weeks before the autumnal equinox. Only between rocks and in narrow ravines do they grow a little taller. There they benefit from the extra protection from winter winds that tend to blow away the thin blanket of snow.

A few members of the heath family (Ericaceae) show adaptations that let them be circumpolar. One, called flowering moss, is a lowly plant that retains its small needle-like leaves under the snow. Each leaf remains

A mother harp seal and her pup on an ice floe off the Labrador coast. (Fred Bruemmer)

evergreen, tough and waxy, ready to make food as soon as the spring sun reaches it. Often snow remains all around when the flowering moss pushes up its pink buds and opens its nodding bells on individual short stalks. The arctic heath, which grows also above timberline on mountain tops in the temperate zone, follows a similar schedule.

Precarious Life on the Lichen Ranges

Where the winter snow and ice are thin on land, exposed rocks are usually coated with lichens barely thicker than paint. Between the rocks, and protected by them, a bushy lichen grows to a height of two to four inches.

112

Forming a gray mat that is spongy when wet, and brittle when dry, it provides nourishment for lemmings and caribou.

Lemmings are particularly important in the Arctic because these short-eared, short-tailed, short-legged rodents convert plant materials into meat upon which predators can survive. Collared lemmings change color with·the seasons from brown in summer to white in winter, whereas common lemmings remain brown. Both show fascinating adaptations that help in getting the most from their inhospitable land. As winter approaches, each collared lemming grows longer, stronger claws on its third and fourth toes. Common lemmings do so only on their first toes—the ones that correspond to our big toe and thumb. With these sturdy claws the lemmings dig efficiently through snow and frozen soil, both to reach their food and to escape from storms and predators. Pelts of collared lemmings in their white phase are so thick and warm that people on the tundra have long made them into gloves, as well as into fur coats for their children's dolls.

Lemmings interact with their food supply, their predators and their diseases to go through explosive increases and sudden "die-offs" of population. Seldom is the peak of the cycle reached in the same year in adjacent regions, but excessive crowding and a change in habits usually come in the fourth year after the previous local die-off. The collared lemmings rarely emigrate, but the common lemmings have made themselves famous by responding to crowded conditions in a single way: they start running downhill. Showing little of their normal timidity, they expose themselves to additional predators, which soon are so replete they can eat no more lemmings for a while. Lemmings that reach the sea swim until they drown. Only a few stay-at-homes, plus the survivors among the collared lemmings, are left as a breeding stock, ready to repopulate the tundra.

The lemmings are the principal prey of arctic foxes, which come in two colors: a white ("white foxes") that turns brown in summer, and a bluish gray ("blue foxes") that is pale in winter and darker in summer. When lemmings are few and wary, the foxes must subsist—if they can—on carrion, such as stranded whales and the remains of the kills of polar bears. Or they dig for edible

"Northern divers" are generally called loons in North America. Several species are common to the New World and the Old. All nest within a foot or two of a deep lake into which they can vanish. (National Film Board of Canada)

frozen plant materials below the snow. The lemmings sustain the population of snowy owls as well. In good years, these big white birds seem to have no difficulty catching enough of the rodents to tide them over the dark winter months. Almost invisible while perching on a snow bank, the owls wait for a lemming to show itself, then glide to the kill inaudibly on broad wings as much as sixty-six inches tip to tip. When lemmings are scarce, the snowy owls fly south, sometimes as far as to the British Isles and the state of New Jersey.

An abundance of lemmings causes a serious depletion in the supply of lichens, affecting the welfare of caribou (or reindeer). The commonest lichen is known as "reindeer moss," because these deer-like animals paw for it through the winter snow. Caribou, which have branching antlers in both sexes, seem unable to survive where reindeer moss has been depleted by lemmings or by a late-summer fire.

The wild caribou of the barren grounds, from which Laplanders domesticated the reindeer strain, are confirmed nomads. By moving on, more or less at random, they rarely destroy their lichen range. Lapp herdsmen have learned to be equally nomadic, letting the arctic plants recover before returning with their herds of

reindeer. This regrowth is extremely slow where the growing season is only a month or less each year. A burned-over lichen range may need 50 to 100 years to become productive again.

Life and Death Along Polar Coasts

During the long summer days, the caribou and musk-oxen often wander toward arctic coasts. There they risk encountering a polar bear that is ashore eating the leaves and fruits of the tundra plants. These big white carnivores will attack a solitary muskox, if one wanders from the herd, or catch a caribou, which the bear can outrun over short distances.

At other times of year, the bears range widely on the ice, and travel southward to stay close to open water around the normal limit of the shifting floes. From these a bear can dive for fishes in open water, catch seals as they come up to breathe, or find seaweed to eat if all else fails. Polar bears are the most agile of all bears in the water, swimming as fast as three miles per hour when they need to do so. They also are the greatest travelers, making the whole shore of sea and ice in the north polar regions their habitat.

The growing season is a little longer along the coasts because the saltier and deeper water melts sooner each spring and stays open after severe frosts have struck the land. In the salt water, seaweeds of microscopic size abound. They benefit from the greater availability of carbon dioxide for photosynthesis in cold water, and from the mineral nutrients brought to them by the ocean currents. Vast numbers of minute animals eat these drifting sea plants, only to serve in turn as food for fishes of many kinds and sizes. Bivalve mollusks partly embedded in the bottom muds filter out the microscopic life as nourishment.

The fishes support roving populations of squids, and also several kinds of seals. Ringed seals, harp seals and hooded seals all take their toll, either remaining in arctic waters all year or coming in regularly, timing

Left: A long-tailed jaeger, smallest and least parasitic of these birds, picking up food from the surface of Hudson Strait, near Killinek Island, Canada. (Fred Bruemmer) Above right: A Canada goose watches alertly while her young feed along the edge of a marsh. (Richard Robinson) Right: Spring brings green to the saltmarsh along the Finnish coast, but causes little noticeable change in the dark evergreen forest that is characteristic of taiga country. (Teuvo Suominen)

their visits to match the extra abundance of food in summer. The shellfishes, on the other hand, sustain most of the world's walrus population. These big animals use their tusks to pry the bivalves loose, and also as ice axes to haul themselves up on ice floes to bask or breed. Some male walruses weigh nearly 2800 pounds, yet they drag their bodies out of the water in the same way.

Evergreen Forests of the North

When the combination of low temperatures and winter winds makes the tundra unbearable for caribou, these big deer of polar lands take shelter farther south where trees grow upright. The essential difference in climate is just enough to thaw the ground completely each summer; permafrost is absent. Yet the trees are shallow-rooted, getting water from the soil that thaws first in spring (when the days are long), even though this is also the soil that freezes first in autumn (when the days are already short).

The topsoil, with its humus matter, is shallow too. Partly this is because there has been too little time for a thicker soil to accumulate during the 10,000 years or less since the glaciers of the Ice Ages melted, leaving boulders and an almost sterile gravel in their wake. Partly it is due to the paucity of life that can tolerate the long winters. Only living things can interact to produce a soil.

Among the evergreen trees of the taiga, the caribou are never left alone to spend the winter. Wolves follow them, in little family groups of three or four, or larger packs numbering as many as twenty-five individuals. Whenever they see a crippled animal or a sick one that cannot defend itself well, they move in and make a kill. Such opportunities come irregularly. Between times the wolves often go hungry while seeking smaller prey. They pounce upon the various voles that reproduce prolifically in the taiga, and upon other rodents such as the wood lemmings of Eurasia and the lemming mice of North America.

The rodents take shelter among the lower branches of the trees, especially when the snow is deep. They find crannies for entry and passageways downward close

Mosses from windblown spores cover the ground, the rocks, the fallen trees, just as they must have during the Carboniferous period, before the earth was left bare by the retreating glaciers of the Ice Ages around the North Pole. (Teuvo Suominen)

to the tree trunks, all the way to ground level and perhaps to burrows that they have dug. Meanwhile the fallen snow holds down the evergreen branches all around like the hem of a big skirt, effectively fencing out the wolves. Only the big weasel, known in Europe as the glutton and in America as the wolverine, is likely to force its way through the branches in pursuit.

In the New World, the taiga has an additional predator, the lynx, and some larger prey animals, the polar hares and snowshoe hares, both of which undergo seasonal changes in fur color, from white in winter to brown in summer. Lynxes prey principally on the snowshoe hares and share their cyclic misfortunes, as is shown by the records of the Hudson Bay Company over the past two centuries. The snowshoe hares, logically known as varying hares, increase in abundance for about eleven years and then die off. The lynx population crashes almost at once, not to rise again until hares are numerous.

Similarities in the Old World and the New

Some of the same kinds of birds frequent the spruce forests in Eurasia and North America, but they often are known by slightly different names. The Siberian tit is the gray-headed chickadee in Alaska and northern Canada, where it frequents the boundary between taiga and tundra and thus avoids competition with the boreal chickadee, which is widespread in the American band of northern conifers. The crossbill of Eurasia is called the red crossbill in America, to distinguish it from the white-winged crossbill of the New World. Both species have the same strange shape of beak, ideally suited to spreading the cone scales of spruce and other conifers to reach the nutritious seeds. The Bohemian waxwings seen most often in the coniferous forests of the American Northwest are just waxwings in Scandinavia and the Siberian taiga. No change in name is needed for the pine grosbeaks or the arctic warbler. But American birders speak of the three-toed woodpecker (*Picoides tridactylus*) of Eurasia as the northern three-toed, to avoid confusing it with the black-backed three-toed woodpecker (*P. arcticus*) in the New World. All three-toed woodpeckers have only one of their four toes directed toward the rear, whereas other woodpeckers hold two in each direction.

Only an expert botanist is likely to notice specific differences between the Old World and the New in the coniferous trees that grow so densely in the taiga. The same genera are represented in both hemispheres. The dominant tree in North America is generally white spruce, with black spruce and balsam fir, named for

the blister-like swellings on its trunk, each blister filled with balsam resin. Boggy places are sites for the North American larch known as tamarack or hackmatack. Although a conifer, it sheds its needles in autumn and does not leaf again until spring.

The European taiga is evergreen with Siberian fir, Siberian spruce and Norway spruce. Bogs have a similarly deciduous European larch, whereas sandy soil favors Scotch pine. Asiatic forests have an evergreen Siberian larch in place of the deciduous one, and Siberian pine in place of Scotch pine. The Siberian fir is usually the dominant tree.

North Americans often speak of the taiga as the "spruce-moose" zone, for it is the favorite habitat of moose. These biggest members of the deer family (Cervidae) occupy the same habitat in Eurasia too, but are known there as elk. They are well adapted for crashing through the shrubby undergrowth, and for wading in the dark pools of the swampland. We have watched them hold their breath for minutes at a time, while they reached beneath the surface for submerged vegetation. At various depths and around the margins, the moose can vary their diet of browse and bark with the roots and other parts of many herbaceous plants.

Throughout the taiga the ponds are full of pondweed, water-milfoil and bladderwort, which are genera of circumboreal distribution. In spring the shorelines are bright with golden-yellow flowers of marsh marigold, white spathes of water-arum, and other plants that are the same species in both the Old World and the New. Over the peat moss, bogbean (or buckbean) often spreads its three-parted leaves; for it too these latitudes on all northern continents are much alike.

These uniformities are most evident in the taiga toward its northern fringe, where the proportion of families, genera and species shared by Eurasia and North America is so great that even an experienced field biologist would have difficulty knowing at a glance whether he was in the Eastern Hemisphere or the Western. Farther south, where the solid stands of conifers give way to mixed forests containing the northernmost representatives of broad-leaved hardwood trees, this uncertainty disappears. There the differences, rather than the similarities, stand out, identifying the Old World as distinct from the New. No longer are the living things primarily those de-

Watchful walruses swimming past Coats Island in Hudson Bay raise their heads high for better vision above the water. (Fred Bruemmer)

scended from kinds that tolerated conditions during the Ice Ages along the southern limits of the great glaciers. Instead, they show characteristics of the subtropics and the tropics, partly in being less adapted to harsh physical conditions and partly in developing more complex social relationships.

Plant Invaders from the South

A few plants of southern origin have spread far north into the taiga. One that is easy to recognize is the round-leaved sundew, which is sometimes used in Scandinavia to thicken milk in making cheese. In bogs and on moist acid soil, it is common all across Eurasia

The European curlew (Numenius arquata), *the largest of Eurasia's wading birds, finds nest sites along coasts, on sand dunes, in meadows, moors and marshes. (Gerhard Klammet)*

and in North America from Alaska to Newfoundland, south to northern Florida and down the mountains of western states to California. Along the southern edge of the taiga it is often mixed with the narrow-leaved sundews, which are equally circumboreal. Still farther south, generally where cold bogs persist or cool shady hillsides have the right soil and amount of moisture,

120

The bull of the grey seal is sometimes rough as he maintains his harem, seizing any female that begins to stray. (Fred Bruemmer)

additional kinds of sundew are found, representing a genus present on all continents and most abundant in Australia. All members capture insects on glandular hairs, digest the nitrogenous compounds and absorb the products as a regular feature of nutrition.

In America, additional insectivorous plants of southern affiliations live in the taiga. They are pitcher plants —the official flower of Newfoundland—whose vase-shaped leaves in whorls hold water. Insects fall in, drown, are digested and yield nitrogenous nourishment to the plant. The only related pitcher plants are native to Florida, California and tropical South America.

As though to compensate for the invasion of the taiga and tundra by southern plants, some of the larches and spruce, peat moss and the sedge known as cottongrass are found hundreds of miles south of the northern coniferous forests, in the Alps, the Caucasus, the Rocky Mountains, and even the low Pocono Mountains of Pennsylvania. Scientists think of them as relics from the Ice Ages, little communities of plants that survived for thousands of years despite warming of the weather. Generally they indicate a chemical situation that is tolerable for them but not for other vegetation. In their distinctive way they perpetuate the acid soil, keep the bog water charged with tannins and protect themselves against competitors.

All boundaries are subject to slow change and interchange. Where laggards among the living things become isolated, sooner or later they generally disappear. Rarely do they possess the versatility required to escape or to become pioneers in their own right and take advantage of opportunities in an environment that has become new to them.

6

Life

in the

Middle Latitudes

Across Eurasia

From the British Isles to Asia Minor, the plants and animals are those of Western literature, of the Bible, of Shakespeare, of Goethe and the poets. To a surprising extent they are the life of the Far East, and of North America too. Some are in all of these places because people took them along to use, or to admire, or just by accident. Many have such a wide distribution because at various times in the past, terrestrial life could spread east and west, encountering similar climatic conditions in temperate parts of the New World and the Old.

For the same reasons, the north coast of Africa, the shores of Asia Minor, and the southern tip of the Crimean Peninsula in the Black Sea resemble the European coast of the Mediterranean Sea. They are isolated from the tropics by broad dry deserts, which extend all across Africa from Morocco to Egypt, and from Sinai and the Negev and Jordan to the shores of the Persian Gulf. In northern Afghanistan and to the north of Burma, the mountains of the Hindu Kush, the Karakorams and the Himalayas rise like a barrier. They cut off the monsoon winds in the basins of great rivers from the Indus to the Irrawaddy, causing chronic drought to the north and producing additional forbidding deserts. Nor do these highlands disappear at the border of China. With scarcely a break for rivers

such as the Mekong, they continue eastward south of the Yangtse almost to the coast opposite Taiwan (Formosa).

Between these southern boundaries and the broad taiga of fir and spruce, Eurasia has impressive forests of deciduous trees, expanses of grassland, desolate arid wastes, and areas where empires have developed around major civilizations. The distance across this immense continental mass from Gibraltar to Tokyo is 7000 miles. Along its southern flank are seventy-five mountain peaks higher than 20,000 feet, whereas all the rest of the world has fifty-one. Both the highest land on earth and the lowest are Eurasian: Mount Everest at 29,028 feet above sea level, and the shore of the Dead Sea at 1286 feet below it. Less than six miles of altitude separate these two extremes. Other differences are more significant in determining what living things inhabit Eurasia's wide middle latitudes.

Instead of a pattern of national boundaries or the conventional division of Europe from Asia, the biogeographer sees temperate Eurasia as a land with four interlocking areas. To the northwest, where a deciduous forest was formerly almost continuous, meaningful patches of woodland remain from Rumania and Hungary through Austria, northern Italy, southern France, to the Basque region of Spain, and northward through Germany and Poland to the boundaries of the taiga. A second area fronts on the Mediterranean and the Black Sea, where the waters act like a balance wheel for climate, keeping it warmer in winter and cooler in summer than would otherwise be the case. This thermal inertia and humid onshore winds give plants and animals around the coasts a "Mediterranean climate." It extends along the north, east and south coasts, and for some distance into Egypt along the banks of the Nile. The tip of the Crimean Peninsula shares to a great extent the Mediterranean types of life.

North of Crimea and into southeastern Poland lies the westward prolongation of the third area, a great steppeland, most of which is in central Asia. It extends southward into the scattered salt flats and deserts of Arabia, upward into the lofty mountains north of India—the so-called Trans-Himalayas—and ends beyond Mongolia. In the Far East it is pinched off where the taiga to the north intergrades with the final area, a deciduous forest sustained by greater rainfall. The rain comes from winds blowing across the Sea of Japan, the Yellow Sea and the East China Sea, and is utilized today for growing crops to feed an enormous human population. Yet enough of the native plants and animals remain in this Far East portion of the Old World

for scientists to recognize its extraordinary wealth of ancient kinds. Living things there seem to have been amazingly unaffected during the Ice Ages by the extensive glaciation elsewhere in the Northern Hemisphere.

Farms and cities now restrict the native flora and fauna to scattered woodlands, parks, ravines and thickets. Yet enough remains to allow a convincing reconstruction of the past, including the changes made by man during the last twenty-five centuries. Within the 120,000 square miles of Poland, for example, the soil and the topography still record the effects of living things and of glaciers. In the southeast are black soils of high fertility that evolved under grassland—the steppe country. The rest of the south in Poland has gray-brown soils where, until less than ten centuries

The unique Père David's deer is extinct in the wild and treasured in captivity. Its large, spreading hoofs seem adapted for walking in marshes, perhaps those of northern China. (Wolf Suschitzky)

ago, deciduous forest stood. In the north are dark brown soils, exposed where coniferous forest of the taiga was felled to increase the space for agriculture. Often it degenerates into gravels and other debris left by the melting glaciers of the Ice Ages, where heaths and other lowly plants of moors and fens now grow. So poorly drained is this part of Poland that it has a special beauty, with dark tarns and lakes that are the natural home of the world's only large colony of wild mute swans. These white birds with black knobs above their orange beaks mature to a weight of forty pounds, the heaviest of all birds that fly. Since World War II, conservation measures have allowed the number of wild swans in Poland to rise above 2200 for the first time since records were kept.

Forest Land, Moors and Marshes

The largest forest in modern Europe is in eastern Poland, preserved near Bialowieza, close to the border of the U.S.S.R. Its 485 square miles compare closely with Grand Teton National Park in Wyoming, or Sequoia National Park in California, which are of medium size for sanctuaries in the world. Yet the Bialowieza Forest is justly respected for its primeval character. More than a dozen kinds of trees rise in impressive stands, but not the European beech because the forest is too far north. Hornbeam is the dominant tree in many areas of it, while in others ash and alder contrast with Scotch pine. Much of the forest is more mixed, great oaks being intermingled with tall elms (English elm and wych-elm), medium-sized Norway maples, massive limes (or lindens), with broad leaves, Eurasian poplars with fluttering foliage, shining white birch and dull gray birch, with here and there a wild apple or a dark Norway spruce.

World of the Wisent

For many years the Bialowieza Forest was the last stronghold of the European bison (*Bison bonasus*), called the wisent. Slightly larger than the American kind, these animals have shorter, thicker, blunter horns, and a heavier cape of long hair over the shoulders and front legs in winter. Once the wisent was as widespread as the deciduous forest. But it was hunted for its meat and gradually exterminated from most of the continent and Britain. By the 1500's it was rare, and by 1900 extinct except for the herd in the Bialowieza Forest, some scattered through the Caucasus Mountains, and a few in zoos.

Since wealthy Polish and Russian sportsmen operated the Bialowieza as a hunting preserve, they kept a rough census of its game animals. Between the 1850's and 1914, their records show that the number of bison shrank from 1900 to 700. They were less favored as trophies and could not compete well with about 7000 big red deer, 7000 smaller, spotted fallow deer, and 5000 roe deer whose reproduction in the forest was encouraged. The bison further decreased to 150 by 1918, mostly as a result of off-duty hunting by military men. The remainder (and the deer) were shot in 1921 by Russian revolutionaries. The Caucasian bison vanished by 1925.

Members of an international society began in 1923 to attempt to save Europe's bison from becoming extinct. They made a census of the animals in zoos and found only 66 alive in 1925, not all of them thoroughbred. A program of reestablishment in the Bialowieza Forest was disrupted by World War II, but has succeeded to the extent that about 100 of the descendants from purebred animals now roam the forest, and zookeepers around the world are well aware that *Bison bonasus* is still on the danger list.

Ecologists are trying to measure the effect of the bison's browsing in the forest, for the balance is a precarious one. If the animals are too few, the seedling trees survive in large numbers and choke out the undergrowth upon which bison feed. If the browsers are too abundant, there are few seedlings left to grow up and replace the old trees as they die. This interaction affects the welfare of the goshawk and buzzard that nest and hawk among the trees, of the woodpeckers, and of the owls that need the nest holes of woodpeckers second hand, of the hazelnut (or filbert) bushes and the wild hogs that root for food in the leaf litter and soft forest soil. The number of bison and deer affects even the lowly world of glowworms and fireflies, whose luminous signals have inspired poets and composers for millenia.

Above right: The Sahara Desert is most famous for its vast seas of wind-blown sand, called ergs. (Pierre A. Pittet) Right: The European salamander spends much of its adult life in the humid litter of fallen leaves in forests, where insects and worms can be found from spring until autumn. (Manuel V. Rubio)

Overleaf: Above left: Spoon-winged flies, found across Mediterranean Europe and Africa to Australia, are close kin to lacewings and antlions. Their grotesque larvae live in loose sand and eat other insects. (J.-F. and M. Terrasse) Far left: The Eurasian bee-eater generally nests in a river bank. Its prey includes dragonflies, bees, wasps and many other insects, which it pounces on from a perch on some bare branch. (Walter Fendrich) Center: Purple loosestrife, native to wet meadows across Eurasia, has been introduced in other lands. (Emil Javorsky) Right: Gardiner Peak (11,175 feet high) and The Dome (11,287 feet) in the Pelvoux Massiv area of the French Alps collect snow early in the autumn, while deciduous trees on lower slopes are taking on bright colors. (Gerhard Klammet)

Relics of the Ice Ages

Marks of glaciation are easy to find in Britain, northern Europe and around the Alps. Rolling land has rounded outcroppings that the ice smoothed and scratched. Conspicuous boulders and gravelly ground remain where the glaciers left them. Because much of this land is unsuited to farming, it has been left reasonably unchanged and harbors many of the same kinds of plants and animals that colonized it at the end of the Ice Ages. Shrubby *Cladonia* lichens and mosses cover sandy and gravelly soils that are physiologically dry because water that falls on them drains away quickly. Peat moss produces miniature bogs where the rocky basins cup the water. Between these extremes are great areas clad in stiff Scotch heather (or ling) and European whortleberry (or bilberry). On sunny slopes the cross-leaved heath straggles among other plants, its stems ending in clusters of deep-pink, barrel-shaped flowers. Scotch broom and dwarf gorse become bright with golden-yellow blooms shaped like those of sweet pea. Common juniper grows in columnar or pyramidal form, its evergreen needles sharp, its seeds covered by a bluish-black flesh coated with a whitish powder just as in North America.

Farther south and west in Europe, the moors and the wetter, marshy areas ("fens"), appear much the same, although the gorse is the more widespread, taller kind that has been introduced into the Western Hemisphere and to New Zealand. The European mountain-ash (or rowan) grows higher, raising its heavy clusters of orange-colored fruits beyond reach of animals standing on the ground. The common hairgrass, and several kinds of bentgrasses that are circumboreal, may grow three feet tall and wave above the heathland. Often they are dripping wet from the mists the wind propels.

In heathlands of Wales, Scotland and Ireland, we have flushed the 15-inch red grouse more often than the larger black grouse or the smaller partridge (known in America as the Hungarian partridge), both of which are common native game birds on the European continent. Meadow pipits which walk briskly or run, wagging their tails up and down whenever they pause, drew our attention frequently in all of these places.

Native Animals that Live Close to Man

Neither history nor imagination is needed to tell us that many of the small mammals and birds in western

Half-wild horses on Eurasian plains swim and wade to reach greener grass. (Richard Harrington)

Europe benefited when people began to cultivate the land, planting fields of crops where previously there had been forest or grassy steppe. Even if these animals did not eat the crop itself, they found insects, snails, worms, seeds and other food to their liking. The brown hare now seems equally at home in deciduous woodland and among field crops. The common hamster has spread westward and northward out of southern Russia and Asia Minor until the present boundary of its distribution is the 60-degree circle of latitude, along the southern coast of Finland. The mouse-sized field voles and the harvest mouse, which are two to three inches long and weigh only about a quarter of an ounce, attack the crop plants and are numerous enough to do much damage. The seeds and insects attract birds of many sizes, from corn buntings and yellowhammers to larks, linnets, partridges, rooks and kestrels.

The larger animals that hunt at night prey on the mice, voles and hamsters, and also the shrews and moles that live on insects, snails and worms. Smallest of these predatory mammals is the weasel; its size allows it to follow easily a mouse or mole down an underground burrow that would require the larger stoat to do some digging. Foxes pounce whenever they can, and dig only as a last resort. Badgers dig better than any of these others. Today, they are generally the largest predator around, for the lynx, bear and wolf have been exterminated over much of the continent and all of Britain. In darkness, the owls are alert to compete for the same prey: tawny owls, barn owls and the well-named "little owls," which stand less than eight inches tall. By day, the buzzard flies over the fields to catch whatever small creatures are careless or desperate enough to become visible.

The brown hare will find a hiding place in a patch of bramble. The fox may crouch for the day in a rather shallow hole. The badger quickly digs itself a snug den in a thicket, perhaps under a hedge where linnets are nesting sociably. Most of the small birds disappear easily for their rest periods in some tree top. Hedgehogs often take refuge merely under a twist of fallen grass, secure in their own spiny coats.

We can expect a barn owl to perch or nest in an outbuilding with an open door or window, anywhere within the circumboreal range of this bird. But tawny owls and little owls in Europe have now taken up the same habit, apparently because tree holes have become scarce and unoccupied badger holes hard to find—as well as damp and dangerous. We have found stock doves taking shelter underground; sometimes they nest in these situations. Admittedly they are then close to the fields in which they glean for seeds. Their larger kin, the domesticated pigeons, show no hesitancy in roosting or building nests of straw on rafters in old barns. Wild birds of this species, known as rock doves, limit themselves to precarious rock shelves on vertical cliffs overlooking the Mediterranean Sea, or similar sites inland.

House martins and swallows (the barn swallow of North America) build nests of mud and straw under eaves, on rafters and ledges of man's sheds and barns, needing only minor adjustments in behavior to keep up with changes in construction. By day they perch on wires, as though these strands of metal had been strung to support their tiny feet.

Transmission lines for electric power, telephones and telegraph, and changes in architecture have a completely different significance for the white storks, which return from wintering in Africa to nest in the European countryside and towns. The white storks, once so

The European beaver builds no conspicuous lodges, but nests in river banks near suitable food. (Wolf Suschitzky)

numerous from Spain to the Baltic countries, east of France and north of Italy, are now scarce. From many areas they have disappeared altogether during the last two decades. Sharp-eyed as these birds are while catching frogs and insects in wet meadows, they seem unable to dodge wires while soaring above roof tops or in flapping flight. Nor can they nest on the flat stone cap over a chimney if a vertical metal rod extends upward there to a television antenna. Even the haystacks upon which they used to nest are vanishing with increased mechanization. Hay-baling machines now make neat packages of hay, which are then stacked in dry barns. Bulldozers level the land, filling in wet places where storks used to find frogs and small fishes.

Beyond the cities, use of farmland has not yet become sufficiently efficient to clear away stone walls and hedgerows. Until this is done, many of the smaller animals and inconspicuous plants will have a refuge. Among these is the adder, Europe's most widespread poisonous snake. Persecution of adders for centuries has had no obvious effect other than to decrease the number of harmless and beneficial snakes; they are killed often and needlessly in the belief that they are adders. The common grass snake (or water snake) is one of the frequent victims. Only one of these rodent-catching snakes of the deciduous forest in southern Europe is known to have a wider range today than formerly, because it is admired, protected and regarded as a living symbol of health. It is the aesculapian snake, whose northern limit now extends through southern Russia, Poland, central Germany, to France, with outposts in Swiss forests near former Roman baths, where it was deliberately introduced.

Undrained Wetland Sanctuaries

Custom and economics have slowed human encroachment in four large areas of marsh and swamp in southern Europe. Today in these places important remnants of the continent's former forest plants and animals can still be seen. One is in Spain, close to the orange groves of Andalusia, between Seville and the Gulf of Cadiz. Known as "Las Marismas," it consists of the broad old delta of the Guadalquivir River, where winter rains soak a moorland and inundate sand dunes, creating a maze of shallow lakes each year. Ducks and geese from breeding grounds in the Far North flock to this soggy area to feed and spend the cold months. Soon after they depart in spring, the sun dries out Las Marismas until it is too arid for plants upon which cattle might feed. Conservation-minded people are trying to set aside 385 square miles of this waterfowl haven as a sanctuary, to keep the game birds flying.

A second marsh continues to enlarge in the delta of the Rhone, west of Marseilles in southern France. Known as "the Camargue," it has been famous for centuries as a mysterious place where cattle and horses roam half-wild, and where extensive rice fields are irrigated with water from the river. Many of Europe's mammals, birds, reptiles, amphibians, insects, crustaceans and rare wetland plants continue to live there, and often can be seen from the roads. European beavers burrow in the margins of the waterways without building any dams or lodges, just as they do in smaller numbers along the Elbe River in Germany, in Poland, Scandinavia and the U.S.S.R. The most spectacular birds of the Camargue are the 25,000 greater flamingoes that nest there almost every year, finding the privacy they need somewhere in the 156 square miles of marshland. Generally they and the other wild animals in the marsh manage despite the frequent storms with strong wind (the "mistral") that blow unimpeded from the north, often for more than a week at a time.

Where the Danube, Europe's second-longest river, finally reaches the Black Sea in Rumania near the border of the U.S.S.R., it too has a vast network of meandering channels and a marshy delta attractive to waterfowl. Its most famous birds are the pelicans that nest there: the more numerous white pelicans and the larger Dalmatian pelicans. The white species otherwise is an African bird and nests around lakes near the Equator. The Dalmatian pelicans are now restricted to the Danube marshes and a few wetlands in the Balkans; in prehistoric times they ranged northward as far as England. Along the Danube they have open waters with a mild current in which to fish and many isolated low islands on which to raise their young.

From the air the full extent of the open water does not show in the Danube delta because, beyond the extensive beds of reed, the surface is so often covered by a seemingly solid stand of floating plants: water soldier and water-chestnut. The borders of the deeper channels are overhung by sallow, which grows to be a tree, intermixed with white willows and alders. On higher ground, woodlands of oak and other deciduous trees attract wildlife of many kinds. Fortunately, the Rumanian government is protecting most of this great marshland. Even the harvesting of reeds, which supports a valuable industry, is regulated to avoid interfering with either the wintering waterfowl or the birds that nest in summer.

The delta of the Volga River, where it flows into the Caspian Sea, is by far the largest of its kind, with an area of almost 4700 square miles. But it is the least attractive marsh to either wintering or nesting birds because its climate varies from hotter than the steppeland in summer to colder in winter. It does have nesting pelicans of both kinds, and large numbers of both the mute swan and the Eurasian coot. But its commonest birds are cormorants, which nest on islands and in trees. The Volga marshes serve wildlife best as a stopping place, for rest and feeding, for birds on migration. Many of the ducks, especially mallards, gadwalls, pintails and shovelers, stay long enough to go through their seasonal moult. They do so without being much disturbed by the fishing activities in the many channels; three-fourths of Russian caviar comes from the sturgeons of the Volga delta. Yet diversions of Volga water for irrigation and the pollution of the remainder with industrial wastes are diminishing the value of the marshland and its channels for any kind of life, with no hope for an immediate change in sight.

The Mediterranean Coasts

Around the Mediterranean Sea, recognizing the native plants and animals grows ever more challenging. For more than forty centuries, living trophies have been introduced from distant lands—by Phoenician traders in their little ships, by special envoys accompanying the armies of Alexander the Great, by returning Crusaders and by explorers to the Spice Islands of the Far East. Later travelers brought novelties from Africa south of the Equator, from the Americas, and still more recently from Australia and the South Pacific. Yet many of the newcomers must be pampered consistently to survive. Living conditions suitable for them are limited by brilliant sun and scanty rain. They grow in patches on the best soil available, while the well-adapted native kinds cling to neglected corners, ready to displace the newcomers at the first opportunity. We can search out the natives on steep rocky hillsides too barren to yield a crop.

According to the old legend, while Hercules was striving to find the mythical oxen of the monster Geryon, he marked the gateway from the Mediterranean to the Atlantic Ocean with two colossal rocks—Gibraltar ("Calpe") on the European side and the Jebel Musa ("Abila") opposite it in Morocco. These "Pillars of Hercules" are fine landmarks, but the actual channel of the Strait of Gibraltar is narrowest at Spain's Point Marroqui—just twelve miles wide and 1100 feet deep at the middle. During the Ice Ages and for a while afterward, sea level was enough lower than today that

the separation of the two continents must have been significantly less. Yet only some animals and the reproductive parts of some plants spread across frequently enough to produce similarities in the life of the two coasts.

Where mists and rains from the Mediterranean support a generous variety in plant life, northwestern Africa and southwestern Europe have distinctive burrowing rodents: the Barbary ground squirrel, which is Africa's only squirrel north of the Sahara Desert. They are daytime feeders, which reduces conflict with the guineapig-like gundis that emerge from separate burrows so regularly to feed on vegetation at twilight that the Arabs refer to this time of day as "the hour when the gundi comes out." Actually, gundis sun themselves in the evening and morning, and resemble the American opossum in going into a cataleptic trance ("playing dead") if suddenly seized and handled.

We cannot feel confident that the Barbary apes, which

Snow-white spoonbills find places to feed and breed in the Danube delta and in marshes farther up the river in Hungary. The same species shows slightly different habits in eastern Asia. (György Kapocsy)

are tailless macaques, ever colonized the European shores without human help. Those that are pampered and famous on the rock faces and fortifications of Gibraltar do not reproduce fast enough to keep up with natural mortality, and fresh recruits must be freed there from time to time to maintain the tradition. Wild ones roam freely in the forests and over rocky ground on the African side of the Strait from Morocco to Algeria. They are the westernmost of the Old World monkeys (Cercopithecidae), and the only ones that may have roamed free in any part of Europe.

From northern Morocco to western Tunisia, as in Spain and Portugal, the magnificent forests of cork oak are carefully tended. At intervals great blankets of valuable corky bark can be separated from these trees without harming them. The need for corks as stoppers for bottles has diminished, but ground cork now has many applications in industry, for flooring, sound-proofing and thermal insulation. Under the spreading oaks on the African coast, the Old World rabbit finds low herbs to eat, and thrives without becoming a pest. The warrens of this sociable burrow-maker often cover an acre or more, but it rarely destroys the soil as it has where introduced in Great Britain, the Ukraine, Australia, New Zealand, and the San Juan Islands in Puget Sound. Domesticated, it is the animal known as the "Belgian hare," a pet or a source of meat.

Seemingly because the European southwest coast has no mountains to compare with the Atlas Mountains of Morocco and Algeria and no deserts as enormous and old as the Sahara, animals of the heights and of the desert fringes live south of the Mediterranean but not north of it. The wild sheep indigenous to this region are the aoudads, which live on grasses, herbs and stunted bushes on rocky, waterless terrain. Apparently they get all of the moisture they need from the sparse vegetation and the dew that forms before dawn where radiational cooling through the clear air produces a sharp chill on the desert. Large antelopes are represented by the scimitar-horned oryx and the spiral-horned addax. Both are well adapted for traveling great distances to find food in even more arid areas in the Sahara. But although both sexes wear horns and can use them effectively for self-defense against predators, these animals are dwindling toward extinction—a result of being gunned down by hunters in motor vehicles and light aircraft.

So far the hyraxes have saved themselves by burrowing or hiding among crannies in rocky hillsides—the habitat that would be occupied by marmots in other parts of the world. These are the "conies" mentioned several times in the King James translation of the Bible, for they find refuge and food all across North Africa and along coastal parts of Asia Minor. Their anatomical details have puzzled scientists for many years. Like rodents they have two prominent cutting teeth in the upper jaw. But, unlike any rodent, a hyrax has four small incisors in its lower jaw. It works its jaw continuously from side to side as though chewing a cud, but produces no cud to chew. Its feet appear hoofed, with four flattened nails on toes of the front feet, and two similar nails on toes of the rear feet. These and other features suggest that hyraxes are closer kin to elephants and hippopotamuses than to members of any

Barn owls have accepted man's buildings as suitable substitutes for tree holes, from which they hunt at night. (Wolf Suschitzky)

other mammalian orders. Yet the soles of a hyrax's feet are unique in having special naked pads and lubricating glands that act as suction cups, letting the animal climb almost sheer rock to reach its burrow or a hiding place. Despite these adaptations, hyraxes seem never to have crossed to southern Europe.

The Rickety Cradles of Civilization

On the western shore of the Arabian Gulf, within the city limits of the capital of the oil-rich sheikdom of Kuwait, stands a national monument in the form of a wooden sailing ship—a dhow fitted out with huge tanks to hold fresh water. Until less than a decade ago it and its sister ships plied between the thirsty community and the confluence of the Tigris and Euphrates Rivers more than 150 miles away. Now energy from surplus gas is used to desalinate all the water needed from the adjacent sea.

Most of Kuwait is a desolate desert because it receives less than five inches of rain annually, most of it between December and March. This is the season when the city of Basra receives six inches and Bagdad seven, in the Euphrates Valley close to the sites of the Sumerian and Babylonian civilizations of 4000 to 5000 B.C. Ever since those ancient times, the land in Asia Minor has had more people than fresh water to supply their needs.

133

An annual rainfall of twenty-five inches on Jerusalem gives no cause to rejoice after the six-months drought begins in May. Fifteen inches on Athens and on Carthage can support no respectable civilization. Even the thirty-five inches of precipitation on Beirut and Rome give a distorted view of the arid conditions a few miles inland from the sea.

Irrigation water from the rivers let the early civilizations flourish in the valleys of the Tigris and Euphrates. But from the chapters of Genesis we can visualize the Semitic patriarch Abram as a stockman feeling dissatisfied with conditions in 1500 B.C., taking his flocks and his family westward in search of greener pastures. He found what he wanted among the foothills of Canaan, to the east of the Mediterranean. But he soon encountered a drought (Chapter 12) and then conflict over grass to feed his flocks as well as those of his nephew Lot (Chapter 13). Perhaps he lived to see the sheep and goats destroy the plants with soft leaves, letting inedible thornscrub take over.

The scrublands may have been more widespread 3000 years ago than they are today. They were the wild country in Asia Minor where Samson met his lion (*Judges*, 14), and that hid David during his days as an outlaw (I *Samuel*, 20). Low branches of scrubby trees, such as pistacia (source of pistachio nuts) and Turkey oak, could catch the long hair of a mounted man, as they are recorded doing with Absolom (II *Samuel*, 18). Scrubland trees include the holly oak, one of the hardiest of evergreen oaks, and the Judas tree, which legend identifies as the kind on which Judas Iscariot hanged himself (*Matthew*, 27). Like its close relative, the redbud of eastern North America, the Judas tree opens a profusion of purplish-pink flowers in early spring, before its leaves appear. In coastal regions around the Mediterranean, these flowers are often gathered and mixed with salad because of their pleasant, acetic taste.

Bible Animals

No lions have roamed the thornscrub around the Mediterranean for many centuries. Probably the last ones were trapped around 400 A.D. to serve in Roman gladiatorial shows. Without this big predator, life must become a little safer for the scattered herds of wild

Above left: The Eurasian bittern stands 30 inches tall among the reeds of marshes and lake shores, where it nests and finds foods such as frogs and small fishes. (Gerhard Klammet) Left: The white-winged black tern ranges from Jugoslavian lakes and western Russia to Siberia. (György Kapocsy)

asses and feral goats. We can suspect, with no way to prove it, that jackals decreased in numbers too, from having to depend on small prey that they caught themselves instead of sharing the lions' kills.

Close to the deserts of the Near East and North Africa, jackals still compete with little fennec foxes which live in underground burrows and hunt at night. With their low build and huge ears, fennecs are the Old World counterpart of the kit foxes of America. As favored prey in the desert they even have the jerboa—a rodent that is equivalent to the kangaroo rat of the New World.

Asia Minor has an extraordinary number of different kinds of rodents, some shared with southeastern Europe, some with Asia across to Turkestan, and others with at least the eastern part of North Africa. We should never forget that the Crusaders, returning from the Holy Land, brought back to temperate Europe inadvertently some of the black rats that are native to Asia Minor.

These coastal parts of the Near East have golden hamsters, rat-like hamsters and common hamsters, all of which are plump-bodied burrowers with short tails and sociable, prolific habits. Asia Minor's quick-hopping gerbils are like those of all North Africa. It has its own Palestine mole-rats, as well as Old World wood mice and many different dormice. Its rocky ground, near grassy areas or crop lands, appeals to ground squirrels called susliks; they are close kin to those of western North America.

The larger animals, such as the chamois and wild goats, have become expert at staying out of sight in the mountains of Asia Minor and southeastern Europe. They may have increased in numbers because fewer eagles now harass them; hunters have shot these birds of prey consistently for centuries. When these mountain-dwelling mammals die, from accident or disease or old age, the vultures—particularly the Egyptian vulture—pick their bodies clean. Then the gigantic bearded vulture, with narrow-angled wings and long pointed tail, may spiral down and feast on the marrow left in the bones. It nests in caves on precipitous slopes and compensates for the weakness of its feet, which are unsuited for helping it tear flesh, by dropping the bare bones it finds on rocks and breaking them into manageable fragments. It could never carry off a lamb or a child, as has been claimed.

Familiar Plants in a Ravaged Land

As we boarded an eastward-bound jet at the Athens airport, we noticed that the craft's tail bore a large red circle emphasizing a stylized picture of the famous cedar of Lebanon that appears also on the Lebanese flag. It brought to mind the biblical story of Solomon accepting the invitation of Hiram, king of Tyre, to send 80,000 hewers into the mountains of Lebanon to cut these trees for the building of the temple in Jerusalem. We marvel that there were so many men skilled in using an ax, but not that the cedars of Lebanon no longer are a resource in the country of their name. One small grove has been spared, but the only real forests of this tree are now in remote parts of southern Turkey.

We think of the descriptions by historians and poets in the Athens of Aristotle and the Rome of Ceasar, mentioning the sounds made by woodcutters as they felled the forest close to urban centers. Some of these men were aware, too, how bare the land became where

The common land snail of northern Europe is a vegetarian that creeps about in woodlands when the relative humidity is high, but retires within its shell to escape dry air. (Ingmar Holmåsen)

sheep and goats were allowed to seek out the last remnants of the native vegetation. The spiny thickets that are left are known in southern Europe as "maqui." Largely evergreen, they often rival in density the chaparral of western North America.

In season, the spiny thickets of gorse attract bees to large golden-yellow flowers, each shaped like that of a sweet pea; the roots bear nodules from which nitrogen-containing substances spread into the soil, enriching it with a natural fertilizer. Globe daisies and rockroses benefit doubly by associating with the gorse, once by getting their roots close to the source of added nitrogen, and then by raising their flowers right through the thorny thicket. Globe daisy heads are spherical, with tiny white blossoms, whereas rockroses open yellow blooms that last just one day. Rockrose stems yield

The Cretan race of the bezoar goat, found wild only in the mountains of Crete, has become a rare animal. An old one would have horns twice as long as this. (Dimitrios Harissiadis)

viscous droplets of an oleoresin called labdanum, which during biblical times was mixed with myrrh. Today it is harvested for use as a fixative in perfumery.

The maqui on mountain slopes and rough hilltops that are too rocky to farm with modern tools remain sanctuaries for native herbs and shrubs. Seeing some of them in flower in late winter, toward the end of the rainy season, we find it hard to believe that they are wild because so many are favorites cultivated all over the world. This region is the original home of crocus,

136

snowdrop, grape hyacinth, narcissus and daffodil. It provided the civilized world with hyacinths, the madonna lily of Easter season, and the Christmas rose that often pushes up through winter snow.

These Mediterranean coasts gave us the cyclamen, the peony and fragrant mignonette; one of its lowly perennials is widely known as "English" daisy. Its decorative shrubs and trees have been transplanted to many lands: firethorn for orange-red fruits in clusters; oleander for fragrant flowers in abundance, despite poisonous leaves; and true laurel, whose evergreen aromatic leaves have been woven into crowns and wreaths ever since athletes began competing in Olympic games. Its contributions to pleasing odors and flavors include lavender, rosemary, marjoram, parsley, chives, leeks (later introduced to Wales and chosen as the national emblem) and garlic.

The ravaged coasts are the original home of the olive tree, the artichoke, alfalfa, the apple, and both the garden pea and the edible chickpea. Where natural underground streams of fresh water created a coastal oasis (as they did until recently at Tripoli in Libya), and inland even in the deserts of Africa and Arabia, the date palm (*Phoenix dactylifera*) provided so much that appealed to man that a way of life could be built upon it. Phoenicia, which flourished along shores that now are Lebanese and Syrian, took its name from the date palms there.

Long before Phoenician times, perhaps around 4000 B.C., the Egyptians learned how to make a translucent kind of paper from the sedge *Cyperus papyrus* that still grows along the shores of southeastern Europe, Asia Minor, and in parts of the Nile Valley south of Egypt. Formerly this plant, the paper reed, was common along the Egyptian Nile. Demand for it may well have exceeded the rate of regrowth, leading to its local extinction. This was the favored material for burial records to be interred in the tombs of the Egyptian kings and for the great library of books on scrolls that scholars studied in Athens and Alexandria. The famous Dead Sea Scrolls are on papyrus paper. It remained the prime bearer of written history until the second century A.D., when parchment made from sheepskins was perfected. Parchment could bear writing on both sides, not just on one, and could be cut into pages to be bound into books that were more convenient than scrolls. But papyrus paper was not really supplanted until after 751 A.D., when Arabs defending the city of Samarkand learned from Chinese who unsuccessfully attacked their position the Oriental method of making opaque paper.

Human history, both written and unrecorded, led the people of Mediterranean coasts to look for centuries toward the site of ancient Babylon and southward into the nearer parts of the Nile Valley as sources of useful plants and animals—the true basis of civilization. Only later, in Caesar's day, did they extend their familiarity northward beyond the Alps, into the forested parts of Europe. Still nearer modern times, and by sea rather than overland, trade with the Far East was broadened. Even today the remote areas north of the Himalayas and west of the great rivers of China remain in need of scientific exploration. In them the temperate zone of the Old World may still conceal some genuine surprises.

Mysterious Heartland of Western and Central Asia

In 1299 A.D., Marco Polo returned to his native Venice and began to tell about his adventures during a fabulous trip across the plateau of Pamir and the Gobi Desert to the court of Kublai Khan at Shangtu. Ever since, men have tried to reconcile his stories with what little was known about this part of the Asiatic world. It seems to have been the place of origin for the most numerous race of mankind, the Mongoloid people. Their ancestors managed to survive during the Ice Ages while walled in by the arctic glaciers to the north and west, by mountain glaciers to the south, and by the Pacific Ocean to the east. As adaptations to long exposure on the cold dry steppes, these people evolved their characteristically stocky build and special pads of fat around their eyes.

Asiatic people were far from casual in developing their resources. To their efforts the world owes the domesticated horse, camels, probably sheep and the house cat. The steppes yielded bread wheat, barley and buckwheat, as well as other familiar plants of lesser value. Yet over most of this vast territory, rainfall is so scanty that the few rivers simply end in a dry lake of salty sand. Even where the growing season is long enough for well-adapted grasses to reproduce, the greenery on the steppe can support only a few animals each year. Except where trading centers have been built, the human population rarely reaches twenty-five per square mile. The best maps show the territory as "unsettled," or as having an average of less than one person per square mile. Few scientists from the East have described in detail the community of plants and animals on the steppes. Few from the West have been permitted to explore there.

Biogeographers marvel that bordering nations have

fought to push their frontiers far into so unproductive an area, or that countries such as Mongolia could maintain their independent identity within it. The area is bounded on the north by the taiga forests of Siberia and deep Lake Baikal. On the west, it suggests a big ameba about to engulf the Georgian S.S.R., for one lobe extends north of the Black Sea to the wheat fields of the Ukraine while another projects into Iraq, Syria and central Turkey. The steppe includes all except the southern part of Iran, as well as northern Afghanistan, northeast Pakistan (but not Kashmir), all of Tibet, the northwestern half of China and western Manchuria.

Nomadic Herdsmen

Somewhere in this Asiatic grassland, probably in its southwestern area, domestic sheep are believed to have

Woolly with hairs, the leaves of the common mullein arise in a rosette from which the flower stalk grows. (Emil Javorsky)

originated; so far, no wild ancestors are known even from fossils. However, related sheep range freely: the Laristan sheep in central southern Iran; Asiatic mouflons (or urials) from mountains of southwestern U.S.S.R. to Kashmir; and Marco Polo sheep (or argalis) with enormous curled horns in the eastern U.S.S.R. and adjacent western China.

Taking advantage of the scanty grasses, particularly the feather grasses that grow in clumps, nomadic herdsmen have driven their domesticated sheep over Asia's heartland for centuries. Each summer they have gone with their animals high into the foothills of the Caucasus, of the forbidding Hindu Kush ("killer of Hindus"), of the Karakorums, and of the lofty Himalayas. Generally they have avoided the great deserts: the Dasht-i-Kavr south of the Caspian Sea, the Kara-Kum to the east of this big inland salt lake, the Takla-Makan in northwestern China and, to the east of the Altai Mountains in Mongolia, the famous Gobi Desert where Roy Chapman Andrews discovered dinosaur eggs.

Mongoloid people in this area domesticated the horse, probably around 3000 B.C., first as a convenient source of meat and beast of burden, then as an animal they could ride faster and farther than a person can walk or run in a day. Other advantages were soon recognized. From the steppes of Central Asia, mounted warriors began riding on wild charges against which men on foot were no match. To defend his Chinese states against these horsemen, the first emperor of China—Shih Huang Ti—had a great wall constructed in the third century B.C.; it extended for 1250 miles along his northern frontier.

Wild horses still roam the plains of Mongolia on each side of the Altai Mountains. They are Przewalski's horses, which differ from the domesticated kind in their stockier build, smaller size, the presence of an erect mane, and lack of a forelock. The familiar horse is a more graceful animal from the same part of the world, derived from ancestors called tarpans that are now extinct in the wild. Once they were widely distributed on the Asiatic grasslands.

In the high country and near the driest deserts, the early Mongoloid people found other strong animals to domesticate. In Tibet and adjacent lofty lands close to the Himalayan peaks, they subdued the long-haired yak (or grunting ox), a type of cattle that tolerate conditions at 20,000 feet elevation better than below 5000. Yaks are still the principal source of power for pulling plows through the slanting fields of introduced potatoes, and serve as beasts of burden, sources of milk, meat, clothing (leather), and fuel (dung). A similar role is filled at lower elevations by the two-humped camel in Afghanistan, between the Oxus River and the Hindu Kush, and the one-humped camel (or dromedary) in the Arabian Peninsula and across North Africa.

Zoo Animals in the Wild

Some of the larger mammals of the Asiatic steppes and mountain slopes, particularly the sheep- and goat-like ones, are better known to science from living specimens in the zoos of the world than from observations in their native haunts. Often the names of these creatures —milus, chirus, saigas, gorals, and takins—are unfamiliar because so few zoos have been able to procure the animals for exhibit.

The milu, known also as Père David's deer, is unknown in the wild, and is probably extinct except for about 400 individuals in zoos around the world. It was discovered originally in 1865 by the Abbé Armand David, who bribed guards at the Imperial Hunting Park in Peking in order to gain knowledge of the herd kept secretly inside the high walls. About 1900, the Duke of Bedford managed to procure some breeding stock from this herd, and it is from these few that all of the existing animals have been bred. The remainder of the original herd was destroyed during political disturbances later in the first half of the century, but in 1960 the Chinese government arranged to reintroduce the milu and build up a new herd on the animal's native land. This deer has large, spreading hoofs which seem suited to wading through marshy lands along the edge of the steppe in northern China. Presumably the milu came originally from this region.

Chirus (or orongos) are Tibetan antelopes of high plateaus and mountainsides, where they nibble at the vegetation along glacial streams or take shelter from the wind in shallow holes they dig for themselves. Since prehistoric times, the Mongol and Tangutan people have regarded chirus as sacred. The long, tapering, almost-straight horns on these animals may have inspired the story of the unicorn for, when viewed from the side and at a distance, the one horn can conceal the other.

Saigas are even stranger antelopes of Mongolia and Kazakhstan, where they show a preference for treeless plains. Their huge Roman noses contain nasal bones with extraordinary convolutions. Their downward-pointing nostrils lead to a sac lined with mucous membranes such as are found in no other mammal except whales. Biologists believe these adaptations relate to the saigas' keen sense of smell, and their ability to survive without drinking by finding vegetation that holds enough moisture.

The gorals are goat-like, hiss when frightened, and bear horns in both sexes. They blend with the rocky mountain slopes of Tibet and most of western China. Takins are shaggy, with spreading hoofs and strong backward-pointing horns in both sexes. They suggest an imaginary cross between a muskox and a Rocky Mountain goat, as they haunt the upper slopes of the Himalayas and lesser peaks in western China. Takins graze on grasses and soft vegetation above tree line in summer, but descend far enough to reach willow shoots and mountain bamboos in winter.

More famous but still rarer animals emerge from the bamboo forests in summer to seek other foods on the mountain meadows of western China and Tibet. They are the pandas, which are kin to the American raccoons. The lesser panda has a long bushy tail, and rarely weighs more than 10 pounds. The comical giant panda is short-tailed, bear-like, and white except for black eye patches, ears, legs, and a band around its shoulders; it weighs as much as 350 pounds. Occasionally the pandas show that they are members of the order Carnivora by forsaking their plant diet and catching fishes from mountain streams, or eating small rodents, or pouncing on the mouse-hares which, in the Old World, are lagomorphs like the pikas of rocky mountainsides in western North America.

Hunters and the Hunted

After the pandas have retired to their caves, rock crevices and hollow trees to sleep, the handsomest of the cat family—the snow leopard (or ounce)—begins hunting above tree line. It frequents mountains from Tibet to Russian Turkestan in the one direction and the Altai Mountains in the other. Snow leopards seek particularly the wild sheep- and goat-like animals, the musk-deer between 8500 and 12,500 feet elevation, the mouse-hares and rodents, and the Himalayan pheasants called monals, which live even higher on the slopes.

The wary Pallas' cat stalks rodents on rocky arid areas of steppe in central Asia. It is about right in size and form to be descended from the same ancestors as the domesticated Persian cat. The fur on its tail and undersurface is about twice as long as on its sides and back. This adapts it well to crouching motionless on snow and frozen ground, waiting for prey to come along. The Chinese desert cat of northern China and Mongolia is larger—as much as three feet long—with indistinct markings, shorter fur and elusive habits.

These predators must compete with the wide-ranging

Old World badger and the striped hyena for burrowing rodents of many kinds. A few, such as the ground squirrels (or susliks) and marmots, are active by day. The rest are nocturnal and include the widespread pest rat, an incredible number of different kinds of jumping jerboas, and desert dormice. The inadequacy of scientific study on the life of the steppes of Central Asia is shown by the fact that desert dormice remained unknown until 1938, when the bones of some were discovered where vultures had dropped them. Soon afterward the live animals themselves turned up—just where they had been, unnoticed, for thousands of years. It is even possible that the wild ancestor of the house mouse will someday be found in this region, where it originally moved in with man and began to share his resources. Its nearest relatives are the wild steppe mice and another of similar adaptability in rocky deserts of North Africa.

In darkness on the Asiatic steppes and desert fringes, short-tailed shrews and piebald shrews hunt silently for insects. Desert hedgehogs, both the short-eared and the long-eared species, can be more obvious in their movements since, at a moment's notice, they can roll up into a ball with stiff sharp spines pointing in all directions. No owl is likely to attack them. No other birds will be active at night.

Birds are scarce in Central Asia, except along mountain streams that flow into the deserts and around scattered marshes. In these local populations, the number of waterfowl that can be seen from a low-flying airplane is often impressive. So is the distribution of plant life, in concentric areas that differ in color around each wet place, each salt lake, each sterile sandy waste. Along the streams the vegetation forms parallel streaks, showing how the various plants are adapted to the harsh environment.

The marsh vegetation is chiefly cattails, which Europeans call club reed or reed mace, and the cosmopolitan true reed. Of this material the great crested grebe and the little grebe (known as the "dabchick"), build their floating nests, just as they do elsewhere, from Australia to Scandinavia. A few of the waterfowl are more limited in range, such as the Baikal teal, the falcated teal, the tufted duck and the red-crested pochard.

Asiatic spoonbills choose to nest among the marsh reeds even where tall trees grow, differing in this behavior from the European spoonbills that are supposed to be the same species. Each old tree with a nest hole is taken over by rollers and hoopoes. This subdivision of available space leaves the bee-eaters and bank swallows (called sand martins), to nest in every sandy cliff. Small birds and young ones are in frequent danger of being preyed upon by a miniature falcon known as the hobby.

Poisonous and Other Plants

In the salty soil along the edges of the Turanian and Mongolian deserts, and over tremendous areas of Russian Turkestan, the principal tree is saxual, which is low-growing and shows no leaves for most of each year. Other members of the same goosefoot family (Chenopodiaceae) are equally tolerant of prolonged drought and salt around their roots. They include the widespread prickly Russian thistle that we have found thriving as an introduced weed around alkaline lakes in Nevada, and tree-like saltworts, as well as many types of herbaceous plants.

The adaptation that saves most of the broad-leaved plants of the steppelands from being eaten in quantity by hungry animals showed up soon after a herbaceous relative of saxual was introduced accidentally into Nevada. Within three decades this plant spread over more than 16,000 square miles of semiarid grazing land in eleven western states. Sheep that were already close to starvation from having been allowed to overgraze their range died when they ate the foliage of the newcomer. The deadly poison proved to be soluble oxalates, which constitute as much as a third of the dry weight of the plant.

In less arid parts of central Asia, people searching for edible foods discovered spinach, also a member of the goosefoot family. It too contains soluble oxalates in its leaves, but not in toxic concentrations. The oxalates will react, however, with any calcium from milk or other source included in the same meal, forming insoluble calcium oxalate, which is unabsorbable.

When we think of oxalates as plant poisons, we remember how dangerous rhubarb can be to the uninitiated. It is another native plant of the arid steppes of central Asia, but concentrates its oxalates in its underground stems and its green leaf blades, not in its long edible leaf stalks (petioles). On the steppes, it grows among members of the vast genus *Astragalus* —the milk vetches—one of the largest groups in the plant kingdom. Milk vetches are legumes, contributing to soil fertility but also forming thickets through which it is difficult to go, because their spiny leaf stalks remain and harden after the short-lived leaflets drop off. Several species, particularly in Iran, yield gum tragacanth. This material is still used in Eurasia as a sizing for cloth and paper, and as an adhesive for envelopes. Buying and selling it is an important aspect of market activity in Baghdad.

Granite boulders left by glaciers have protected softer rocks from erosion in the Transylvanian Alps of southern Rumania, producing strange columns to which fanciful names are given. (Gerhard Klammet)

The familiar words, "Beneath the bough, a jug of wine, a loaf of bread," from the verses written by Omar Khayyam, the Persian mathematician and poet who died in 1129 A.D., strike but one heretical note: alcoholic beverages are forbidden in a Mohammedan community. Out of the "Wilderness" of the steppe country came plane trees, the shade of whose leafy boughs is appreciated today over much of the Old World. Travelers brought seedlings to Europe from the great area between central Turkey and the northern flanks of the Himalayas, where these trees are native.

From these same semiarid parts of western Asia the world gained barley, bread wheats and buckwheat. Barley was man's first cereal. Egyptians were already cultivating it on a useful scale in 5000 B.C. It still retains its popularity for use in porridge and flatbreads in Mohammedan areas of western Asia and North Africa, just as it did in biblical Asia Minor, in classical Greece, and throughout the old Roman Empire. Elsewhere barley-eaters changed to wheat, to which about a fifth of the world's cultivated land is now devoted; this is four times the area used for growing barley, and twice as much as for rice. Yet no single ancestor for modern wheats has been discovered. Their ancestors may include wild wheats, goat-grasses and quack grass (or witchgrass) through a series of complex successful crosses. Wheats require less water to bear fruit heavily than any other cereal raised in quantity by man. They also contain more protein, including two relatively simple ones (gliadin and glutenin) that react in bread dough to form the gluten that keeps it porous when raised with yeast.

We seldom think of grasslands and desert fringes as harboring handsome flowers suitable for introduction in temperate gardens around the world. Yet from Asia's steppes have come some favorites, whose origin is often forgotten. Some tulips originated there, as did oriental poppies; the crown imperial, a spring-blooming herb of special beauty; and the mock-orange, a spreading

141

shrub with white flowers of delightful fragrance. Even the botanists, who should be better informed, have sometimes been misled about the source of the plants they named. One that seems destined to bear indefinitely the scientific name *Molucella laevis* is shellflower, or balm-of-gardens, a mint from the steppes that appears exotic enough to have come from the Moluccas instead.

The Far East's Enigmatic Life

Until the late 1600's, the western world knew almost nothing about the flora and fauna of the Far East. Nor did the first plants and animals sent to European centers of culture give any scientist the idea that the native plant life would include so many representatives of genera familiar in Europe and North America. Early shipments tended to be of kinds that seemed strange to western visitors. They were understandably impressed by the ancient oriental culture and the values, both esthetic and practical, placed upon living things in the Orient.

The Jesuit missionary Pierre d'Incarville began sending plant specimens home from Peking (then Peiping) between 1740 and 1757. Among them was the Chinese tree-of-heaven, now planted so widely in temperate countries. North Americans think of it as the tree that grows in the most crowded parts of Brooklyn. It came from the most crowded parts of the Far East, the area where currently nearly half of the world's people live on a small fraction of the world's land.

Esthetics and practicality often seem close together in the Far East. At least as long ago as 1500 B.C. the Japanese chose their single chrysanthemum as a national symbol, to be stylized and included among the royal insignia. Even earlier and independently, the Chinese discovered the insecticidal properties of pyrethrum powder from a related plant, the painted lady. By the time samples of the Chinese product traveled along the trade routes to Europe, the material gained the popular name "Persian insect powder," which is still in use. A similar error is perpetuated in the West in the name "rice paper," a writing material from the Far East that has no connection with rice. When the actual source of rice paper was discovered to be a tree, it was easier to call it the rice-paper tree than to correct the original mistake. The tree itself is a close relative of ginseng of the same warm southeastern part of China and Taiwan (Formosa), from which Orientals obtain a medicine that is highly esteemed as a cure-all in China.

Despite pressures from so many people crowded in so little space, a wealth of native plants has been maintained in gardens, around religious shrines and in public parks of the Far East. From these places a succession of ornamental trees and flowering plants has been obtained to give pleasure in distant lands. Some have already become old-fashioned favorites, such as bleeding-heart. It was brought to the West by the Englishman Robert Fortune, a traveler who visited China four times between 1843 and 1861. His name is honored for a citrus tree, the kumquat (*Fortunella*).

Horticulturalists have praised as the "most gorgeous of flowering shrubs" the Japanese *Magnolia kobus* and one called *M. soulangeana*, which is a hybrid between two Chinese species. Often the source of these and other plants from the Far East tends to be forgotten. They enrich the beginning of the growing season with winter jasmine and goldenbells, brighten the house with the white-trumpet lily (or Bermuda lily) of Japan as the "Easter lily" when Easter is early—long before the flowering time of the traditional madonna lily of southern Europe. Multiflora rose offers an abundance of bloom and then a harvest of red fruits for birds on autumn migration. The weigelia shrubs and the common hydrangea with its huge balls of blossoms generally open and fade before the camellias, the primroses, the balloonflower (or Chinese bellflower), the Japanese laurel, and the golden-banded lily of Japan come into flower. Plantain-lily (or funkia) and Chinese asters bloom late in the growing season, just before the nandina bush shows scarlet in its decorative berries. Outdoors, the vine called Boston ivy drops its leaves and then its leaf stalks for the winter, while indoors the old-fashioned cast-iron plant survives despite neglect.

The commonest of fan palms in hotel lobbies and similar places is a Chinese tree. The tea in the coffee shop and tea garden is from a tall shrub or low tree (*Camellia sinensis*) native to the whole area from Japan to India. Outside the building may be an imperial tree from China and Japan, introduced to many other temperate areas for its handsome large leaves and loose clusters of purplish blue flowers. Often growing forty feet tall, it is one of the few members of the figwort family (Scrophulariaceae) to attain tree size. The katsura tree is even more of a curiosity, for it belongs to a little family of eastern Asia with only five members; its inconspicuous flowers open at the same time as its leaves in early spring, whereas the blossoms of its near relatives (*Euptelia*) are handsome and bright red.

Trees of Far Eastern Woodlands

Walking about in public parks and patches of woodland in the Far East, a naturalist from Europe or eastern North America feels much at home. Most of the mixed hardwood trees belong to familiar genera. But the species are usually distinctive. Maples and poplars are particularly numerous. There are native walnuts (such as Japanese walnut), hornbeams, beech, chestnuts, oaks and elms, witch-hazels, lindens, dogwoods, rhododendrons, and low-growing ash trees. Yet the next grove encountered may be distinctly Asiatic: an orchard of camphor trees which are native to Japan, southern China and Taiwan. They are even more valuable now for the crystalline camphor that can be steam-distilled from chips of the wood. This is not used as a repellent for clothes moths, as when our parents were young, but as a plasticizer for commercial motion picture film and other technological products.

Canadians, who know the maple leaf as their national emblem, are particularly surprised to discover the maple trees of the Far East. Of 150 species known, 130 are native to this small part of the Old World; western Asia and Europe have seven kinds, and North America thirteen. Maples are conspicuous components of the mixed forest between 6500 and 9000 feet elevation on the island of Taiwan, and at lower altitudes on the islands of Japan, on southern Sakhalin Island, and on the mainland from eastern China and Korea north to the southeasternmost U.S.S.R.

Until the patterns of glaciation during the Ice Ages were understood, no one could explain why the Far East had so many trees of the class Gymnospermae—woody plants with naked seeds. During the Age of Reptiles these were the dominant members of all forests. But except in eastern China, Japan and nearby areas, few kinds survived to the present day. These few were able to reproduce progressively closer to the Equator during the Pleistocene and then move back again when the glaciers retreated, as they did farther west in Eurasia and in North America. The Far East seems to have remained relatively undisturbed because, to the north, too little precipitation fell for snow to accumulate into glaciers and spread southward from the Arctic. Ancient types of trees could remain where their ancestors had been for millions of years.

The Far East has its share of native pines, larches, arborvitae, cypresses and junipers. But it also has several genera of gymnosperm trees found nowhere else. The Japanese cedar yields good timber, and grows as a decorative evergreen. Many a western home has a frog, a turtle or other curio carved from *Cryptomeria* wood, in which the alternating grain of pale spring wood contrasting with narrow bands of dark summer wood gives an unusual, artistic pattern.

The parasol-pine (or umbrella-fir), which bears its branches in handsome whorls, and the Hiba arborvitae (or akeki), which is often more irregular, are native to Japan but often cultivated elsewhere for their ornamental forms. Taiwan has a unique tree, *Taiwania cryptomerioides*, somewhat resembling the Japanese cedar, and also one of the two known kinds of *Cunninghamia*, the other being China-fir of southern China.

Human care, or lack of it, has much to do with the survival of these ancient trees. One, the maidenhair tree (*Ginkgo biloba*) is unknown in the wild, but people have saved it from becoming extinct. The ancestors of all ginkgoes alive today were trees held sacred for uncounted centuries in monastery gardens of the Far East. Now it has become a shade tree, whose strange fan-shaped leaves and spreading roots show a spectacular tolerance for the polluted air and paved environment of city streets along which it is set out. The dawn redwood (or water-fir) of China may have been equally on its way to extinction in 1941 when a few groves were discovered—totalling about 1000 trees—where woodcutters were felling every one of useful size. *Metasequoia* is among the few types of plants known first from fossils, and then sought by botanical expeditions in areas where a few might have survived—and had.

Along the Great Yangtse

Two-thirds of a million square miles of the Far East lie in the drainage basin of the Yangtse River, which reaches the China Sea at Shanghai. The river is not only a waterway for human commerce, but a haven for native animals that manage to survive close to crowded people. Least appreciated is certainly the brown rat, which came from East Asia and probably Shanghai, reaching Europe by ship about 1553, only to be named the norway rat. It became a major pest —the sewer rat—larger and more aggressive than the black rat. Yet an albino strain of the norway rat is one of the most useful animals in experimental medicine and psychology.

The margins of the Yangtse are the principal habitat for a uniquely Asiatic mammal, the little water-deer, and it is among the few places in the world with a giant salamander, an alligator and a paddlefish. Water-deer are social creatures, rarely more than twenty-one inches tall at the shoulder or heavier than twenty pounds. The size of a large dog, they bear no horns in either sex, but the upper canine teeth of mature males grow long enough to project like tiny down-pointing tusks. Koreans maintain that water-deer have a fatal bite, although no unquestionable instances can be cited as

143

proof. Perhaps superstitions have allowed these little cud-chewers to survive to the present day despite their proximity to so many hungry people.

Giant salamanders seem to be good eating anywhere. The Chinese kind (*Megalobatrachus davidianus*, named for Père David who discovered it) grows to a length of 41 1/2 inches. Like the Japanese giant salamander (up to 60 inches long), it wears sensitive fleshy barbels around its mouth. The closely related and smaller giant of eastern North America, the hellbender, measures less than thirty inches long and has no barbels.

In Yangtse waters, the Chinese alligator grows to be five feet long, generally hibernating through each winter.

The giant panda of western China resembles a bear more than it does its close relatives, the raccoons. (Wolf Suschitzky)

Its only near relative is the alligator of the southern United States. Similarly, the Chinese paddlefish has its sole counterpart in the Mississippi Valley. Both of these strange fishes, which are distant kin to sturgeons, swim about with their enormous mouths agape, straining small crustaceans and other minute animals from the water before it passes out through their gills.

A goodly number of different animals beyond the

Yangtse and its wetlands are obviously related to North American kinds or to those in the deciduous forest area of Europe. The Asiatic black bear finds food and refuges in the mountains of Taiwan and Japan, as well as across Southeast Asia to the Himalayan foothills where it was first discovered. The Asiatic birch mice may have come from the taiga, and the Chinese jumping mouse from the steppes, although both show preference for the woodland borders of mountain streams. The Japanese flying squirrel shares its genus with the widespread Old World flying squirrel of the northern coniferous forest from Finland to Siberia. The biggest of these gliding rodents is the giant flying squirrel of Taiwan, a distinctively red-bellied animal measuring as much as four feet and weighing up to three pounds.

The Far East has pit-vipers much like those of the New World, although none are rattlesnakes. The species with a large head shield, a close relative of the North American cottonmouth, ranges from the steppes of the southeastern U.S.S.R. to the islands of Japan. Several members of the genus *Trimerosaurus*, which closely resemble the fer-de-lance and kindred snakes of the Americas in having separate scaly plates atop the head, commonly have red tails in southeastern China and on Taiwan. Their venom is like that of rattlesnakes, but fatalities from them are seldom reported.

Far Eastern Animals and Men

In parts of the Far East where human ingenuity has been able to provide irrigated fields and ponds for rice culture, people and some of the native animals have evolved together a productive integrated community.

Ducks and geese, carp and goldfish live among the human population. The ducks are mostly domesticated white varieties of the widespread mallard, while the geese, equally white, are derived from the swan-goose that winters in northern China and breeds normally on the Siberian taiga and tundra north of Lake Baikal. The birds swim about, fertilizing the water with their droppings, eating the weeds and excess rice plants culled from the crop. The same foods maintain carp, whose soft flesh is highly esteemed, and decorative goldfish as much as eighteen inches long. Often esthetic pleasures from irrigated land are increased by having a few of the outstandingly beautiful mandarin ducks among the white ones. Mandarins, which are native to eastern China, Japan, Korea and Taiwan, tame easily, as might their only near kin, the North American wild wood ducks.

All across the middle latitudes of Eurasia, the living things have outlived more conflicts among people than we like to count. We marvel that so many native plants and animals have held a place while human populations have grown, declined, and swelled anew to claim the land. Each new civilization in succession has reduced the amount of forest cover, the area of open grassland with a naturally rich community of plants and animals, and the numbers of most kinds of animals for which a use could be found. It has led to spring floods and summer droughts, and loss of the topsoil which is both the product and the requirement of many kinds of life. Despite these odds, the living things have shown incredible tenacity in holding on to their old ways, or have evolved new ones in a hurry.

7

Life in the Middle Latitudes Across North America

The plants and animals of Canada and of the United States show far more similarities to the living things of Eurasia's temperate latitudes than to those of the American tropics. Yet the differences in climate over this great area in the New World are too great for many kinds of life to be familiar north to south and east to west.

A few special kinds of plants and animals give the middle latitudes of North America a distinctive character. In this one area are almost all of the ninety known kinds of goldenrod; just one grows in Britain, and a few others have been found in the grasslands of Eurasia and South America. Its show of asters is particularly splendid, for again Britain has just one native kind, and the few in Asia, Africa and South America comprise only a token among the 600 different species. Choke-cherry and box-elder (or ash-leaved maple) are widespread, quick-growing pioneer trees.

Only this part of the New World has rattlesnakes whose rattling tails distract prey while the reptile gets into position to strike with its venomous fangs. Skunks are almost as distinctive, although a few species range southward to the Straits of Magellan and thrive in the altiplano of the Andes. Fortunately, the rattlesnakes are far less common than the skunks. Skunks that pass unseen in the night often leave an odorous marker where they have encountered danger. There are few places in southern Canada and the conterminous United States where a person with a discerning nose can live long without detecting skunk, or encountering an aster and a goldenrod.

One reason why so many people in the United States and Canada watch the wild birds and delight in seeing them around is that certain ones are to be met, like old friends, in most parts of the middle latitudes. The European house sparrow and European starling, which have spread so far in this region since their introduction to the New World, are no more nearly universal than a score of the native birds. Indeed, a checklist of the two dozen commonest and most widely distributed birds in temperate North America would include just these two as naturalized citizens:

Pied-billed grebe	House wren
Sharp-shinned hawk	American robin
American bittern	Bluebirds (3 species)
Killdeer	Starling
Spotted sandpiper	Yellowthroat
Common snipe	House sparrow
Mourning dove	Red-winged blackbird
Screech owl	Brown-headed cowbird
Belted kingfisher	American goldfinch
Hairy woodpecker	Rufous-sided towhee
Downy woodpecker	Juncoes (5 species)
Common crow	Song sparrow

Although many of these American birds migrate great distances, only one nests also in the Old World; the snipe lives in Norway and east of the Baltic Sea.

Other living things in America seem more affected by differences in the amount and seasons of rainfall, which follow a fairly simple pattern on the two sides of the mountain chain that divides the narrow Pacific drainage area from the broad Atlantic. West of the mountains, precipitation decreases from north to south; it supports a rich cordilleran forest from coastal Alaska well down into California, then a semiarid thorn scrub inland through an area called the "Great Basin," and finally the Arizona-Sonoran desert, which extends into Mexico and has some of the strangest plants and animals in the world. East of the continental divide, the rainfall increases progressively southward and eastward. Accordingly, the living things have produced a great natural grassland—the prairies—grading into a huge deciduous forest eastward, and then a broad coastal plain extending from Virginia to eastern Texas.

Giant coastal redwoods of great age and height form forests near the Pacific Ocean. (Martin Litton)

146

Southward through Mexico, Central America and South America, the mountains continue. South of the Tropic of Cancer they provide a great many strips and high islands of temperate climate into which life has emigrated from the middle latitudes. At intervals in the past, unusual weather has favored this southward spread of northern plants and animals. Often, in isolation, they have been able to evolve independently into new species without losing the generic features that help in tracing their origin. In this way the currants and gooseberries of the North Temperate Zone have extended their range to Tierra del Fuego. Tailed amphibians (salamanders) in the New World show a similar distribution. The Andes have blueberries and cranberries and the spectacled bear—the only bear in the Southern Hemisphere. We think of all of these outliers as "belonging" in the North Temperate Zone.

North of the Tropic of Cancer there are no corresponding islands of tropical climate. Except for a few versatile animals, such as the nine-banded armadillo and the marsupial mammal called the Virginia opossum, tropical life has been unable to spread northward at a matching pace. Both of these creatures have taken advantage of man-altered environment, including cultivated fields and bridges, to pass the barriers that previously held them back. For most animals and plants adapted to the tropics of the New World, the change from low-moist-warm to higher, drier and cooler environment to the north exceeds the tolerances they have inherited.

The Cordilleran Forest of Pacific North America

Onshore winds from the Pacific Ocean bring mists and rain to some of the world's most valuable timber trees on slopes rising into the mountains. The extraordinary extent of this forest, for hundreds of miles north and south, is due to ocean currents. These arise through a great eastward flow of water toward the coast of Oregon, where the stream splits. A northward current brings

Above left: The commonest of nesting thrushes in the deciduous forests east of the Great Plains is the wood thrush, which has particularly large breast spots and a rust-colored head. (Jack Dermid) Left: The female black-chinned hummingbird of western mountains in the United States and Mexico works alone to build her nest, incubate her eggs and tend her two young of the year. (Robert Leatherman)

148

unexpected warmth as far as to Juneau, Alaska, while the southward counterpart chills the coast even beyond the Mexican boundary of the United States. We have experienced the astonishing change in just a few miles of travel eastward from the Pacific coast at the latitude of San Diego, California, beginning near the coastal current under a fog bank and an air temperature in the low 60's F., getting quickly beyond its influence to a cloudless sky and a temperature close to the 100° mark. The chill along the shore was still there upon our return. A similar change in Alaska matches travel from the temperate forest along the shore, through a taiga, and then to a tundra a short distance inland.

The tallest and the oldest of the world's trees grow in the forests of northern California, southern Oregon, and along the California-Nevada boundary. Some of the coastal redwoods within thirty miles of the Pacific Ocean measure more than 369 feet tall and 65 3/4 feet

The dappled shade under the spreading boughs of an avenue of maples becomes brighter as the leaves fall. (Emil Javorsky)

in circumference at a man's chest height. Other genera of conifers have their giants too: western hemlock to 259 feet; western red cedar to 250 feet; sugar pine on higher and drier ground to 245 feet, with cones up to 18 inches long.

The Douglas fir, which sometimes grows 300 feet tall, outranks the rest in production of marketable timber. Mostly it is in the same size range, between 150 and 200 feet, as other valuable trees such as western yellow pine, lodgepole pine, and western larch.

Gnarled bristlecone pines, surviving higher on the mountains than the lodgepoles can tolerate, form a timberline of their own. Quite a number of them have

been found to exceed 4500 years in age. One that was felled for study embarrassed the scientists who cut it down by proving to be more than 4900 years old. These trees, and the dead carcasses of others still higher on the slopes of the White Mountains in eastern California, have let dendrochronologists extend their study of ancient weather to about 7000 years before the present day.

In many of these cordilleran forests, an understory of broad-leaved trees develops. Among the coastal redwoods, it may prevent seedlings from perpetuating the forest—a condition that presumably was corrected in former times by an occasional fire that cleaned out the lower vegetation and exposed the ground. Pacific dogwoods as much as eighty feet tall become especially

The newborn young of the horned lizard of Arizona deserts often remain for a few days with their mother, climbing on her back or scampering off to eat tiny ants. (Lorus and Margery Milne)

handsome in spring when they display their compact flower heads, each advertised by about six glistening white, petal-like leaves. Madrone trees of the heath family tower even higher near the Pacific shores. Inland they are shorter, sometimes mere shrubs, but each is clad in smooth red bark that peels in thin layers, and holds glossy evergreen leaves. Common undershrubs include the pink rhododendron and the salal, whose leaves have a lemonlike fragrance. Most of these woody broad-leaved plants produce edible fruits, which animals eat and then disseminate the indigestible seeds.

Where the cordilleran forest grows high on the mountain sides and meets climatic conditions suitable for trees of the taiga, deciduous groves of aspen flourish. Their new growth and suckers provide a favorite browse for mule deer, which otherwise seek a varied diet including lichens, nuts, fungi, herbs and grass.

Many of these same foods appeal to a unique type of short-tailed burrowing rodent called the sewellel or mountain beaver. These heavy-bodied animals grow to be about eighteen inches long; they emerge from their tunnels only at night and then travel for short distances, chiefly along the banks of mountain streams or over wet soil where they can find the foods they like. While looking for signs of sewellel activity, we have often come across great mounds of woody cone scales left by another denizen of the forest: the Douglas red squirrel. These western squirrels cut off, drop and later collect the unopened cones from various pines and other conifers, stuffing them into hollow stumps or under logs, or piling 150 to 200 in a single heap on the moist soil near a stream. Natural dampness keeps the trophies from drying out. During winter, the squirrels burrow through three or four feet of snow to reach these caches and sit there feasting, tearing the cones apart and eating the seeds.

The red squirrels use their teeth to carry cones, and remain active all winter. Golden-mantled ground squirrels of about the same size have capacious cheek pouches, and fill them with a great assortment of dry fruits, seeds, mushrooms and even insects for transport to underground storage chambers. Yet the ground squirrels rely upon their body fat during their deep hibernation, and do not feed on their stores until they awaken, lean and hungry, in early spring. The smaller western chipmunks, by contrast, have little fat; they sleep lightly and visit their food reserves at frequent intervals until the weather warms up. They too use large cheek pouches to carry whatever surplus nourishment they find.

Larger rodents live where the cordilleran forest is open, or along its upper boundaries in the mountains.

Above right: The giant saguaro cactus often grows 40 feet tall in the Arizona-Sonoran Desert. (Martin Litton) Right: A paler race of the timber rattlesnake is known as the canebrake rattler (Crotalus horridus atricaudatus) *along the Gulf Coast and around the lowlands of Florida. (Jack Dermid)*

151

The western, yellow-haired race of the American porcupine shows a clear preference for the young twigs and inner bark of the yellow pine, whereas the marmots (*Marmota*) restrict their feeding to the plants that grow on rocky slopes where there are plenty of crevices in which to take shelter from predators.

These forested western mountains are now the principal refuge for mountain lions (or cougars, or pumas), which prey particularly on the deer. Formerly these big cats ranged across America to the Atlantic coast, but they have been progressively eliminated east of the continental divide. They still find enough to eat and avoid man over more territory than any other mammal of the New World—from northern British Columbia to Patagonia—often attaining a weight of more than 200 pounds.

Grizzly bears and black bears are far heavier when mature, but rely upon plant foods when unsuccessful in finding meat. They use their great strength and formidable claws to dig out ground squirrels, and sometimes pull huge rocks aside in an attempt to reach the hideaway of a marmot.

Porcupines are the favorite prey of the large weasel-like fisher (or pekan), which ranges eastward through forest country to New England. Hunting mostly at night, it seems able to flip a porcupine on its back and kill its victim by a quick bite through the unprotected under surface. Fishers will climb trees to reach the porcupines, ascending and descending headfirst with a dexterity that seems extraordinary in an animal weighing up to 20 pounds. They pounce also on sewellels, squirrels, marmots, mice and seldom overlook even a dead fish, such as a spawned-out salmon. A smaller relative, the pine marten is even more agile, relentlessly pursuing squirrels, catching them by day or night, in the treetops or on the ground.

Both the marten and the fisher are ready to raid the nests of the great gray owl, a two and a half- to three-foot bird of the same forests and those of western Eurasia as well. But the owl is quite capable of turning the tables; it is a giant among owls, with a continual need for meat.

So thoroughly do the pine martens explore the cordilleran forests for food that we marvel at the number of bird nests they overlook. The black-billed magpie needs big trees to support its massive platform of twigs, and ordinarily builds near the forest edge so as to be able to forage more freely in the open country beyond. Barrow's goldeneye ducks incubate their eggs in large tree holes, such as old woodpecker nest sites, within 100 yards or less of a pond, a lake or a river to

which they can lead their ducklings on hatching day. The varied thrush and Townsend's solitaire, like the flycatchers and Steller's jay, build their nests among the branches, following the commonest of all patterns in bird behavior. The blue grouse and the little Oregon junco nest on the ground; the former leads off its precocial chicks as soon as they hatch, while the other continues to bring insects to feed the young birds until they are old enough to fly. Only the dipper (or water ouzel) seems to find an inaccessible place in which to incubate its eggs and care for its offspring. It builds its oven-shaped bulky nest under the curtain of a waterfall, or on some ledge overhanging a swift mountain stream through the forest. This little gray bird is unique too in the way it goes after insects: it walks right into deep pools and swirling rapids, and uses its wings to propel itself while completely immersed, to find caddis-worms and stonefly naiads, blackfly larvae and others that cling to stones or burrow in the bottom.

Scrublands of Western North America

From a low-flying airplane, changes in color and in the height of vegetation follow contours more complex than the cuts in any jigsaw puzzle. Those that mark the boundary between the cordilleran forest and the western scrubland record a difference in the amount of moisture available to vegetation due to a whole array of interacting features: the slope of the land and its direction in relation to the sun and to prevailing winds; the drainage character of the soil, which reflects both the kinds of plants now growing and those that have left their products in the past; the altitude; the latitude; and the distance from the Pacific Ocean. On the ground, these changes are most noticeable along hillcrests and valley floors where, in just a few feet, the community of living things characteristic of a humid land gives way to the community natural to a semiarid region. The thorny shrubs with their leathery leaves are well adapted to a chronic drought which is not quite severe enough to produce a desert.

Spanish explorers pushed through this scrubland, mostly on horseback. They found their animals blocked or hindered by dense thickets of stiff shrubs and low trees. On alkaline soils these are commonly greasewoods and chamiso. Dwarf oaks and spiny acacias constitute chaparral. To protect their own legs from

Where glaciers scoured valleys, lakes now reflect snow-clad mountains and evergreen forests. (Martin Litton)

being scratched and torn, the horsemen invented leather protectors called "chaparajos" (now abbreviated by cowhands to "chaps"). The word chaparral now refers also to scrubland itself, and includes great areas from Mexico to British Columbia and eastward toward the plains, where the commonest plant is the tall sagebrush. Although the blossoms of this shrub are small, yellow and inconspicuous, Nevadans have chosen it as their state flower. Sagebrush leaves give off a spicy fragrance that to us is the surest sign that we are in the semiarid West. When we smell it, we can count on soon seeing a long-eared jack rabbit (actually a hare) dash across the road in front of us. Generally it will jump three or four times in flat arcs scarcely higher than the sagebrush and then take a great leap as much as five feet into the air, apparently to see better whether it is being pursued.

We are likely to hear a coyote bark at night, but not to notice this wily member of the dog tribe, for it blends almost perfectly with the sagebrush and benefits from keeping out of sight. Coyotes dine frequently on jack rabbits, and also on the Rocky Mountain cottontails that are common enough to be called sage rabbits. The rabbits feed on many different plants, such as the lupines, whose flower clusters often give the sagebrush country a purple hue.

The most conspicuous bird may be a sage thrasher singing loudly from a perch in a scrub oak or an acacia. If we pause a while to listen and enjoy its song, we may notice a little flock of sage grouse foraging among the sagebrush, and actually eating the leaves and fruits. The sage sparrow is more wary, and flicks its dark tail nervously as it hops around, searching for seeds or for insects to feed its young.

Where the oak trees grow taller, particularly the Californian live oak, the acorn woodpecker is often abundant, sociable and noisy. Its small red cap contrasts sharply with its dark blue back, wings and tail tip and with its white markings as it flies in an undulating

The pine barrens tree frog, native to bogs and swamps in New Jersey, is heard more often than seen. (Manuel V. Rubio)

Above left: The dog-day cicada, or harvester fly, of deciduous forests in eastern North America spends two years underground, feeding from roots. Males produce a loud, prolonged whine on hot days. (Emil Javorsky) Far left: The golden-fronted woodpecker, with a distinctive golden spot on the nape of its neck, inhabits deciduous woodlands from Texas into central Mexico. (William J. Bolte) Left: The big-eared white-footed mouse of eastern woodlands converts fruits, seeds and insects into flesh attractive to owls, foxes and other predators. (Jack Dermid)

course from tree to tree. More than any other woodpecker, it specializes in preparing small holes into which it can hammer an acorn, choosing for its storage sites not only the bark of oak trees but also the treated wood of telephone poles. A square foot of surface may hold two dozen acorns before the woodpecker goes elsewhere to continue its obsessive hoarding.

Attempts to discourage the acorn woodpecker from attacking telephone poles in this way have been scarcely more successful than efforts toward clearing the scrubland and using the area for raising wheat or livestock. The growing season is too short and precipitation too erratic unless supplemented with irrigation water brought from great distances and at great expense.

The North American Deserts

When scientists who are familiar with the truly arid lands of the Old World visit the deserts in the North American Southwest, they generally exclaim over the amount of vegetation present. Every few feet they see another plant growing and, if they dig down through the bare soil between the plants, they usually find sturdy roots a few inches below the surface. Deserts

155

elsewhere rarely show this much life, either because rain comes only a few times in a century or because man and his domestic animals destroy each pioneering plant for fuel or fodder.

Much of this vegetation is adapted to put forth root hairs within hours after moisture penetrates the soil, and to store water underground or in thick stems coated with an impervious cuticle. Leaves may be lacking altogether, as on the many members of the cactus family (Cactaceae). The giant saguaro cactus has accordion-pleated stems as much as seventy feet tall and three in diameter, capable of expanding to contain tons of water. Barrel cacti hoard moisture in the same way. The jointed flat stems of prickly pear get thicker, still protected by long spines amid clusters of short brittle bristly ones.

Coachwhip (or ocotillo) opens small green leaves after a rain, only to drop them as soon as drought returns. Its stiff spiny stems remain slender and never develop into woody trunks. Mesquite, on the other hand, follows much the same schedule with its foliage but becomes a tree of medium size, mostly because it extends roots to a depth of sixty feet or more where subterranean moisture is available while other plants can reach none at all. Creosote-bush can tolerate a 50 per cent loss of moisture from its thick leathery leaves without harm, surviving droughts that would kill other plants with foliage exposed above ground.

After a rain, seedlings spring up all over the desert. New leaves spread out, stems branch and grow, and bright flowers appear. Day after day the desert changes color as one kind of plant after another comes into flower. Representatives of the daisy family are usually among the first to open. The goldpoppy, which Californians have chosen as their state flower, blooms, sets seed and dies in just a few weeks. Sand verbenas may be the last of these ephemeral annuals, for aromatic resins in their leaves allow them to resist desiccation particularly well, letting them benefit from a few extra days of hot sunshine. The main body of the plant generally dies while its fruits continue to ripen. Then life remains only in the seeds, which may wait for months or years for another rain to help them germinate.

At the other extreme, the desert palm grows tall and old, a few fan-shaped leaves dying each year and an

The cacomistle, or ringtail, hunts at night for prey on deserts and along thornscrub edges in the American Southwest. It eats chiefly rodents in winter and spring, insects and an occasional bird in summer. (Willis Peterson)

equal number of new ones taking over their duties, where steep walls of rocky canyons shield the trees from wind. The Joshua tree branches grotesquely, each branch ending in a cluster of bayonet-shaped green leaves around the bud of a flower cluster; the bud will develop and its flowers open only after a generous rain.

When the desert plants are in flower, bugs and beetles, wasps and bees, flies and other insects flit about by day, sipping the nectar, attending to pollination, and laying eggs. These hatch promptly into young, for the food supply will not last. Hummingbirds join the insects at the flowers, coming for many miles. Birds of other kinds arrive almost as promptly to dine on the insects, or to eat the fruits and seeds that soon will be ripe. From deep in the soil, amphibians and even snails come up wherever water collects in shallow pools. Frogs and toads and salamanders may all be swimming around on the day following a rain. By then the spade-

Above the level of thornscrub and below the level of coniferous evergreen trees, each mountain in western North America has a band of aspen trees whose foliage turns gold in autumn. (Bill and Mary Lou Stackhouse)

157

foot toads will have mated and laid their eggs. In one more day, their eggs will hatch. Their tadpoles feast on microscopic plants growing in the warm water, and increase in size at an amazing rate. If some of the desert pool remains a month after the rainstorm, all of the surviving tadpoles will not only have transformed into adult toads, but each will have attained sexual maturity. If the pool dries up sooner, the young toads merely burrow into the soil and wait until the next rain gives them a chance to complete their growth.

At other seasons, almost all animals in the deserts of the Southwest restrict their activities to late evening, nighttime and early morning hours. Most spend the day underground, either buried a few inches below the surface of the soil or in definite burrows. In this way

Long after the young of the Virginia opossum are too big to get back into their mother's pouch, they use their sharp claws and prehensile tails to cling to her as she travels in search of food. (Jack Dermid)

they escape the deadly heat of the midday sun and minimize their loss of precious water by evaporation into the dry air. At night, in fact, radiational cooling often drops the temperature on the desert so far that the relative humidity rises far enough for dew to form. Animals that venture forth in these dark hours can lap it up, gaining enough moisture to last them several days in their nocturnal way of life.

Lizards far outnumber the snakes. Most, like the

various kinds of horned lizard, are insect-eaters that feed in morning and evening hours, finding their food by sight. The chuckwalla and the desert iguana prefer leaves and flowers, even those of the creosote-bush. The beaded lizards seize smaller lizards, eat the young of burrowing rodents, and whatever bird eggs and nestlings they can find on the ground. The Mexican beaded lizard, which has a long tapered tail and grows to be thirty-two inches long, and the Gila monster, which has a thick, short tail and attains a total length of only twenty inches, are the only poisonous lizards in the world. They are active primarily at night and for only those few months each year when the rains are most frequent. Whether in seizing a victim or biting in self-defense, the beaded lizard merely holds firmly and grinds its short teeth into the wound while the poison in its saliva takes effect—attacking the nervous system. Although these lizards should be treated with respect, only eight people in the last four centuries are known to have died within a few days after being bitten by one—and even in these cases other causes of death cannot be ruled out.

The rattlesnakes that hunt on the desert at night for prey that is warmer or colder than the soil are far more deadly. Even though they rarely grow to more than five feet long, they inject their venom more efficiently and, if alarmed, can stab their poison fangs into a person several times in a few seconds. Their principal food consists of desert rodents and sleeping birds, each victim struck, poisoned, and followed at a leisurely pace until it is inert enough to swallow whole. The most highly-adapted of these snakes is the sidewinder, which can climb loose sloping sand by throwing loops of its body as though it were a helical coil; its track is a series of parallel straight lines slanting up the slope.

The productivity of the plants in the deserts of the Southwest can be judged from the extraordinary numbers of nocturnal rodents that reproduce prolifically there. By spending each day in burrows eighteen inches deep or more, they have a natural equivalent of air-conditioning—a temperature of about 60° F. and almost 100 per cent relative humidity, while the surface of the desert above reaches 150° and moisture is scarcely measurable. Often these animals bring into their burrows the air-dried seeds and fruits they find and leave them for a few days to absorb water from the humid air. Then, when they eat these foods, the animals

Black bears, when young, often climb trees to escape suspected danger on the ground. (Martin Litton)

From northern Mexico to southern Canada, raccoons with black masks explore for food. (Wolf Suschitzky)

get the absorbed water as well as the nourishment. They convert some of the carbohydrates into metabolic water and, by having remarkably efficient kidneys, get enough moisture in these ways to satisfy their needs for life without ever having to drink.

Best known of these desert rodents are the attractive little kangaroo rats; they are so nocturnal that they stay hidden on moonlit nights. In complete darkness they hop along on their hind legs, balancing with a long tail. If pursued, they can make ten two-foot leaps per second—which is more than thirteen miles per hour—and zigzag as they go. If a predator gets too close, they stop this fast hopping gait and, instead, kick sand into its eyes. Yet they get caught and eaten by persistent coyotes, by kit foxes, which earn their other name of "swift foxes," and by some of the desert-visiting owls. Badgers sometimes dig up kangaroo rats, destroying the whole system of burrows the little rodents have made.

These same beasts and birds of prey catch also the pack rats that weigh three times as much as a kangaroo rat—nearly a pound when fully grown. So big an animal, with its preference for plants as food, has to be abroad for many hours each night to eat enough. During all of this time it is in continual danger, and

must be ready to dash for home. Its refuge is not a burrow but a formidable fortress built of spiny cactus pieces, with plenty of hiding places in twisting passageways among them. No one knows how this soft-furred rodent keeps from being hurt on cactus spines while eating cactus, let alone while it is constructing its shelter or going into and out of it in darkness. Predators make no attempt to tear apart the pack rat's home, and snakes (which might slither through the crevices) are unlikely to come by day when the owner is in. At that hour, the rodent's greatest danger may be from intense heat. But this it tolerates, crouching motionless in its shady mound.

The badgers of the desert have no difficulty digging deep and spacious quarters underground, and they retire to these as soon as the day becomes uncomfortably warm. The slender-bodied ringtails (or cacomistles), which are related to raccoons, find deep crevices in rocky cliffs or stretch out on the shadowed surface of high limbs on trees. Coyotes and bobcats commonly dig lairs for themselves on the north side of a mesquite or other shrubby tree that gives some shade. The bobcats climb well and are quite ready to take refuge in a taller tree, high above the surface of the ground which, by mid afternoon, is the hottest place of all. From any of these refuges the predators can reach the desert floor toward sunset, and take advantage of anything edible that comes their way.

The desert birds are almost equally divided among those that are directly dependent upon the plants—whether for nectar, fruits or seeds—those that eat insects and other small animals, and those that scavenge for whatever carrion they can find. Hummingbirds of several kinds make regular rounds of the flower clusters on century plants. Little Inca doves with long square-ended tails run on short legs along the ground, pecking at seeds. Larger white-winged doves come to visit the saguaro cactus flowers, and actually serve as the principal pollinators.

The Gila woodpecker eats insects but cuts a nesthole in a big saguaro. Unable to fill so large a hole with sap, the cactus merely produces a viscous coating against the air in the nest cavity. The coating hardens to become the part of the saguaro most resistant to erosion. Long after the cactus has died and disintegrated, this lining of the nest will lie on the ground and be known as a "desert shoe" because of its shape and single opening—the place where the woodpecker went in and out.

The smallest of American owls, the elf owl, often nests in the Gila woodpecker's cactus hole after the woodpecker leaves. The owl stands barely six inches

160

tall, and may sit in the doorway facing out all day, only to fly to the ground at dusk. It specializes in catching scorpions, seizing each by the tail and nipping off its stinger; then it swallows the scorpion's body, pincers and all. The cactus wren is a bigger bird, with a heavily-spotted breast and throat. It commonly builds its cup-shaped nest in the middle of a cholla cactus, where the large needle-sharp spines pointing in all directions give almost perfect protection. Like other wrens, it is an insect-eater, helping significantly in the control of bugs, flies and caterpillars. A much bigger nest in a cactus, more open to the sky, is likely to be that of the road-runner (or chaparral cock), a unique bird that is reluctant to fly. Actually a large kind of cuckoo, it has a long tail, a distinctive crest, and a taste for snakes and lizards instead of other food. It dashes after its

Preying principally on rabbits, hares and small rodents, the stubby-tailed bobcat, or wildcat, matures at a weight of 40 pounds, compared to the 200 pounds of the long-tailed cougar. (Ed Park)

prey, and runs off with it to the shade of a bush or a clump of cactus.

Every few days a vulture soars over each area of desert, examining the ground below. Some mammal may have died of thirst, or disease, or accident. Part of a carcass may remain from a predator's kill. In summer, the soaring bird is likely to be a black vulture with a short tail and a white mark below each wing. During other seasons, it generally is a turkey vulture with longer tail and unmarked wings—wings so broad

161

that the bird seldom needs to flap them to maintain its altitude. In winter, when all turkey vultures have shifted to more southern territory, they are usually closer together and each keeps watch on what its neighbors do. If one vulture spirals down to feed on a dead animal, others come from all directions, ready to share in the meal.

The Great Plains

Until a century ago, these scavenging birds could find far more food over the grasslands that extended east of the Rocky Mountains from the Prairie Provinces of Canada (Alberta, Saskatchewan and western Manitoba) to northern Texas. Immense herds of American bison and pronghorns grazed as nomads, attended by packs of wolves and the numerous coyotes. The dog-like carnivores kept the grazers alert, and feasted frequently on unprotected mothers that were giving birth, or unattended young, or adults that could no longer keep up with the herd because of injury, disease or age. Between feasts, the wolves and coyotes stalked or dug for rodents, hunted the sixteen- to eighteen-inch grouse known as prairie chickens, and pounced on mice and grasshoppers. The most numerous rodents were burrowers: prairie dogs, ground squirrels and pocket gophers, all largely vegetarians. They thrived, as did the mice—grasshopper mice, harvest mice and prairie deer mice—where the grazers came frequently, keeping the grasses and herbs low. The rodents then could see a predator before it pounced.

Black-footed ferrets took refuge in the extensive warrens of the prairie dogs, and seemed to prefer their hosts as food. Long-legged burrowing owls nine inches tall and prairie rattlesnakes shared the underground quarters of the prairie dogs but usually ate elsewhere. The owls caught grasshoppers by day and mice by night, whereas the rattlers hunted primarily in darkness when mice were most active.

Today the mice and grasshoppers still find a place on the Great Plains, but most of the other animals are gone. The ferrets and the prairie chickens have become so rare that they are listed among the world's most endangered wildlife. Prairie dog towns are scarce enough to need protection. The surviving bison are in parks and special refuges. The pronghorns have moved to the edges of the desert and the foothills of the mountains. The coyotes have spread elsewhere in America where they can find more food.

The perennial short grasses of the drier western plains—the buffalo grass and the bluegrasses—have been plowed under to make space for introduced wheat and other grains. So have many thousands of square miles where formerly the Junegrass and needlegrasses grew two feet tall, and each stream bank was bordered by thorny buffaloberry, prickly prairie rose and leathery-leafed wolfberry.

Farther east, where the rainfall is greater and the native grasses formerly attained a height of six to eight feet before late-summer drought ended their growth for the year, other introduced crops have replaced the bluestem, the switchgrass and the Indian grass. This was the vegetation to which the Plains Indians set fire each year, deliberately extending the area over which the bison and pronghorns would roam. The fires encroached upon the natural forests to the east, increasing the spread of the plains grasses and the other living things that benefited where the land was open to the sky.

The common sunflower, which sometimes grows fifteen feet tall and is topped by a massive head of flowers nearly twenty inches across, appealed immediately to colonists from the Old World. They largely ignored the age-old Indian method of parching the thin-shelled seeds, then grinding the meats into meal. Instead, they used the sunflower leaves as fodder for cattle and used the seeds in their shells as food for domestic fowl. Today this plant is cultivated in Eurasia as well as temperate North America for the sweet yellow oil that can be pressed from the seeds, and for the oilcake residue which, together with the stem and leaves, can be converted into nourishing fodder for livestock. And some gardeners the world over raise the common sunflower as an ornamental.

North America's Deciduous Forest

From eastern Manitoba to the coast of Maine, south to Virginia and, on the mountains, to Georgia too, a forest of deciduous trees grew as an almost solid stand until the European colonists began to fell it. Enough remnants can be seen today to let us credit the statement that an American gray squirrel might once have traveled through the interlacing treetops from the Atlantic shore-

On a wet bank in the Smoky Mountains, fire pink and wild stonecrop attract different animals to pollinate their flowers. From the book Appalachian Wilderness: The Great Smoky Mountains *by Eliot Porter. Copyright © 1970 by Eliot Porter. Published by E. P. Dutton & Co., Inc. Reprinted with permission of the publishers.*

line to the Mississippi River without ever descending to the ground.

On the north, substantial amounts of this hardwood forest survive as second growth, adjacent to the spruces and firs of the taiga. Many a valuable grove of eastern white pine grows along the irregular boundary. Hemlock forms an understory of the deciduous forest wherever no fire, to which it is particularly vulnerable, has come for many years. Arborvitae and eastern red cedar provide other touches of evergreen foliage, especially where the soil is thin and rocky outcroppings show how recent was the last extension of the Ice Age glaciers.

One of the tall hardwoods that gave character to the deciduous forest—the American chestnut—has virtually disappeared, the victim of fungus disease introduced accidentally from the Orient into New York state in 1904. A second hardwood—the American elm—is dying out today, seemingly doomed by another disease, which came from Holland. Black walnuts have been logged out, first as a source of attractive wood for cabinetwork, then for material with which to make propellers for airplanes in World War I, and most recently as the preferred wood for gunstocks, both for military and civilian use. Black locust and honey locust, like most kinds of maple, oak and birch, have lost most of their former popularity; the supply of large straight trunks has shrunk too far for these trees to provide raw materials for industry, and substitutes have been found to reduce cost, generally eliminating the hand work required by the grain of wood.

Sugar maples are still tapped for their sweet sap in Quebec and New England to provide the solution that can be boiled down to yield maple sirup and maple sugar. Pecan trees are planted in orderly rows to obtain commercial nut meats, but mostly in the southeastern states and seldom in their original home: the Mississippi Valley forests, from Illinois to eastern Texas. Sycamores and basswoods have become shade trees for city streets and suburban lawns, often replacing ash-leaved maple, which is short-lived, and poplars, whose roots too frequently invade sewer lines and clog them. The catawba tree (or Indian bean) and the flowering dogwood tree are planted for their floral displays wherever the climate is favorable.

The great horned owl stands twenty inches tall. Except where killed off, it finds forest homes and prey from coast to coast and from the Arctic to the tropics. (Manuel V. Rubio)

The wealth of woodland wildflowers in eastern North America has diminished as the trees were cut or forest land turned into public parks. Hikers and picnickers must be asked not to pick the handsome orange woodlily, the dog's-tooth violets, the bloodroots and the mayapples. Naturalists make little pilgrimages in season to admire the remaining untouched stands of marsh marigold, skunk cabbage, hepaticas, lady's-slipper orchids, and jack-in-the-pulpit. Cardinal flower and wild bergamot have been moved beside treasured exotic plants as ornamentals in flower gardens.

For some animals, the changes in the deciduous forest region during the last three centuries have been opportunities. Never before have there been so many woodland borders and cleared areas where, through neglect by man and pioneering by plants, new growth is available as cover and food close to ground. White-tailed deer find hiding places and forage that are lacking where the trees are tall. Cottontail rabbits take refuge among the stiff thorny canes of wild raspberry and blackberry, escaping from both the red foxes of more open country and the tree-climbing gray foxes of the forest itself.

Other woodland animals, especially the larger birds, have lost much of the habitat they need for survival. Hawks find fewer places with tall dead trees and solitude, suitable as nest sites and lookout points. Dead trees may be scarce, because the forest is managed for maximum production; woodpeckers go elsewhere, searching for bark beetles and wood borers to eat, and for trunks in which they can make a nest cavity. Owls that would inherit the woodpecker nestholes have few available. Raccoons and black bears need den trees too. Quite possibly the passenger pigeons, which nested in incredible numbers between the Appalachian Mountains and the Mississippi River, would have dwindled to extinction early in the present century for want of forest tracts to nest in, even if market hunters had left the birds alone. The sole survivor of this species that died in 1914, a captive in the Cincinnati zoo, may actually have outlived her wild habitat.

The Southeastern Coastal Plain

From the eastern lowlands of Virginia southward and around the Gulf of Mexico to eastern Texas, the most distinctive plant community is often swampland dominated by tall baldcypress. This deciduous conifer generally supports ghostly gray streamers of epiphytic Spanish moss, a member of the pineapple family (Bromeliadaceae) that has spread northward out of the American

tropics. Marshes often prove to be almost solid stands of a sedge called sawgrass. Both types of wetland are winter havens for waterfowl from nesting grounds on the arctic tundra and around prairie potholes—birds that regularly follow the Atlantic coast or the Mississippi River on seasonal migrations south and north. Many of these birds need never reach Florida or the Gulf of Mexico because they find inland extensions of the coastal plain up river valleys, particularly in the Mississippi basin.

These are the home areas for the brown pelican, the American alligator, and the wetland pit-viper known as the water moccasin (or cottonmouth). Each of them catches fish in its own distinctive way, and competes to some extent with the Louisiana heron and the much larger great white heron. In the baldcypress swamps, the huge ivory-billed woodpecker has dwindled until now it is on the verge of extinction, and is known only in eastern Texas.

Slightly higher land, often with a sandy soil, supports a southeastern forest of evergreens. Tall slim pines, particularly the longleaf whose needles are twelve to fourteen inches long, and the slash pine of Florida and Cuba, are now tapped regularly for turpentine and other naval stores. Shortleaf pine and loblolly provide valuable timber. Often the saw palmetto and prickly pear cactus grow low over the thin soil under the pines. The stiff fan-shaped leaves of the palmetto, each on a stalk armed with spines as sharp and regular as saw teeth, are assumed erroneously to be the crown of foliage on a young cabbage palm—another member of the palm family with similar foliage. Cabbage palm grows erect from coastal Georgia through the Florida Everglades and near the Gulf of Mexico to eastern Texas. It tolerates frost somewhat better than the royal palms native to southernmost Florida and Cuba, where they are often planted along avenues, seeming to add dignity with their smooth columnar trunks and shade-giving crowns.

For millenia the Colorado River and its tributaries have carved into the plateaus of the American West, creating cliffs and gorges. (Bill and Mary Lou Stackhouse)

Two evergreen plants of this region that are often overlooked have special interest to scientists because they represent relics from millions of years ago. The coontie (or Seminole bread plant) is the only kind of cycad native to America north of Mexico. From a thick starchy subterranean stem it sends up tough leaves as much as two feet long, each dark green and divided to the midrib like those of simple ferns. But its conelike reproductive bodies show it to be closer kin to the conifers than to ferns or to palms; its ancestors were prominent among the forest trees of the Age of Reptiles, and some of them are not much different from their forerunners. The other relic is the whisk fern, which consists merely of repeatedly forking slender stems, with no roots, no leaves and only inconspicuous spore cases for reproductive organs. Similar in these ways to the earliest known of land plants, more than 350 million years ago during the Age of Fishes, it grows today from the rotting stumps of baldcypress trees and the debris that collects among the broken leaf bases of cabbage palms, all across the southern part of Florida.

Among the most unique plants of the coastal plain is the Venus'-flytrap, which now takes advantage of the abundant light and occasional watering on golf courses near its native Wilmington, South Carolina. Its leaves extend in a flat rosette, each with an enlarged blade at the tip, fringed around with stiff bristly appendages. If trigger hairs on the upper surface are jostled by a fly or an ant, the two halves of the blade quickly fold together, their bristles interlocked; if they capture the insect, the leaf presses against it and secretes digestive agents. After absorption, or if the prey is missed, the leaf opens again and waits like a trap.

In the thickets, where better soil supports broad-leaved trees and shrubs, the whistling of the cardinal and the endless imitations of the mockingbird (which now ranges in the South from coast to coast) are often the most conspicuous sounds. They conjure up a scene replete with huge live oaks, handsome magnolias, sweetgums with star-shaped leaves, and persimmons for whose fruits birds must compete with squirrels, raccoons and opossums—the only marsupial mammal that has spread into the nearctic realm on its own.

8

Australia and New Guinea

Of all the distinctive land areas in the world, only Australia is both a continent and a single country. Its 2,940,599 square miles make it the smallest of the seven continents, and sixth in size among the countries of the world. No other continent or country has so many unique native animals and plants. It is a biological showpiece, harboring some of the strangest life on earth.

Fully a third of Australia is desert, in a broad central expanse that grades into semiarid areas of sparse grass, low thornscrub or scattered trees. The chronic drought requires living things there to be highly adapted to survive. In the northwest, the west and the south, grasslands and savannas extend to the coast, but less than 1,000,000 square miles of Australia receive from rain, snow and underground streams enough moisture to support a forest. Rain forest is even more restricted, mostly because the mountains are so low. The highest peak, Mount Kosciusko in New South Wales, is only 7316 feet above sea level.

Australia's forests contain almost all of the world's 600 different kinds of eucalypts (*Eucalyptus*), which are evergreen trees with narrow aromatic leaves that generally hang vertically downward. Both surfaces receive the intense sunlight at some time of day, while the sun's rays pass through layer after layer of leafy branches. Except in the rain forests, eucalypts are the dominant trees. They account for fully a tenth of the continent's variety in flowering plants.

Australians call their eucalypts "gums," and recognize how well these trees are fitted for growth where the air is dry and the soil has only a little moisture most of the year. They talk of "sclerophyll forests," meaning that the leaves are hard-surfaced, tough and gray-green, or reduced on many of the shrubs in the undergrowth to stiff waxy needles or mere lumps on green branches that remain all year in this seemingly leafless condition. Under the gum trees the ground is littered with fallen leaves of a tan color and willow-leaf shape, that are scarcely drier than while they were alive on the trees above.

The Tropic of Capricorn divides this continental country into a tropical half and a temperate half. But drought, rather than seasonal cold, determines the distribution of living things. The farthest extremity of Australia's outlying state—the little island of Tasmania (26,383 square miles)—is at a latitude south of the Equator comparable to that of Marseilles, Istanbul, Chicago or Boston in the Northern Hemisphere.

Close to Australia on the north is the gigantic island of New Guinea (317,000 square miles), separated by only a narrow shallow strait in which small islands offer the biogeographical equivalent of steppingstones. New Guinea has a tropical climate, rainfall in excess of sixty inches annually, towering mountains and lush forests. In these forests are plants and animals of Australian origin, some peculiar native kinds, and many that are characteristic of the East Indies and the Orient.

Australia and New Guinea are so famous for their egg-laying mammals (monotremes) and their pouched mammals (marsupials) that the native mice, rats and bats (placental mammals) are overlooked. Actually, there are more kinds of placental mammals than of the primitive ones that receive the publicity. Admittedly they are smaller, and yet important in the web of life.

Australia's Southeastern Forest

Anyone with a sensitivity to plants recognizes at a glance how distinctive are the sclerophyll forests of southeastern Australia. Naturalist-explorer Joseph Banks, who accompanied Captain James Cook on his

The shining white bark of the ghost gum makes it a distinctive landmark in the semi-desert parts of Australia, whether in the extreme north of the continent or much farther south. (J. Allan Cash)

celebrated circuit of the world, convinced his commanding officer to give the name Botany Bay to the harbor where they landed on the southeast coast. The bay itself is only six miles across. Overlooking it is a public reserve called Kuringai Chase, just north of Sydney, where much of the sense of wildness Banks found in 1770 is still maintained. From the headlands and cliffs of weathered sandstone, tall eucalypts of several kinds tower above low trees and shrubs of the protea family known as "Australian honeysuckles" and named *Banksia* to further honor Sir Joseph. One of the common shrubs with fragrant pink flowers is named the native rose, although it actually belongs to the citrus family (Rutaceae). The strange grass-trees of the lily family are often called "blackboys" because of a fancied resemblance in silhouette to the outline of an aborigine in a grass skirt—an appearance given by the short black trunk under a great cluster of down-curving narrow leaves, and a slender flower stalk rising vertically.

Whenever we explore Kuringai Chase or similar forested areas, we can imagine ourselves in a forest of sycamores or plane-trees because the trunks of the gum trees are so mottled in their patches of peeling bark. Their narrow leaves and hard, dry fruits tell us otherwise; they are members of the myrtle family. So are many low-growing shrubs and small trees, such as the bottlebrushes, the tea-trees, which often hold sand dunes in place, and the bread-and-meat plant. Small herbs with gray leaves and upright flower heads surrounded by flannel-like floral bracts prove not to be edelweiss, which they resemble, but flannelflowers of the carrot family. Great clumps of green flattened upright stems are not a spineless gorse, but bossiaeas with a brilliant display of butter-yellow flowers in spring, so much like those of gorse that we know them to be members of the pea family (Leguminosae). Shrubby spider flowers bear near the ground the same peculiar type of blossoms as the silk oak holds among its silky leaves; silk oak, a valuable timber tree, is just another member of the diverse protea family. Other members are all around in the Australian sclerophyll forest: shrubs known as wooden pears because of the woody inner tissue in their pear-shaped fruits; hakeas, such as the sea-urchin whose crimson flowers are followed by a woody fruit; and the handsome waratah whose heads of small red flowers are cupped by ranks of attractive red floral leaves.

Forest Birds

Active birds from sparrow-size to jay-size visit the flowers of the gums and other trees, the banksias and various shrubs, sucking nectar and eating the small soft fruits. About ninety kinds of these perching fliers are primarily nectar feeders, but they are important in pollination. Most efficient in inadvertently insuring that fruit will form and seeds develop are the honeyeaters (Meliphagidae), which generally have slender downcurved pointed beaks and a long tongue with a brushy tip. In use they curl the tongue to form a tube through which they suck the nectar. Particularly varied and numerous in Australia, these birds are represented too on New Zealand, many islands of the Southwest Pacific, as far east as Hawaii and west as the Moluccas.

The New Holland honeyeater, which perpetuates the old name for Australia, has a melodious call that makes the bird easy to recognize from a distance, despite its dull olive-green plumage. Helmeted honeyeaters travel in small flocks, bickering noisily with each other over the flowers, but joining forces to drive birds of other species from the particular tree in which they are feeding. The painted honeyeater shows such a fondness for small fruits that it regularly visits different trees at appropriate seasons. The closely related wattlebirds and friarbirds, which have colorful wattles behind the eye or featherless areas on the head, time their arrival in each forest area so regularly to match the ripening of fruits that they are regarded as true migrants rather than nomads.

Australia is the center of distribution for wren-warblers, a peculiar group of Old-World warblers that carry their long tails cocked up almost vertically as they dart among the low shrubs, pursuing insects. One is the red-tailed heath wren, which sometimes abounds in the low shrubs known as Australian heaths. The little emu wren, which has a bright green tail longer than its head and body combined, is one of the few birds known with only six tail feathers—the absolute minimum.

Several of the native birds that haunt the forest undergrowth for insects have become well known to Australians through their willingness to venture into pastures and suburban gardens. Conspicuous among them are certain members of the family Muscicapidae (the Old-World flycatchers) that are so unlike ordinary flycatchers in plumage and action as to be known as robins and fantails. Yellow robins are particularly numerous in southeastern Australia. Other robins (five species of *Petroica*) are boldly patterned in red-brown, black, white and blue. They hop along the ground, instead of dashing after insects on the wing.

One of the fantails, affectionately known as the willie wagtail, resembles in many ways a towhee of North America. Blue-black above and white below, with a white eyebrow mark, it perches on low branches, clotheslines, cow's backs, and all sorts of vantage points from which it darts down to seize small prey upon the ground. Except for its boldness, it might be compared to an American bluebird. No matter where, willie wagtails are seldom still or silent. Almost constantly they swish their tails from side to side, or up and down, or fan out the feathers and then close them together again. Sometimes they seem to forget that they are less than nine inches long, for they pursue far larger birds and drive them away, much as a kingbird might a crow.

Marsupials of the Night

The marsupial mammals that take over the forest in darkness are often highly selective in their diet. The greater gliding possums climb out of treehole nests they have lined with stripped eucalyptus bark and leaves, and peer around with prominent eyes. Using their long prehensile tails, which are hairless only underneath near the tip, they work their way along branches to reach young leaves and blossoms on the gum trees. Or they spread the membranes between their legs on each side and plane for as much as 300 feet from one tree to another, like flying squirrels except that they weigh four pounds or more and measure eighteen inches from nose tip to base of tail. Among the foliage they may meet a lesser gliding possum half as big, hurrying along a branch with a load of leaves securely held in curls of its flexible tail; if frightened, the lesser glider drops its load and planes away.

The least timid of these marsupial climbers is the brush-tailed possum, which often descends from eucalyptus trees in the midst of cities throughout eastern Australia. It is a slow-moving animal, seldom releasing hold with its prehensile tail before getting a good grasp with all four feet on some new support. It is more versatile than most possums, supplementing its diet of vegetation from many kinds of plants with insects, snails, and an occasional bird too young to get away. After a brush-tail baby is too big for its mother's pouch, it generally rides along on her back for a month or more. One mother that we found while exploring by flashlight seemed reluctant to move, even when we touched her pink nose with the glass covering of the shining bulb. At other times these animals click loudly if disturbed, or hiss, or rise up to full height and scream repeatedly.

Neither of the spotted marsupial cats—the smaller (fourteen to seventeen inches long in head and body) or the larger (up to almost thirty inches)—seems common enough in the gum forests today to give much trouble to middle-sized mammals. Indeed, they feed extensively on cold-blooded vertebrates, insects and crustaceans, as well as on birds and marsupial mice. Occasionally they attack a bandicoot that is using its long, pointed snout to reach insects, worms, and other delicacies among fallen leaves and rotting logs.

The forester (or great gray kangaroo), which is the largest living marsupial and may weigh 150 pounds, is too big and powerful to fear any native predator. It browses throughout the night on a wide variety of low foliage, and grazes on many of the grasses that grow in the shadeless forests. By day these animals rest inconspicuously, always ready to leap away on their extraordinary hind legs, balancing themselves with their long muscular tails.

Higher humidity in the cooler gum forests farther south along Australia's eastern coast favors the growth of tree ferns and tall fan palms. The strange Illawara lily (or beefsteak lily) of the amaryllis family produces large rosettes of coarse green leaves and then a flower stalk eight to fifteen feet tall bearing a cluster of dark red blooms. This is lyrebird country, although the repertory of mimicked whistling calls produced by the concealed male may suggest every other bird in the forest. Only rarely is a naturalist lucky enough in the early morning or late afternoon to see a male pursue a potential mate into an open place and there raise his amazing tail-covert feathers up over his back and forward until he is peering through the display to see how the hen bird is responding. Lyrebirds feed mostly on insects, crustaceans and worms uncovered by scratching among the litter on the forest floor.

The confused "woo-oo-om" calls of a flock of wonga pigeons led us to stalk them where they were feeding on the ground. But these big birds were too alert, and flew off with a great whirr of wings. We could not tell what fruits and seeds they had been finding, for, between the shrubs, the only plants we could recognize were terrestrial orchids. Greenhoods in flower reminded us a little of jack-in-the-pulpits in a rich woodland of eastern North America. Some of the shrubs on sandier soil included the native fuchsia and waxflowers, both of them among the handsomest of spring wildflowers in this part of Australia.

High on Low Mountains

Although the peaks along Australia's continental divide are only 4000 to 6000 feet above sea level, the climate near the crests is cooler enough to hold snow through the winter and to give alpine wildflowers a special advantage in spring as meltwater becomes suddenly

While courting, the male lyrebird throws his fan of display feathers forward over his head. (L. H. Smith)

available. More than fifty kinds blossom in the wake of the melting snow. A big white buttercup thrusts up among the granite boulders. Snow daisies produce pink buds that open flower heads suggesting those of giant fleabanes or small daisies of the Northern Hemisphere. And on all sides the snow gum grows as an evergreen, its twisted trunk smooth and mottled brown, cream and red where the bark has peeled away.

Later in the summer season, golden everlastings expand their dry floral display, and the spiny-leaved mountain parsleys produce flower heads that display flat-topped umbels. Among the strangest are the trigger plants, whose relatives live at lower elevations in southwestern Australia and in tropical Asia. Small bees are attracted by the petals of the pink or white flowers, but as soon as they alight to seek nectar, the central column of the blossom bends over smartly, picking up whatever pollen is on the insect's back and then applying another load of the golden dust from a pair of stamens. Trigger plants need about fifteen minutes to reset the mechanism.

Water Birds of the Eastern Coast

The mountains are so close to the Pacific coast that the rivers flowing east are short and swift. Yet each of them maintains a broad estuary where birds of swamp and marsh find places to nest and feed. The Hawkesbury River in New South Wales is the southerly limit for the tropical lily-trotter (or jacana), and near the boundary of tolerance for the jabiru stork and white ibis. Australian snakebirds swim under water in pursuit of fish, and then join the fish-eating cormorants in drying their outstretched wings while perching on a tree branch overhanging the water.

Among the reeds and cattails, small moor hens quack sociably. Sunlight glitters from the iridescent blue chest, bright beak and red legs of swamp hens. The native reed warblers sing melodiously. The swamp hens are native to New Zealand too, but the others are like the black swans in being Australian representatives of genera that are found world-wide. So are the red-billed silver gulls that scavenge along the coasts.

In winter of the Northern Hemisphere, about thirty species of migrant birds arrive from arctic tundras to feed along the eastern shoreline of Australia. They include golden plovers, Mongolian dotterels, little whimbrels, and the smallest of sandpipers, the red-necked stint. At the same time of year, other individuals of the same kinds search out places to rest and feed in New Zealand and other islands of the Southwest Pacific realm. By traveling so far they gain a summer at each end of the migratory route, but rarely become so attached to the Southern Hemisphere that they nest there instead of going north again.

The Temperate Rain Forests

Southeastern Australia (Victoria) and adjoining South Australia receive up to sixty inches of rain each year, just as does the western half of the adjacent island of Tasmania. In all of these places, some of the mountain-ash trees (*Eucalyptus regnans*) tower 350 feet above the ground in dense forests that are true temperate rain forests. Lesser, broad-leaved trees include valuable hardwoods such as the lilly-pillies, which have soft fleshy fruits although they belong to the myrtle family. Below their crowns is an understory of tree ferns. Still lower are wiry maidenhair ferns and many different terrestrial orchids.

Giant earthworms ten feet long and as big in diameter

The narrow, pendant leaves of gum trees cast only modest shade in Australian forests. (Emil Javorsky)

as a man's thumb burrow through the rich loam, making gurgling and bubbling sounds in wet weather as their bodies displace water in their passageways. Smaller worms are present too, and insects of many kinds. Apparently these edible creatures and the habitat in which they thrive have led to the evolution and survival of the world's second kind of lyrebird, which has a less spectacular tail development than in the more northern relative, and the famous platypus (or duckbill), the egg-laying mammal with a beak like that of a duck. Platypuses swim freely in the small streams, and dig burrows as shelters and nests in the bank. But at night they crawl out or seek food along the stream edges, finding small animals by smell and touch.

In these rain forests the black cockatoos show a strange dietary habit for members of the parrot family. Instead of fruits and seeds, they seek grubs, caterpillars and other insects, getting these by tearing away the bark on trees and the dying giant fronds of tree ferns. In a sense they take the place of woodpeckers, of which none are present in any part of Australia or New Zealand. Woodpecker-like in another way, they nest in tree holes high above the ground. No doubt this brings them into competition for space with the nocturnal marsupial mammals which, in Victoria, include a mouse-sized dormouse possum, which grows fat in autumn and hibernates, and also Leadbeater's possum, which is extremely active. It was believed extinct until 1961, when numbers of these insect-hunters were discovered in dense forest.

Tasmania—The Island State

Rising as it does from a southern extremity of Australia's continental shelf, the island state of Tasmania is separated from the mainland by open water about 150 miles wide and less than 250 feet deep. The gap between the two—Bass Strait—was dry land during the Ice Ages and for a while afterward because the oceans were lower in relation to the land. Living things could spread across the bridge in both directions, producing similarities in the flora and fauna on both sides.

The island is about the same size as Ceylon or Scotland, but shows three different climates, east to west. The center is a high plateau, with mountains around and within it, the highest (Mount Ossa) rising to 5203 feet above sea level. The alpine meadows have characteristic little cushion plants of the family Stylidiaceae, including the rock cushion and the alpine cushion, which grow in similar situations in New Zealand and Tierra del Fuego.

The rainfall of thirty to sixty inches yearly on the plateau itself maintains two beautiful lakes and supports an open savanna. The trees include a tall gum, middle-sized she-oaks and various shrubby wattles. In the south, the cooler climate favors the growth of a small tree despised as the "horizontal." Its spiny stems arch over and take root, much as briars do in the Northern Hemisphere, producing an obstructive platform of vegetation above the ground. It is a member of the Southern Hemisphere family Cunoniaceae, to which the favorite Christmas bush belongs.

The eastern slopes are too dry to bear trees, but sixty inches or more of rain fall annually on the western slope of the island—enough for a temperate rain forest. Much of the original timber has now been cleared, particularly the tall gums. Antarctic beech grows 100 feet tall. Distinctive conifers include the endemic huon-pine, a celery-pine and the 100-foot King Billy pine—another endemic. In boggy places, a sedge called button grass forms great tussocks as much as six feet tall, just as it does amid the sclerophyll savanna of the plateau.

Tasmania's "Devil" and its Neighbors

Most famous of the purely Tasmanian animals are the marsupials known as the Tasmanian wolf and the Tasmanian devil. The "wolf" is teetering on the edge of extinction, unable to defend itself against introduced dogs or the stockmen who claim it attacks their herds. It is a strange predator. As much as fifty inches long, with a twenty-five-inch tail, it prefers the blood from its victims, or blood-filled tissues such as nasal membranes and liver; it rarely eats muscles or returns to its kill. This habit leaves carrion for the smaller "devils," which are usually less than twenty-four inches long with a twelve-inch tail. The "devils" can crush bones in their powerful jaws, but combine hunting with their scavenging. They attack even the poisonous black tiger snake and the copperhead, which are Tasmania's principal dangerous reptiles.

Biogeographers generally assume that if a species on an island is the same as a distant neighbor beyond the water barrier, then it arrived in recent times. On Tasmania, the emus, the forester kangaroos, the wallabies and bandicoots, the broad-footed marsupial mice and marsupial cats would all fit this assumption, because others just like them live north of Bass Strait. By contrast, kinds of life without close relatives nearby are likely to have come from ancestors that reached the locality far back in the past. By this criterion the most antique creatures on Tasmania would be two kinds of primitive, shrimp-like crustaceans found no-

where else in the world. So far, no members of the order to which they belong have been found on other continents, and there is no way to tell where the crustaceans came from originally. They swim or crawl over the bottom of streams, pools and lakes in parts of Tasmania where erosion by glaciers has left the mountains jagged and spectacular, the rivers swift or cascading from one rocky ledge after another. The crustaceans feed on algae and on smaller animals, and sometimes swim so rapidly that they leap out of the water.

Another explanation may account for animals that are limited to Tasmania, such as the pygmy possum and the largest of the spiny anteaters, or echidnas. The island may simply offer them more security than the continent across Bass Strait. The same species may have lived along the north shore of the strait until the European red fox was introduced into Australia. The fox finds similar native mammals to be easy prey, and has all but exterminated from the southern parts of Australia one of the scrub wallabies (or pademelons). These

two used to be as common there as they are today on Tasmania.

Barriers in Bass Strait

The animals that are present on the north side of Bass Strait but missing in Tasmania tell something of the past barrier to their spread. Tasmania has no lyrebirds or yellow robins or willie wagtails. It lacks pythons and snake lizards (pygopods), geckoes and goanna lizards, tortoises, and among marsupial mammals, the glider possums and beloved koalas. Trees (particularly the eight kinds of gum trees on whose foliage the koalas feed) may simply have been missing along whatever

175

land bridge connected the coast of Victoria to Tasmania in the past. The true barrier is always a lack of suitable habitat between two areas. This view could be extended to explain why the coarse-haired wombat is found on both sides of Bass Strait, whereas the soft-furred wombat is missing from Tasmania. The coarse-haired species prefers to eat grass and roots, but sometimes forages among the storm debris on sea coasts; the saltwater margin of a land bridge would be suitable for it. The soft-furred wombat eats grass too and hides for the day in a burrow it has dug; however, it usually remains on drier, higher inland areas, or among cliffs of sandstone and limestone—and these may well have been lacking along the bridge to Tasmania.

In Bass Strait itself are several islands fringed about by leathery kelps. To one of these, appropriately called Albatross Island, the shy albatross comes to nest—its only home territory except for isolated Mewstone Rock off the south coast of Victoria. It lays its eggs and raises its young on the flat top of these havens, while short-tailed shearwaters dig nest burrows into the shallow soil. Australians call the shearwaters "muttonbirds," and know that they spend most of their lives far at sea, circuiting the vast Pacific Ocean, feeding on fishes at the surface but returning to breed where they themselves were hatched. Many of these islands are almost overrun by poisonous snakes, including tiger snakes up to six feet long that eat young small shearwaters and grow fat. For the rest of the year the snakes must catch fast-moving lizards, such as skinks, and live on their fat.

A few of the islands in Bass Strait offer sanctuary to the small-headed Cape Barren geese; the others live on equally isolated bits of land along the south and west coasts of the continent. Unable to fly to any safer place, these geese survive only where predators are lacking and where they have grassy pastures in which to graze. The total population of these unique birds varies between 4000 and 6000, and there is always the danger that they will dwindle into extinction.

Australia's Scrublands, Grasslands and Deserts

The settlers who sought to homestead the center of Australia gradually developed their own way to evaluate the hostile climate. For each place they counted the months in which the average rainfall would leave a surplus for the plants to use—at least three times as much rain as would evaporate and drain away in the same period. A region with this much water for vegeta-tion in nine months of every year could be cultivated profitably for almost any crop. Areas with between five and eight months of adequate moisture might produce grasses and forage for livestock, but nothing more. Areas with less than five months moisture were deserts.

Occasional storms do reach Australia's desert heartland. They may come in January from the north as extreme disturbances of the summer monsoon system, or in July from the south as parts of the winter storm system in the Antarctic. Yet no water may fall for year after year on any particular area of desert. And the rivers that flow into the heartland from higher ground, particularly from the western slopes of the continental divide, simply end in dry lakes that shine in the bright sun with white salts left where the moisture evaporated.

Billalong, Mallee and Mulga

Only two major rivers actually reach the sea: the Murray and the Darling, which join and empty near the city of Adelaide. In dry season they often cease to flow and, for much of their length, are represented only by a series of ox-bow lakes called billabongs. Between one stagnating lake and the next, river gums and coolibah trees mark the channel edges, just as cottonwoods or willows might in the Northern Hemisphere.

The billabongs have their own native life, based upon the fishes that inhabit the river water, with adaptations to match the current in times of flood and the stagnation during recurrent droughts. Under these conditions, the Murray cod shows that it is an olive-green relative of the big sea basses by growing to a length of six feet and a weight of 180 pounds. Australia's unique blackfish becomes especially abundant and about twelve inches long. The perch-like yellowbelly is intermediate in size. The young of all of these native fishes and other small fishes too attract three different kinds of cormorants, four of egrets, two of bitterns, and many kingfishers. At least six kinds of ducks find water plants to their liking. Black swans cruise about and nest. Australia's maned goose (or wood duck) breeds if, close to the billabong, trees with nest holes are available.

Productive pastures and grain fields now cover the cooler and better-watered parts of the Murray-Darling

The famous koala "bear" of eastern Australia lives only in the sclerophyll forests where it can find its favorite kinds of eucalyptus foliage. On Australian highways, signs warn "Koalas cross here at night." (J. Allan Cash)

176

Basin. Formerly the region supported a savanna of native grasses studded with dwarf gum trees referred to generally as "mallee" or mallee scrub. In places it formed dense thickets. In others the trees remained isolated because the rainfall was too unreliable to favor germination of seedlings.

Mallee, like billabong, is an aborigine word adopted by the white settlers. So, in a sense, is "budgerigar"— the abundant, small and slim parrots of mallee savanna. The "budgie" that is so popular as a cage bird got its name from a common misunderstanding. When an aborigine was asked, "What are those birds?" he answered in his own way *Boogeree gar* ("Good food") instead of giving the distinctive name. The response could have applied equally to about half a dozen other kinds of parrots that frequent the same Australian territory: the white cockatoo with its sulfur crest, the galah with its pink breast and general gray appearance, the blue-bonnet parrots, which are red underneath, the red-rumped parrots, the yellow rosellas or the daintily crested cockatiels.

Strangest among the birds of the mallee scrublands are the chicken-sized mallee fowl, or lowan, the only one among the ten different mound-builders to nest in

such dry country. At nesting time the male mallee fowl digs a great crater in the soil and fills it with a mixture of dry, sandy material and the freshest vegetation he can find. Soon the plant tissue begins to ferment and produce heat. The male probes with his beak into the mound he has built, apparently guided by its temperature in his automatic actions—opening up the mound to let heat out or scratching more earth over it to keep the heat in. He produces in this way an incubator in which his mate lays her eggs, and continues to tend it until the little birds hatch out and run off to fend for themselves.

To the north and west of the mallee is more arid territory, where the dwarf gums are replaced on the savanna by the mulga shrub. A mulga grass grows well in some years and poorly in others, making this area marginal for ranchers with sheep and cattle. They call their land simply "the mulga," and have difficulty utilizing it efficiently because the domestic animals on a private holding can never be as nomadic as the emus and red kangaroos that are the native grazers.

Australia's Desert Heartland

With so much desert to challenge them, Australians notice small differences as the basis for distinctive names. The Great Sandy Desert in the northwest is neither as sandy nor as barren as the Sahara. The Great Victoria desert, which occupies so much of the southern half of the state of South Australia, is partly the amazing Nullarbor Plain where a level layer of porous limestone 900 feet thick prevents the growth of trees by letting whatever rain falls percolate downward quickly beyond reach of roots—hence "Nullarbor" ("No trees"). The Simpson Desert in eastern central Australia is largely a level tableland too, with conspicuous pebbles and small stones ("gibbers") rolled in all directions by the winds; the area is often called "the gibber desert." In the midst of all this aridity, some low mountains (the Macdonnell and Musgrave Ranges) and a famous rounded sandstone rock (Ayers Rock) focus attention on a small area of grassland near the town of Alice Springs.

These mountains in the desert offer a refuge at elevations between 4100 and 4950 feet, where half a dozen kinds of gum trees grow. They include the ghost gum,

Wedge-tailed shearwaters tend underground nests in darkness on Heron Island and other isolated land areas on the Great Barrier Reef off the east coast of Australia. (J. Allan Cash)

178

which has a white trunk; and along the dry river courses the river gum grows in much the same situations as mesquite trees would grow in the deserts of the American Southwest. Corkwoods of the protea family provide a shrubby undergrowth, their main stems gnarled and twisted. A desert palm spreads its fan-shaped leaves at least 700 miles away from any other palm trees. A palm-like cycad represents a far more ancient group of seed plants.

Below the ghost gums, the palms and corkwoods, and in the shadows under the cycads, a litter of fallen leaves conceals insects of many kinds. The spiny anteaters (or echidnas) of northern Australia and New Guinea are there in the mountains to scratch for the insects each night. By day they hide in crevices among the rocks or in burrows. Rock wallabies with long bushy tails marked by a succession of dark rings, and hill kangaroos (or euros) find small caves in which to hide when neither basking in the sun nor grazing in darkness.

A hill kangaroo will eat even the spiny porcupine grasses that grow like giant pincushions twenty feet across and as much as six feet tall. These cushions get water through deep roots, and provide a refuge for many kinds of insects and reptiles. They also produce quantities of grains that nourish the desert birds, such as the common zebra finches, and the burrowing mice. Often the same burrow serves the Australian hopping mice, which are true rodents, and the Australian marsupial mice. Both types are jerboa-like in their ability to jump and to get all the water they need from their plant foods.

Like other deserts, the Australian ones have their ephemeral annuals in plants adapted to germinate, produce leaves and flowers and ripe fruits within a few weeks, and then live almost indefinitely as seeds until the next time a heavy rain drenches the ground. The grains of *Spinifex* grass seem especially able to wash along with the rain water, but eventually they catch their spiny extensions between the mineral particles of the soil and stick there, ready to sprout. Some perennials grow as herbs, using moisture whenever it is available to send up visible parts from hidden roots or under-ground stems. Handsomest of them is Sturt's desert pea with brilliant scarlet flowers marked with glistening purplish black. Parakeelyas of the purslane family have succulent stems on which large rose-red flowers appear.

Woody plants are more scattered, and often amazingly tolerant of alkaline soils. Outstanding in this respect are saltbush, which generally grows on clays, and bluebush, which favors calcareous sites. Neither of them normally attains more than five feet in height.

Cypress-pines, which are true conifers of Australia, remain mere bushes in regions of moderate aridity, but become timber trees if they receive a little more moisture.

The ephemeral plants have their counterparts in the two-inch shield-shrimps and smaller fairy shrimps that develop rapidly from drought-resistant eggs as soon as a storm leaves a temporary pond in the desert. The crab *Telphusa* and a desert crayfish are almost as well fitted to profit from erratic rains. Desert snails are numerous. A desert frog digs deeply and stays dormant for years if necessary, surviving between rains on an inner ration of water stored in its body.

The perennial lizards and the free-roving birds of the desert are easier to comprehend, as they are more like ourselves. They get what they need from the plants and plant products, or from animals that do. Australia has more of its distinctive dragon lizards (family Agamidae) in the desert than elsewhere, and one—the eight-inch devil lizard—is perhaps the most grotesquely spiny reptile in the world. The devil lizard moves slowly in search of ants on sandy soil, and may eat 1000 to 1500 of them in a day. Never does it rush about on its hind legs as do the Australian bloodsucker and the fringed lizard of Australia and New Guinea. Both the fringed lizard and the bearded lizard have the startling habit of turning on any attacker and suddenly distending membranes around the head and opening the mouth to display the brightly colored inner surfaces. Abruptly the lizard seems a ferocious opponent for any predator.

The deserts have more kinds of monitor lizards and of geckoes than any other part of Australia. The monitors are called goannas, and hunt by day, although they are usually timid enough to race back to their burrows if alarmed. By night they sleep, which restricts their prey to other reptiles, birds' eggs and nestlings on the ground, unwary mice, and large insects such as grasshoppers. The geckoes search for insects at night, and spend the day either basking in the sun with the slit pupils of their big eyes closed to a series of minute pinholes through which to view the world, or burrowing into the soil to escape the heat. Some of the Australian geckoes have extraordinary knobs at the tail tip, and a few the disconcerting habit of standing up on long legs and barking loudly. In the darkness a six-inch gecko can sound many times its actual size.

The Garden Corner—Southwest Australia

Isolated by the deserts from the rest of the continent, the southwest corner of Australia has become famous for its distinctive plants and animals. A generous rainfall favors the growth of great forests of karri (*Eucalyptus diversicolor*), with straight white trunks 300 feet tall, and jarrah (*E. marginata*), which yield valuable timber, and red gum (*E. calophylla*). The red-flowering gum (*E. ficifolia*) is different from most of its close relatives in having bright red flowers instead of yellowish-white ones.

In many ways the flora of southwestern Australia is the richest on the continent. It includes 112 out of the 180 kinds of spider flowers, 106 out of the 120 kinds of tea-trees, 71 out of the 100 of corkwoods, 37 of the 50 kinds of Australian honeysuckle shrubs, and all 52 of the known dryandras. Only in this one corner does a native mistletoe start out being parasitic on the roots of trees, then become independent as a forty-foot tree— the Australian Christmas tree.

Of nearly 6000 different kinds of flowering plants, almost 80 per cent are native nowhere else. A pitcher plant, so different from any other as to require a family of its own (Cephalotaceae), bears lower leaves that are vase-like and hold water in which insects drown, and upper leaves that are specialized only for photosynthesis. A member of the amaryllis family has irregular flowers so similar in shape to the furry toes of a mammal that it is known as kangaroo-paws.

In this garden of Eden, the dangerously poisonous mulga snake of the desert turns up occasionally, along with the equally venomous brown snake and tiger snake. All are members of the cobra family. They prey on mammals and birds, most of which are different from those seen in Australia's eastern forests. The southwest corner has its own native marsupial cat, and its own honey possum that sips nectar from the flowers of corkwoods, dryandra bushes and Australian honeysuckles. It has a rat-sized marsupial anteater called the numbat and, on Rottnest Island off the coast near Perth, a small wallaby that may already be extinct on the mainland. Koalas and Tasmanian devils are lacking, although fossils prove that they once lived in this area. Nor are there any gliders, despite the apparently suitable habitat for them in the gum forests.

No lyrebirds or friarbirds or kookaburras call out in southwestern Australia. The black swans and willie wagtails are the same, and so are the silvereyes except that they have green backs instead of gray ones, as near the eastern coast. The common robins are a separate species, with a gray breast, not a yellow one. The black cockatoo is of a kind with a white tail and not a yellow tail. The honey parrot (or lorikeet) is a purple-crowned one about seven inches long, the smallest of parrots in the southwest but equipped with a brush-like tongue for sipping nectar just like those on various honey parrots in the northeastern forests.

Australia's Tropical Coast

From the vicinity of Darwin on the north coast to somewhat south of Brisbane on the east, Australia is fringed by rain forests with a truly tropical flavor. The Cape York Peninsula points like a finger toward the adjacent island of New Guinea, and clearly has served in the past as a travel route for land life between New Guinea and Australia. South of the peninsula lies the Atherton Tableland where, for 200 miles north and south, 30 miles east and west, Australia has a tropical rain forest closely similar to those on New Guinea.

It is a strange forest for Australia, without a gum tree (*Eucalyptus*) or tea-tree (*Melaleuca*) or wattle (*Acacia*) or beefwood (*Casuarina*) of any kind. Instead, its conspicuous trees belong to families and genera found in the East Indies and on South Pacific Islands. Almost certainly their ancestors emigrated from New Guinea, and not from any other part of Australia. They include kauri-pines and hoop-pine, which are evergreen conifers of the araucaria family—so characteristic of the Southern Hemisphere. Toon, or red cedar, and Australian beech are members of the mahogany family. Australian maple belongs to a little family (Icacinaceae) of trees and shrubs represented in tropical forests all around the world. Often these trees are laced together by clambering fan palms of the Old-World tropics; in the undergrowth the same palms form spiny tangles that are almost impenetrable. Australians know the vine-like growths as "lawyers."

They give the name "anthouse plants" to some of the vegetation that perches on outstretched tree branches and in crotches against the trunk. When jostled, these plants suddenly swarm with small ants that nest inside. Biting at one end and stinging at the other, the ants repel animals that invade their territory. Others of these perching plants are the staghorn and elkhorn ferns that

Above right: Dunes of red sand, patterned into ripples by the wind, form only one of Australia's many types of desert. (Janet Finch) Right: The tiger snake is one of the largest and most venomous in Australia—a continent on which more than half of the snakes are poisonous and where victims of snake bite have about a 50 per cent chance of surviving. (Photographic Library of Australia, Sydney, NSW)

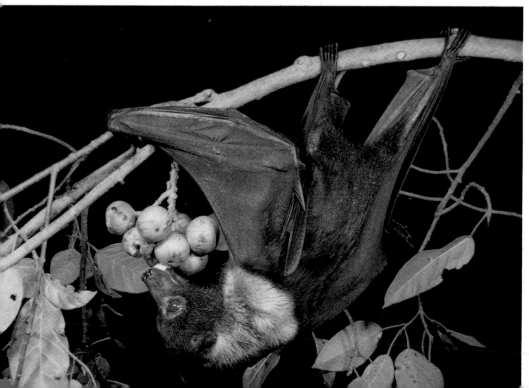

can be grown as curiosities in humid greenhouses; different species are native to Australia, New Guinea and the islands to Malaya, and to the rain forests of tropical Africa.

"Hunter's rope" is almost any flexible, tough tropical vine. In the Australian tropical rain forest it may be any of a wide assortment of aroids related to the monsteras and anthuriums of tropical America, or a climbing shrub of the screwpine family similar to others found from Ceylon to islands of the South Pacific.

Stalking about under the tree ferns and other vegetation of modest height are cassowaries, which are the third largest among birds, after African ostriches and Australian emus. Unlike the others they have a horny, featherless crest and a pair of dangling red and blue wattles at the throat, making them "the birds with the built-in jewelry." The same species and two more live in New Guinea, from which the ancestors of Australian cassowaries probably emigrated less than a million years ago by walking across a land bridge. Further interchange today is prevented by the shallow waters of Torres Strait.

The same rain forests in northeastern Australia have two native mound-builder birds: a scrub fowl and the brush turkey. Shy and reluctant to fly, they use their big feet both to scratch among the leaf litter for fallen fruits and small edible animals, and also to draw together the big mounds of wet vegetation which will generate heat from fermentation and serve as an incubator for the birds' eggs. For these mound-builders the task seems much easier than for the mallee fowl in semiarid savanna country. The other seven kinds of mound-builders are native to islands from Samoa in the South Pacific to the Nicobar Islands in the Indian Ocean. Biogeographers interpret this distribution as indicating that the Australian mound builders were immigrants long ago from the East Indies by way of New Guinea.

Above left: Second only to canaries as a familiar cage bird over most of the world, the Australian budgerigar in the wild travels in immense flocks that seek out shallow lakes from rare rains. Far left: A spectacled "flying fox" feeding on fig fruits near Cairns, along the tropical coast of Queensland. (Both by Graham Pizzey) Left: Rainbow lorikeets, which are slightly smaller and much swifter than pigeons, use their brush-like tongues to get nectar from flowers and soft pulp from ripe fruits. They are found in both Australia and New Guinea. (Lorus and Margery Milne)

Bowerbirds and Birds-of-Paradise

A similar southward spread probably explains the presence in northwestern Australia's tropical forests of seven of the world's nineteen kinds of bowerbirds, and four of the forty-three kinds of birds-of-paradise. The rest live in New Guinea and on nearby small islands, such as those of the Aru group at the mouth of Australia's big northern Gulf of Carpentaria.

The Australian bowerbirds show all the gradations in behavior of their relatives. The green catbird, which is fifteen inches long and named for its mewing calls and heavy beak, builds no bower at all. The tooth-billed bowerbird constructs only a stage on which to display himself in attracting a mate. The golden bowerbird, although only nine inches long, makes a sort of roofed gazebo as much as nine feet tall, and lays a carpet of soft mosses in front of it on which to cavort. The satin bowerbird, thirteen inches long and resembling a short-tailed grackle, is like three other related birds in building an elaborate avenue with two rows of upright sticks in the middle of his display ground. Adding bits of moss and slender vines, he closes over an arching roof until his "bower" becomes almost a tunnel. He decorates its floor and walls with bright pebbles, bits of bleached bone, shells, colorful leaves and fresh flowers (or petals), as an amazing lure—as though to pique the curiosity of a female who might become his mate. To lead her to the entrance, he whistles and chirps, imitating the calls of other birds, or making sounds like grunts and wheezes that somehow prove effective.

Only the least spectacular of the birds-of-paradise live in Australia, where all four kinds are known as riflebirds from one of their most distinctive calls: a high-pitched buzz ending in a dull click, like the sound of a bullet and then its impact. Australians who have not heard this sound sometimes claim that the birds were called riflemen because of a similarity in plumage colors to the green and black uniforms of British rifle regiments.

Bird-wing Butterflies and Pouched Mammals

These rain forests of northeastern Australia shelter a fascinating array of animals, including most of the continent's 350 different kinds of butterflies. The largest of them, known as bird-wing butterflies, are actually giant swallowtails as much as nine inches in wing span. Their broad fore wings are generally marked in green and velvety black, their hind wings smaller, rarely with any tails. These handsome insects visit flowers of trees and vines in the high canopy, then descend to seek out particular plants—woody climbers of the birthwort

The male of the golden bower bird builds an elaborate tunnel in which to approach a mate. (John Warham)

family (Aristolochiaceae)—upon which their caterpillars can develop. Australia is the southernmost extremity of their range, which extends through the East Indies and the Philippine Islands.

Some of the marsupials in these forests are almost as widespread, such as the tiger cat and the bandicoot. Others are more local, such as the musk kangaroos, which dig and turn over rotting logs to get insects and worms. A majority have so many close relatives in New Guinea that there is no reliable way to learn which direction their ancestors took to produce their present distribution.

The striped possum is found on both sides of Torres Strait; others of the genus are in New Guinea. The tree kangaroos, which rarely come to the ground, include three kinds in New Guinea and two in the rain forests of northeastern Australia. The slow-moving cuscuses, which remind us of both the kinkajous of America and the tailless slow loris of Southeast Asia, hold tightly with their long prehensile tails while curled up sleeping by day and foraging for fruits and sleeping birds by night. In Australia their principal danger is from the amethystine rock python that has also invaded Aus-

tralia from the north, to become the continent's largest snake, sometimes twenty feet long.

Magnetic Ant Nests and Man-eating Crocodiles

Downslope to the west of the Atherton Tableland is a flat area of northern Australia extending to the coast. It receives virtually all of its rain between October and March, during the summer monsoons. This change in climate is encountered in just a few miles and with a slight decrease in altitude along the edge of the rain forest. The perching ferns and orchids disappear, along with most of the mosses and lichens. Spaces between the trees increase, and the landscape becomes a savanna with tall gums, particularly the stringy barks, sixty-foot fan palms, forty-foot tea-trees with white bark peeling in paper-thin sheets, and twelve-foot cycads like miniature feather palms as an understory. In some places, the strangest trees are bulky baobabs, similar to those on savannas of eastern Africa.

The most peculiar features in this semiarid landscape are the so-called magnetic ant nests, each one a thin vane of dried mud as much as twenty feet tall, built one sand grain at a time by an Australian termite (or white ant). Oriented with one flat face to the north and the other to the south, they cast a minimum of shade and intercept a minimum of heat from the tropical sunlight. Inside the "magnetic" nests the termites stay reasonably cool while the soil on all sides heats up to dangerous temperatures.

Along the north coast itself, and the lagoons and estuaries of rivers that flow only a month or less each year, mangrove trees of fourteen different genera grow in dense thickets. Through the adventitious roots of the mangroves, the stemless nipa palm often extends its feathery fronds, adding to the tangle. Just to make a passageway through the mangroves far enough to look out over the water can be a major undertaking. It is a strange amphibious world, with fishes that climb out on land and birds that dive or swim or wade. Crocodiles lie in wait, submerged except for their watchful eyes and open nostrils.

East Indian fishes of shallow water swim near the surface. One of them, the famous archer fish, takes aim at insects on mangrove roots and overhanging foliage, then spurts a jet of water that often brings a victim within swallowing distance. Mud skippers with bulging eyes use their fins to creep up slanting roots, but keep their tails in the water; if alarmed, they skip and flip over the mud flats and sand bar from pool to pool. Always the fishes must be wary of mangrove bitterns and jabirus (Australia's native stork) that pose motionless, ready to stab for a meal. Mangrove kingfishers hover over the water, and dive for fishes near the surface.

An endemic Australian crocodile, which rarely grows to eight feet long, is a menace too. But the real monster is the saltwater crocodile of Southeast Asia and the East Indies, which visits Australia's north coast; sometimes this reptile exceeds eighteen feet in length. A ten-footer almost killed a friend of ours who is over six feet tall and muscular. He saw the crocodile rise close to him in a coastal pool, and had the misfortune to slip into the water as he backed toward safety. The reptile seized him by the arms and followed its normal procedure of rolling over and over while dragging its victim toward the bottom in a deep place, where drowning would be inevitable. Our friend felt helpless, but as he struggled he may have kicked the crocodile in the belly, for the animal released him. He pulled himself ashore, his arms and hands badly mangled.

Where the water is fresher, the coasts are more marshy. The handsome blue water lily floats leaves that are nearly two feet across in the water film. The vegetation attracts the semipalmated goose (or magpie goose), just as it does along the south coast of New Guinea. Green pygmy geese are found on both coasts too, but a close relative is on the one side only and known appropriately as the Australian pygmy goose. Along the two facing coasts, the wandering tree ducks (or whistling ducks) of the East Indies and western islands of the South Pacific grow larger and may be a separate subspecies.

The Tropical Pacific Coast

Along the northern part of Australia's Pacific shore line, east of the Atherton Tableland, the native plants and animals are quite different. The most conspicuous trees are often the screwpines and Australian pines, neither of which is a true conifer. In Australia the "pines" are called she-oaks (from the *shee*-ing sound of wind through their jointed green twigs), or horsetail trees (from the scale-like leaves in whorls at the joints), or beefwood and bloodwood (from the red color of the heartwood), or South-Sea ironwood (for the hardness of the timber from many kinds).

Trees with naked seeds—gymnosperms—are represented along this coast by native kinds held in high esteem. The bunya-bunya pine is so productive of two-inch edible seeds that in aboriginal settlements, ownership of the crop from each tree is regarded as a living heritage, a right to be passed to successive generations in a single family. Two different cycads that yield a starchy edible pith are not only endemic, but also the

world's only cycads with fronds that are twice-divided, like those of so many ferns.

Often the plantsman is challenged to recognize the family to which the various trees belong, particularly when they are not in bloom. He may see that the most prominent gum tree (the carbeen) is a member of the myrtle family (Myrtaceae) without realizing that the giant tristanias, which sometimes tower still higher, are close kin. The nettle family (Urticaceae) is represented by the Australian nettle tree, on whose leaves are glandular hairs with a substance intensely irritating to human skin. Firewheels and Queensland-nut trees are of the protea family. The famous wattles, with golden powderpuffs offering nectar, and the tall Moreton Bay chestnut (or Australian bean-tree) are legumes. The Australian flame tree, which is highly regarded for its displays of flowers, and the ornamental kurrajong are distant relatives of cacao in the sterculia family (Sterculiaceae). An antarctic beech with spreading buttresses probably grows closer to the Equator than any other member of the beech family (Fagaceae).

Along this tropical coast, we found ourselves needing constant guidance from naturalists who had spent years, not just weeks, in the Old-World tropics between Ceylon and Samoa. Otherwise we were bewildered by the flying creatures we met and the local names given to some of the strange fishes. "Burnett salmon" and barramunda are names of the Australian lungfish, whose red flesh is edible. Apparently herbivorous, it attains a length of six feet; it has a rounded tail, uniquely paddle-shaped fins, and a single lung. Its eggs and young resemble those of frogs, and are to be found along the Burnett and Mary Rivers of southeastern Queensland, where this ancient fish is native.

Fruit bats known as "flying foxes" hang by day among the mangroves and other trees. The adults are as big as a half-grown cat, have a six-foot wing span, and bicker so loudly over choice branches as supports that a roost of these creatures in a swamp can be amazingly noisy even at midday. At dusk they set out for fruit trees, and often do great damage to orchards. Biting into fruit that is almost ripe, they crush the pulp in their mouths, swallowing the juice but spitting out the residue and any seeds.

Some of the larger parrots and pigeons can be equally destructive and, along the tropical east coast of Australia, both parrots and pigeons of medium and large size are incredibly varied and numerous. Their bright colors are understandable in a land with so few native predators. Protective resemblance to green foliage would have far less value than patterns that could be seen by other birds of the same kind at a distance. Even large size has definite advantages. With a big beak or a wide gape, a bird can eat large woody fruits, either by cracking them parrot-fashion to reach the seed inside, or by swallowing them whole as the big pigeons do. Moreover, small parrots would be at a disadvantage in the tropics because the large parrots are less troubled by the numerous stinging bees and wasps that quickly take over any treehole of a size suitable as a nest site for small parrots. This seems to explain why the principal hole-nesters among small tropical Australian birds are the few parrots and kingfishers whose habit it is to dig cavities in termite mounds. There the termites go about evicting the bees and wasps.

So often the names of the birds along this coast are misnomers coined by pioneers on the spur of a moment or as approximations for aboriginal words. What were we to look for when told that we should see twenty-inch currawongs and Australian butcher-birds (family Cracticidae); frogmouths (Podargidae); drongos (Dicruridae); wood-swallows (Artamidae); magpie-larks (Grallinidae); minivets and cuckoo-shrikes (Campephagidae); babbling thrushes (Timaliidae); honeyeaters (Meliphagidae); white-eyes (Zosteropidae); flowerpeckers (Dicaeidae); waxbills, mannikins and grassfinches (Estrildidae); Australian sunbirds and spider hunters (Nectariniidae)? These unfamiliar families are native nowhere else. To us the words pitta or jewel-thrush (for members of the family Pittidae) seemed to fit hundreds of colorful little birds, and not just a specific few. It was a relief to recognize a parrot or a pigeon, or to get to know the regent bowerbird and the Albert's lyrebird.

Land Life Along the Great Barrier Reef

From about the latitude of the Tropic of Capricorn almost to New Guinea, a succession of coral islands and coral reefs along the Queensland coast form the continent's rampart against the waves of the Pacific Ocean. More than 1250 miles of this Great Barrier Reef lie within the tropical zone, partaking of the life of the vast sea to the east, of the coast fifteen to fifty miles to the west, and of the East Indies to the north. Beyond the reef on both sides, near the edge of the continental shelf, the water is about 180 feet deep. From personal experience we know how rough the wind can make the water even on an ordinary day. On an extraordinary day, when a storm beats in from the ocean, the reef and its islands must prove their sturdiness.

Opportunities for land life are limited, but fully utilized. "One-Tree Island" may have two trees—both alike, or one a she-oak and the other a twisted, brown-barked, broad-leaved *Pisonia* of the four-o'clock family. Among shrubby undergrowth will be a few screwpines, a *Muehlenbeckia* of the buckwheat family (Polygonaceae) that seems to take the place of its close relative the sea grape in the New World, and a *Scaevola* or a *Goodenia* to represent a little family (Goodeniaceae) whose center of distribution is Australia. Under them, on the coral sand, will be succulent green branching stems of the carpetweed known locally as pig's-face. All of these plants tolerate salt spray.

In the shade of the *Pisonia* trees, white-capped noddy terns build their nests of interlacing twigs on forking branches. All night long they call to one another restlessly, until dawn lets them fly away to dive for fishes. White terns of half a dozen kinds divide up the open beach into unmixed colonies, island by island and month by month. Wedge-tailed shearwaters (or mutton-birds) excavate burrows in the earth to hold their single eggs and the parent that is incubating. The shearwaters come and go in darkness, tending their hatchlings until each chick is rolypoly. Then they desert it and let it complete its development alone. It must fledge out, find its way to some open place where it can get a running start, learn to fly on its own, and venture forth for food from the broad Pacific.

White reef herons make big nests of twigs high in dead she-oaks, where they can look in all directions. They stalk for crabs along the beach, choosing the lee shore where waves will interfere least. Small, red-beaked silver gulls stand facing the wind, watching the few that are airborne to see if the scouting birds descend. This is their way of staying alert all day for any food that may wash ashore, any distressed fish or crab caught in a shallow pool by the retreat of the tide, any tern egg or chick left unattended. We found the gulls insistently efficient in catching, killing and picking clean any young green turtle that emerges from the sand and tries to scramble to the sea from the place where its mother laid her eggs. Between 5 A.M. and 7 P.M., the turtles have little chance. The successful ones make the perilous trip in darkness, braving only the land crabs, and swim out on the high tide to deep water.

Mysterious New Guinea

Second only to Greenland in total area and far exceeding it in space for living things, the island of New Guinea long baffled explorers. Among those who participated in many expeditions was the late Dr. E. Thomas

Gilliard. We remember his advice when we mentioned a wish to visit the island for a month or so. "You're over twenty-five," he said, "and that's too old to tackle such dense forests and rugged terrain. Even the botanists hardly got beyond the fringe of New Guinea until the mid-1920's." He knew all about the rivers that are too violent for navigation, the small patches of bare soil inland that are too steep for an airport runway, and the native people who keep so busy with tribal wars they have scant tolerance for strangers.

Helicopters and road-building machinery have changed all this faster than a man can age. Now New

"Magnetic" termite mounds are oriented like thin vanes that intercept most heat from the sun in early morning and late afternoon, but little at midday. They tower ten to twelve feet tall in northern Australia. (Robert Smith)

Guinea has many reasonable roads, rental automobiles, and motels of a pioneer quality. Like a transition on a movie screen, when one scene fades out and a new one takes its place, the mystery of New Guinea is dissolving into an exciting picture of a lush island on the crossroads between Asia and Australia.

187

Knowing how shallow and narrow are the waters of Torres Strait, it is easy to assume that New Guinea is just a big outlying island of the Australian continent, closer and more similar to it than Tasmania. The plants prove otherwise, for most of them are of kinds found in the East Indies and islands of the South Pacific. Few have clearly come from Australia. The eminent botanist Dr. Ronald Good commented on this in the 1964 edition of his fine *Geography of Flowering Plants:* "With the Australian flora out of the picture, the remaining distributional pattern is orderly and comparatively simple; with the Australian flora included where it is, the pattern is essentially disorderly."

Glaciated Peaks in the Tropics

Any explanation for the presence or absence of particular kinds of life on New Guinea must allow for the tremendous geological changes there during the last 15 million years. The lofty mountains, which include many active volcanos, probably rose to their present (or greater) heights within that time. They have permanent glaciers today. During the Ice Ages these ice fields spread to within 6000 feet above modern sea level. The sea then was lower, perhaps by 500 feet, which not only linked New Guinea to northern Australia by a land bridge but also narrowed the open water barriers to the west, north and east.

Presumably the alpine plants would be least affected by the spreading and retreat of New Guinea's glaciers. Cooler weather would force them to grow at lower elevations, and a rise in average temperature permit an upward spread. Today the alpine meadows range from 10,000 feet elevation to snow line (14,500 feet). Below the meadows is a montane forest, including large areas of antarctic beech, others of coniferous podocarps, as well as oaks and forty-eight different kinds of true rhododendrons. Probably the beech and podocarps came from Australia while the weather at sea level was chilled by the Ice Ages. Yet no oaks spread in the opposite direction, and Australia has only a single species of *Rhododendron.*

Between the lower edge of the montane forest (2000 to 4000 feet elevation) and the mangroves along the coasts lies a dense belt of rain forest composed mostly of trees, shrubby climbers and perching plants common in the East Indies, with some from Australia. Wherever this forest has been cut repeatedly, the eroding land is generally covered by a stiff grass known as kunai. Growing six feet tall, it is hard to walk through, but useful for thatching native huts. Otherwise this grass is regarded as a weed all over the archipelago.

More animals than plants may have crossed between Australia and New Guinea and found a place to settle. From his own and other's experiences, Dr. P. J. Darlington, Jr., says in his *Zoogeography:* "Various Australian animals extend to the areas of open eucalyptus woods in southern New Guinea, and various New Guinea animals extend to the areas of rain forest in northeastern Australia." Beyond those areas is a world of difference.

As in Australia, the only native mammals of New Guinea are primitive egg-layers and marsupials, bats and true rodents. The island has no duckbill, but probably is the original home of spiny anteaters (echidnas) since it has three kinds of long-billed anteaters and a short-billed one. Australia has the same short-billed spiny anteater, and an additional kind belonging to the same genus in Tasmania. New Guinea has about fourteen kinds of kangaroos and wallabies (family Macropodidae), of which the tree kangaroos and forest wallabies appear to have evolved there; Australia has about thirty-nine members of this family, only the two tree kangaroos in northeastern Queensland having come from New Guinea. New Guinea's twenty-two kinds of phalangers (family Phalangeridae) include about five different cuscuses, two of which have managed to cross Torres Strait to Australia's Cape York Peninsula. The other seventeen presumably crossed in the opposite direction. New Guinea has nine kinds of bandicoots, two of which reached Australia, in addition to nine others that are native to the continent. And New Guinea's twelve small pouched carnivores (family Dasyuridae) include only six that belong to genera missing from Australia.

The greater richness of the island's rain forest and tropical mountain slopes must be allowed for in comparing New Guinea with a continent where drought and cool climate affect such large areas. The well-watered, mountainous tropical island affords far more different diets and opportunities for privacy. In consequence, although the area of New Guinea is only a tenth that of Australia, the island has more kinds of bats and more different genera of true rats and mice than the continent. It has 568 different kinds of nesting land birds and freshwater birds, as compared to Australia's 531.

A "Zone of Subtraction"

If we look toward New Guinea from Singapore at the tip of Malaya, as so many experienced scientists have done, we see on the map an archipelago beginning with nearby Sumatra, extending to the atolls of the South

188

Pacific. Along the island chain, the number of different families of animals decreases gradually with distance. Biologists speak of the archipelago as a "zone of subtraction." Within it, the freshwater fishes show no overlap at all with those of Australia. The Asian fishes, in fact, are found no farther east than Java and Borneo, with those of eastern Borneo more like those of Java, and those of western Borneo showing many ties to those of Sumatra. These same three islands are as near New Guinea as gibbons and lorises get; or flying squirrels and rodents other than mice and rats; or the Asiatic wild dog; or the leopard, the rhinoceros and the deer-like chevrotains.

Alfred Russel Wallace, while sorting out his personal knowledge of the animals along this archipelago a century ago, saw that Bali was the farthest east for the tiger, the scaly anteater, a tree shrew, two kinds each of monkeys and squirrels. The porcupine goes one step farther—to Lombok. But New Guinea is beyond range of the shrew, the crab-eating monkeys, the palm civet of the coast of Southeast Asia, the wild pigs and deer that are found on Timor, Ceram, the Moluccas or the Philippines.

Wallace drew a line—now called Wallace's Line— between Bali and Lombok. Only a few of the types of birds that are characteristic and widely distributed in the Orient have spread beyond the line toward New Guinea. None of the pheasants, woodpeckers, trogons, barbets, broadbills, fairy bluebirds or true finches has succeeded. Pittas and drongoes go beyond New Guinea into northern and eastern Australia, but hornbills and crested swifts have not colonized beyond Torres Strait. The strange frogmouths show a split in the family (Podargidae), with the nine kinds of *Batrachostomus* in forests of the Orient as far as Java, Borneo and the Philippines, and the three of *Podargus* (the only other genus) in New Guinea, Australia, Tasmania, and the Solomon Islands. In all of these places the lethargic birds appear to sleep all day and perch all night, opening their huge mouths in darkness to display a brightly colorful interior as though it were a flower. Insects that investigate the "flower" are abruptly swallowed.

In the westward direction, the archipelago shows subtraction too. The mammals of New Guinea and Australia have traveled very little. Ceram, just a short distance from New Guinea, has its own distinctive bandicoot—an animal so rare that only four specimens are known. One cuscus inhabits the forests and scrubland on Celebes and the islands of the Moluccas. All of the Australian mammals stop short of Wallace's

The flightless cassowary of tropical northern Queensland, New Guinea and a few adjacent islands moves swiftly through the undergrowth, its head somewhat protected by a large, bony helmet. (Australian News and Information Bureau)

Line. Nor do the distinctive birds-of-paradise, bowerbirds, magpie-larks, bell-magpies, scrubbirds, lyrebirds, owlet frogmouths and cassowaries go westward from New Guinea. The mound-builders (family Megapodiidae) have representatives in the Moluccas and Celebes, and members of the genus *Megapodius* are distributed more widely—to the Philippines and to Lombok, with an outlier in the Nicobar Islands, and additional species in the South Pacific. One of the wood-swallows ranges over much of the Oriental tropics, and another to New Caledonia and Fiji; the other eight are in New Guinea, Australia and Tasmania. The flowerpeckers of New Guinea have a few close relatives in Australia and Tasmania, a few in the Philippines and South China, India and Ceylon. One honeyeater has spread as far as Bali and Celebes. A cockatoo reaches Lombok, but not Bali.

None of the alternatives to Wallace's Line offers much improvement. While the differences between the faunas of New Guinea and Malaya are impressive, any choice

189

of the boundary between the Oriental and the Australian realm is about as arbitrary as choosing a wavelength in the solar spectrum to separate blue from green.

New Guinea's Turtles

One distinctive turtle (*Carettochelys insculpta*) found only in New Guinea has been given a family all its own. Biologists regard the creature as a fortunate survivor, a "missing link" that is not missing, showing how the main stock of softshell turtles in Asia, Africa and North America split away from the line of hardshells that bend their necks vertically to withdraw their heads for safety. The New Guinea softshell lacks a bony shell altogether, and has no horny plates either. Its body is covered by only a thin layer of soft skin. Yet the animal is a powerful swimmer and lives an almost completely aquatic life in the rivers and estuaries, paddling along with webbed feet that have lost all but two claws.

These same rivers are home to some related snakeneck turtles of New Guinea and Australia which seem strangely out of place. Nowhere else in the Old World are there members of their family (Chelyidae); the rest, which are the majority, live in South America, where they include the peculiar matamata. To conceal their heads, these turtles merely loop their necks around and push their snouts into the loose skin in front of their shoulders. How these turtles reached the rivers and swamps of New Guinea remains a mystery.

New Guinea's Birds-of-Paradise

With so few endemic families and genera in New Guinea, and such obvious influence by immigration, it is hard to explain why one group of crow-like birds should have diversified into so many extraordinarily adapted birds-of-paradise. Yet the island has forty different kinds. Four more have moved south into northeastern Australia, and a fifth (called the paradise crow) into the Moluccas.

To show these wonders of New Guinea to readers elsewhere, the distinguished explorer-ornithologist R. Bowdler Sharpe published almost a century ago an oversize *Monograph of the Paradisaeidae, or Birds of Paradise, and the Ptilorhynchidae, or Bower Birds*, with large colorful lithographs by W. Hart. To our delight we found at the Australian Museum in Sydney that we could purchase two plates from his work. One on our study wall reminds us of the black sicklebill, which is the largest of all birds-of-paradise. A quick glance at the female, positioned between two courting males, might suggest that she was a brown thrasher from the New World, with a longer, down-curved beak. But the glossy, dark-blue males are too spectacular to be mistaken for other birds. Each has several long downcurving tail feathers, spreads a comb-like spray of undertail coverts to the sides, and raises enormous black epaulets fringed with electric blue and yellowgreen. In the New Guinea forests below 9000 feet elevation, these males go through acrobatic gestures, meanwhile calling unmusically to draw attention to their iridescent plumage. For each one, success is measured by the number of mates he can entice down from the high branches to the display ground he keeps clear under the tall trees.

A century ago, when the millinery trade in Europe and America supported the collecting and export of trade skins of these extraordinary birds, the native hunters saw no advantage in saving the featherless feet along with the bodies. This led in Europe to the naming of the greater bird-of-paradise *Paradisaea apoda*—"the footless" one—and the invention of the story that these birds kept soaring day and night for life, unable to rest on tree or ground, mating in the air, and even carrying their eggs with them aloft until each youngster hatched and could fly off on its own. For a while, 50,000 skins a year were exported from this one big island. Some species may have been exterminated. Today the traffic is controlled, and visitors to New Guinea have a chance to see these native birds in action. A small colony of this particular species has been established on Little Tobago Island in the West Indies, and there we have watched the males in their seemingly social display, tossing above their reddish brown backs the golden plumes that then seem as fluid as the water jets in a colorfully illuminated fountain.

In New Guinea, two related birds-of-paradise have the strange habit of hanging upside down by their feet from a horizontal branch while performing their mating displays. In this way, gravity helps them spread their lacy flank plumes—bright blue in the blue bird-of-paradise and golden with ivory-white vanes in the Emperor of Germany bird-of-paradise.

The smallest of the family is the king bird-of-paradise, of which the male is white below, scarlet above, with green plumes on the sides that can be opened like Japanese fans, and a pair of extremely long curving tail feathers that appear to be mere wires tipped with iridescent flags.

The seven-inch King of Saxony bird-of-paradise lashes the air with head plumes that may be eighteen inches long, the shafts each bearing thirty-five to forty "flags" of a sky-blue color. On the twelve-wired bird-of-paradise the special feathers are six in number on each side, growing stiffly from each flank and having bare springy shafts as much as twelve inches long beyond the vane part. In the sago swamps along the freshwater portions of the rivers, each male identifies himself with these distinctive plumes and with erectile feathers like a collar whose iridescent violet-black and green-black contrast sharply with his lemon-yellow underparts.

Australia and New Guinea

As we think about our experiences in this land of primitive mammals, rodents and bats, cockatoos and cassowaries, spectacular pigeons and birds-of-paradise, eucalypts and *Banksias*, we realize how many lifetimes we would need to get acquainted properly with all the wonderful kinds of life there. We were warned to avoid all snakes, because so many are both poisonous and aggressive—the bite of a large tiger snake or a Queensland taipan may cause death in a few minutes. The Australian bulldog ants are similarly to be feared, for these inch-long insects leap to the attack, hold on with sickle-shaped jaws and jab repeatedly with an extremely venomous stinger. Left alone, they use these jaws and venom to capture and tear apart small animals as prey, then share the fresh meat with the ant larvae. And swimming along the coast would be risky because of vicious sharks and sea snakes with a venom of the cobra type. Yet, somehow, we explored and came away unharmed.

If the great Alfred Russel Wallace had been able to explore as easily and painlessly as we, his greater store of information would surely have kept him from regarding Australia as central to a broad Australian (or Australasian) realm. Today, nearly a century after the appearance of his famous and stimulating book, *The Geographical Distribution of Animals* (1876), we see Australia mainly as a center of evolution for its own kinds of life. Enough marsupial mammals spread to New Guinea to distract attention from the real progression of plants and animals from Malaya through the East Indies to the South Pacific.

In good conscience we cannot call New Zealand "Australian," since so little came to it directly across the open waters of the Tasman Sea from the big continent 1500 miles to the northwest. New Zealand's living heritage is more from the tropics to the north and perhaps, long ago, through Antarctica to the south. The Pacific islands show far more of New Guinea than of Australia. The biogeographer regards them as the idyllic ends of a long distribution line.

9

New Zealand and the Islands of the Broad Pacific

At Bluff on the coast of New Zealand's South Island, we were about 7250 miles from Los Angeles, 6800 from Lima, and 6500 from Cape Town. Tokyo (6200 miles) was farther away than Tierra del Fuego (4600). Yet the yellow pointed signs clustered above our heads gave the mileages to other places by the most direct route: London, 11,820; New York, 9380; Equator, 3112; South Pole, 2998; Sydney, 1250; Hobart, 1050; Stewart Island, 22. A point on the open ocean 180 miles from Bluff is the farthest a Londoner can get away from home. From that point, whichever way he heads, he is going back to London.

As we stood beside the signs, thinking about their meaning to nonhuman living things, the cold waves surged against the rocky shore. From our ears they concealed the cries of migrating shearwaters flying rapidly on their dark long wings low over the water, and also the purr of diesel engines in oyster-fishing boats returning with their catch to a nearby harbor.

We wondered if we should make the special trip to Stewart Island. Its 670 square miles are bleak, but still New Zealand. The coast of Antarctica is 1700 miles beyond. Half that distance in the opposite direction is North Cape at the far tip of New Zealand's North Island.

Much closer to where we stood were remote islands administered from New Zealand: Campbell Island, and the Antipodes Islands, and Chatham Island, in an arc about 420 miles out to sea. They are all nesting grounds for shearwaters and albatrosses, and have a few kinds of special plants found nowhere else. The tree daisy on Chatham Island has the largest flowers of any among the 125 different kinds of these shrubby plants, which are native to New Zealand, New Guinea and Australia. This island has the largest known forget-me-not, a giant perennial with hairy leaves and, from November until February, big purple flowers in attractive clusters. Its other endemic plants include a little geranium, a shrubby *Senecio* with dense clusters of two-inch daisy-like flowers, and a creeping *Cotula* of the daisy family—a member of a genus common in New Zealand.

Beyond us lay Bounty Island, and nearer at a latitude of 50° 40′ south, the Auckland Islands which bear the southernmost of tree ferns as well as a true subantarctic forest of trees of the myrtle family, known by their New Zealand name of southern rata.

Evergreen Islands the Maoris Found

Rata, kiwi and Rotorua are good examples of the plants, animals and places in New Zealand known by Maori names. The proportion of these in use is amazing in a country where the primary language is English. In many instances, each species has a Maori name; almost invariably the members of different genera are distinguished. Apparently the ancestors of the Maoris brought this exceptional awareness of living things with them to New Zealand when they immigrated from Polynesia between 900 and 1000 A.D.

Fortunately, many of the white colonists who began arriving in 1840 recognized this Maori skill. Rather than adopt the unfamiliar Maori names, some preferred to apply names from the Northern Hemisphere, such as red pine (rather than rimu) for the principal timber tree, which is a podocarp and not a pine; or swamp hen (instead of pukeko) for the handsome big gallinule of marsh margins.

Using the Maori names, it is easy to speak of a northern rata-kauri-tawa forest and, in New Zealand, be understood as well as a person in England mentioning a beech-oak-pine forest. The botanist remains consistent with usage in all countries, perhaps by

Spectacular mountains, green valleys, lakes left by the glaciers, and long arms of the sea delight the eye in Fiordland National Park. It is located near the south end of the west side of South Island in New Zealand. (Michael F. Soper)

192

remarking that *Metrosideros robusta* (northern rata) is unusual in being a sturdy tree ninety feet tall, whereas most members of its genus are dwarfs or woody vines; that *Agathis australis* (kauri) of the araucaria family is one of the world's giant trees, its stout trunk often shining white in the forest; and that *Beilschmiedia tawa* (tawa) of the laurel family grows eighty feet tall, with a fine black bark and reddish-purple plum-like fruits in season for the birds.

Where the forests have not been cut on the two big islands of New Zealand, they have a luxuriance in tree ferns (especially *Cyathea*) that shows how much humid air and generous rain comes in from over the surrounding salt water. Unlike Australia, the islands are largely volcanic, with abundant signs of recent activity in the form of hot springs and steam vents. Those around Rotorua on North Island are exceeded in size only by the thermal areas of Yellowstone National Park in Wyoming, U.S.A., and of Iceland. In neither of these other places have volcanos in the past few million years contributed such spectacular scenery as in New Zealand. From Fiordland in the southwest of South Island, the mountains continue like a backbone, with Mount Cook's summit 12,349 feet above the sea. On North Island, Mount Ruapehu (9175 feet) appears smaller than Mount Egmont (8260 feet) only because it is at the bend in an L-shaped mountain chain instead of rising above surrounding flat lands. Between the shore and the peaks are coastal herbs and shrubs, coastal forests and high forests, scrublands and rocky outcroppings, grasslands and tussock country, and alpine gardens with such extraordinary plants as "vegetable sheep" and

gigantic buttercups. Such varied landscape offers a great variety of living spaces for plants and animals. Yet representatives of only certain types of life have reached the islands from elsewhere and been able to diversify to match the opportunities.

Despite the location of New Zealand's two big islands in the temperate zone, almost all of the trees, shrubs and nongrassy herbs are evergreen. Fall colors never develop, and only a few woody plants are deciduous. Most notable of them are a tree daisy of the south island, and the rewarewa—a tree of the protea family sometimes called New Zealand honeysuckle.

Almost every habitat has a distinctive representative among the daisy bushes, which occur also in New Guinea and eastern Australia, and the fleabane-like celmisias, which are found elsewhere only in eastern Australia, and in the same daisy family (Compositae) some groundsels—a genus that is nearly cosmopolitan. True heaths are almost all snowberries, whose relatives range in America from Tierra del Fuego to Canada. New Zealand's heath-like shrubs are many, among them a whole array of hebes belonging to the snapdragon family (Scrophulariaceae). Stiff, prickly spaniards, or speargrasses, prove by their flower clusters to belong to the carrot family (Umbelliferae). Peculiar grass trees are New Zealand forms quite unlike their Australian relatives in the epacris family (Epacridaceae), since they are not in the least heathlike. The shrubby coprosmas of the madder family resemble those found in eastern Australia, but one of them (*C. foetidissima*) in New Zealand completely justifies its generic name ("dung odor") by emitting a vile odor when merely brushed against, and a stench when broken or crushed in any way.

The Last of the Beak-Heads

We had reason to appreciate coprosmas when we visited some relatively inaccessible islets between North Island and South, where the lizard-like tuataras manage to survive. These animals hide underground or bask in the sun by day, then emerge in darkness to hunt for land snails and big, wingless crickets called wetas. Tuataras live under the coprosmas in burrows dug by shearwaters that fly to these islands to nest. From one nesting season to the next, the tuataras keep the

Using its strong legs to push itself through the undergrowth, a kiwi probes deeply into New Zealand forest soil for insects and worms it can detect by scent in darkness. (Michael F. Soper)

burrows open, and apparently move over to share the space with mated shearwaters when the birds return to breed. The feathers from the shearwaters and the dung from both the birds and the tuataras contribute to the soil just the ingredients needed to make coprosmas grow well. The churning of the earth as birds occasionally dig new nests helps also. Yet the plants contribute too: roots keep the soil from eroding in each rainstorm, and evergreen foliage shades the soil, preventing the sun from baking the earth into a mass too hard for a bird to burrow into. It is after a rain that the mixture of feathers and dung with mineral matter and dead plant material develops the odor that gave the coprosmas their bad name.

Tuataras remain "living fossils" in the animal world because they alone survive from one whole order of reptiles—the beak-heads (Rhynchocephalia)—that was well represented in many parts of the world during the

Steam vents and thermal springs on little volcanic White Island, off the northeast coast of the North Island of New Zealand, indicate how precarious life can be where eruptions are possible at any time. White Island has the look of a newborn island. (Robin Smith)

Age of Reptiles. Their ancestors saw the dinosaurs come and vanish. Tuataras continued, each with a structure atop the brain inside the skull that serves like a third eye, informing the reptile when daylight has faded and the time has come to emerge. Recent research by Professor William Dawbin, a New Zealander now at the University of Sydney, shows that tuataras need about twenty years to reach maturity and may well live (as the Maoris claim) to an age between 100 and 200 years. No other reptile is known to be a juvenile for so long. Yet their fate today is bound up with

195

The gallinule known as a pukeko in New Zealand is reluctant to fly, preferring to run, wade and swim through the marshes in which it finds food and nest sites. (Michael F. Soper)

coprosmas and shearwaters. If introduced goats destroy the New Zealand shrubs that shade the nesting sites of the birds, or the birds fail to return to nest, the future is bleak for tuataras.

Feathered Survivors

During the last thousand years, approximately half of the native birds on New Zealand have become extinct. Only bones and feather capes show that twenty-two different species of moas lived there until the Maoris arrived. Some kinds of moas were no bigger than a domestic turkey. Others were taller than modern ostriches, but somewhat more slender. Their only surviving relatives are the three kinds of kiwis, of which the most widespread (the brown kiwi) is somewhat larger than a domestic fowl. The wings of this peculiar bird are less than an inch long, and bear a curved gray

claw. No barbules hold the feathers into vanes, which gives them a hairlike texture. Captured kiwis try to defend themselves by kicking with their powerful legs, but make no attempt to bite or otherwise use the long pointed beak, which bears the nostrils near the tip. At night kiwis probe the forest soil and seem able to open the tip of the beak just enough to seize insects and earthworms they find with an acute sense of smell. They use the beak also to pick up and eat small animals and fruits found on the surface of the ground.

Until white settlers arrived and brought four-footed mammals to New Zealand, the islands formed a sanctuary for flightless birds. Not only the kiwis, but two of the rails—the weka and the takahe—are grounded by the shortness of their wings. A third rail, the pukeko, seems most reluctant to fly, although it travels long distances once it gets airborne. Pukekos stay around fresh water, where they can retreat quickly among marsh plants such as the cattails known as raupo. Twitching its short tail and flashing the white undercoverts there, a pukeko draws attention to itself. Its orange-red legs and feet and a scarlet beak matched by a shieldlike bare area between the eyes contrast with the iridescent greenish black upperparts and purplish blue underparts and thighs.

Wekas walk along deliberately, flicking their short tails in a similar way. But they rely upon being brown, darker above, streaked and spotted, to blend with their surroundings. If alarmed, they dash off through the undergrowth or swim across streams. While hunting for small animals and bits of edible vegetation, they haunt the dense scrub and the forest edges, or come down to the sea coast to examine flotsam thrown ashore by the waves.

Takahes mature at a size distinctly larger than any pukeko or weka, with a heavier beak, stouter legs and still shorter tail. Originally known only from subfossil skeletons and believed to be extinct, takahes were found to be still in existence when hunters brought back individual dead birds in 1849, 1851, 1879 and 1898, all from the southwestern portion of South Island. In 1948 a few live birds of this kind were sighted and reported by Dr. Geoffrey Orbell, a New Zealand dentist. Subsequent explorations by wildlife biologists led to the realization that about 200 of these rare birds live in tussock land west of Lake Te Anau in "Takahe Valley" of the Murchison Range of mountains. Every effort has been made to protect these survivors, to mark them with numbered metal anklets, and to learn how best to introduce a breeding stock into suitable places elsewhere. Still on the danger list of species close to extinction, the takahes may make a comeback if these measures succeed.

A flightless parrot, the twenty-four-inch kakapo, is even closer to disappearance. Sometimes known as the owl-parrot because of the pattern of bristly feathers around its eyes, it is a denizen of the antarctic beech forests of South Island. There it clambers over branches and fallen trees, or walks along the ground, often following its own tracks through the undergrowth to fruit-bearing shrubs and grasses on which it feeds. The kakapo is a heavy bird, but able to glide for 100 yards or more on its broad wings. Rarely does it take to the air, however, perhaps because it is abroad in twilight and at night when its vision is inadequate to prevent it from colliding with obstacles in the forest.

The kakapo is just one of six members of the parrot family (Psittacidae) native to the main islands of New Zealand. Two others (the kaka and the kea) are big olive-green birds seventeen to nineteen inches in length. The remaining three are brightly colored parakeets less than twelve inches long. Kakas travel in little flocks of about a dozen birds, squabbling over the nectar in rata flowers, eating fruits of many kinds, and ripping off loose bark to get edible insects. Sometimes they raid orchards in broad daylight and cause considerable damage. Keas, on the other hand, are mountain birds of South Island, and feed from the upper limit of the forest down to lower elevations. When they can find food in alpine grasslands, particularly in autumn, they range more widely. Apparently in these places they normally act as scavengers, eating whatever carrion they can find. In some localities, keas have learned to stay close to sheep ranchers' slaughterhouses, where entrails are often dumped in the open. From using this source of food, it was only a minor step for a kea to seek out a sick or dying sheep and use its sharp beak to rip into the flesh. Stories of keas attacking healthy sheep have been credited widely, but the evidence for this is extremely rare.

The presence of six kinds of parrots in New Zealand —a fifth of the country's native land birds—might seem improbable since open seas are on all sides and the climate is temperate rather than tropical. Apparently it is due to the habit of parrots to travel in mated pairs. The arrival of one individual as a colonist is likely to establish a member of the opposite sex as well. Pigeons show a similar habit, and New Zealand has a particularly large and handsome native pigeon. Its back is a rich purple, its wings, chest and face metallic green, its beak and feet red, and its underparts otherwise white.

Another fifth of New Zealand's native land birds are of little families found nowhere else: the New Zealand wrens and New Zealand wattlebirds. The bush wren and the rock wren are still numerous, but the Stephen Island wren was apparently exterminated in 1894—the year it was discovered—by the lighthouse keeper's cat, which brought home dead the only specimens of this bird ever seen. The smallest of New Zealand's birds, the three-inch rifleman, is an elusive forest dweller with a distinctive snapping call but the habits of a wren. Probably the closest relatives of these four wrenlike birds are the pittas (Pittidae) of tropical islands from the Solomons westward to northeastern Australia, southern Asia and south-central Africa.

New Zealand's wattlebirds rarely fly. Instead they hop along the ground, or from branch to branch, the two members of a mated pair usually close together and calling to one another almost constantly. One of the three known species (the huia) is presumed extinct, for none has been seen since 1907. The tieke (or saddleback) is listed among the world's endangered birds. We were delighted to see one at close range, a ten-inch bird whose glossy black feathers on most of the body set off a bright chestnut-colored saddle area and conspicuous orange wattles. The kokako (or wattled crow) is distinctly larger and more numerous, given to gliding down from one tree to the next and then vigorously hopping on its long legs to ascend to levels where it can find fruit. A dark grayish blue bird with a shiny black beak that is almost hooked, it wears ultramarine-blue wattles (North Island subspecies) or rich orange ones (South Island subspecies). The alert movements of these birds support the ornithologists' belief that they are related to starlings, possibly descended from ancestors that flew south to New Zealand from tropical islands long ago.

Even for a gull, the distance is too great for ordinary travel from any point of land to New Zealand. Those that live there are local species and are not found in Australia or New Caledonia or New Guinea. The red-billed gull scavenges along the coasts, whereas black-billed gulls stay around freshwater lakes and streams. Gull physiology prevents these birds from continuing day after day over open ocean, drinking salt water, without poisoning themselves. Even though they arrive where fresh water is available to them, they die in a week or less. Probably other birds that can settle on water and take off again find a similar barrier that keeps them from colonizing distant lands. Those that cannot rest until they reach land have the additional need to make the trip nonstop. Comparatively few have the speed and endurance to succeed.

A few kinds of birds have clearly come to New Zealand across the Tasman Sea. The Australian white-eye arrived in June 1856 in large numbers, and competed successfully for nectar, fruits and insects with the two native honeyeaters, the bellbird and the tui (or parson bird). New Zealanders call the white-eye the silver-eye. As active and noisy a little bird as a chickadee, it generally associates in small flocks and ranges through the undergrowth from the coast to mountain slopes as high as 3500 feet above the sea.

Birds that travel in flocks are more likely than those that fly alone to colonize remote islands. When one arrives and settles in, finding a place for itself among the kinds of creatures already present, it generally can locate an acceptable mate and start a family. By contrast, birds such as herons, which tend to be solitary, may reach an island as a waif or stray without establishing a new population. Even the cattle egrets, which recently have shown extraordinary ability to reach and colonize remote places, came to New Zealand only in 1963 (one bird) and 1964 (many, on both islands).

Islands of Opportunity

Places and ways to live on New Zealand must have gone unused for millions of years, awaiting the arrival of plants and animals to fill them. The newcomers changed to fit the opportunities they found, and in this evolution became native species. Only thirty of them are land birds; fourteen lizards (geckoes and skinks) and three kinds of small uncommon frogs managed to reach the islands and to find niches in the community. There are no native snakes except sea snakes, no turtles except marine ones, no toads or salamanders, and no native fishes. Two kinds of bats have come by air, and the antarctic fur seal by water. One of the bats, the New Zealand short-tailed, is endemic to forests of the main islands and some of the small islets nearby; it has a family to itself, partly because of its special ability to roll up its wings and run about agilely on all fours. Although insectivorous, it does not hibernate, and in this respect differs from the other bat—the long-tailed, lobe-lipped bat, which belongs to a genus found otherwise in Australia and on Norfolk Island.

Sixteen kinds of butterflies visit the flowers of New Zealand. One of them is the wanderer, which somehow reached the islands from North America, where it is known as the monarch. Other insects have lived long enough on New Zealand to have evolved winglessness. This can be advantageous where strong winds carry weak fliers out to sea and certain death. The "Maori bug" is a black cockroach nearly an inch long, found throughout the islands under loose stones and debris, from which it emerges to scavenge at night. Several stick insects, some of them five inches long and spiny along the sides of body and legs, use their protective resemblance to leafless twigs to escape being attacked by insectivorous birds; they eat foliage of many types. And the distinctive wetas, such as the large-headed *Hemideina megacephala* whose body is fully two inches long and heavily built, creep out of tree holes and from under bark as soon as day ends, to eat leaves and fruits anywhere from ground level to the tops of the highest trees. In a way, the wetas take the place of many different kinds of mammals absent from the New Zealand scene: fruit bats, monkeys, forest mice, squirrels, and all the various arboreal marsupials met in Australia or New Guinea. Indeed, the wetas and the trees they live in are now threatened by the Australian brush-tailed possum, which was introduced repeatedly between 1858 and 1920 as a potential source of valuable fur. The commercial gain proved to be an illusion, but the modern destruction of forests by this ten-pound nocturnal mammal can be seen even from a low-flying airplane.

Other introductions seem to have succeeded without disturbing the balance of nature. Life must have become easier, in fact, for New Zealand's kiwis when white colonists introduced earthworms from England. Previously the islands lacked segmented worms of this kind. Now they form an important part of kiwi diet, and are eaten also by a native carnivorous snail, whose flat spiral shell may be as much as three inches across.

Natural Way Stations: Lord Howe and Norfolk Islands

In seeking to understand the past routes by which living things of the land reached New Zealand, a biogeographer tends to look northward toward New Caledonia and westward toward Australia. In between are tiny islands that might have served as natural way stations: Lord Howe and Norfolk Islands.

Lord Howe Island, whose central peak 2840 feet above sea level dominates a land area of only five square miles, lies about 350 miles from the Australian coast, 800 southwest of New Caledonia, and 1200

Mitre Peak rises from the deep waters of Milford Sound in Fiordland National Park, New Zealand. (Virginia Carleton)

northwest of North Cape on New Zealand's North Island. Norfolk Island, with an area of fifteen square miles, is similarly volcanic and wooded, but 1000 miles from Australia and less than 600 from either New Caledonia or New Zealand.

Some plants and animals might be expected to spread between New Caledonia and New Zealand by way of Norfolk Island, while others would go between Australia and New Zealand either with only a single stop on Lord Howe or, more circuitously, by way of both Lord Howe and Norfolk Islands. Being larger, Norfolk Island might be expected to shelter a greater variety of life. Actually it has fewer kinds of plants and animals than Lord Howe, and a smaller proportion of them are found nowhere else.

Lord Howe Island has four kinds of palms, one of them the endemic form that is now cultivated in other countries for its beauty. It shares with New Zealand the pepper tree (or kawakawa) and one of New Zealand's twenty kinds of native brooms. Lord Howe shows ties with both Australia and New Zealand in having a prostrate shrub of the buckwheat family (Polygonaceae), which is found nowhere else, and in having a single species of shrubby sandalwood of a genus (*Exocarpus*) represented by ten kinds in Australia, seven in New Guinea, six in New Caledonia, three in Hawaii, and one each on Fiji and Norfolk Island. Of Lord Howe Island's fifteen native kinds of land birds, eight are now extinct; three apparently came originally from New Caledonia, two or three from Australia, and one from New Zealand.

Norfolk Island is the home of the Norfolk Island pine, a conifer whose symmetrical whorls of branches bearing short evergreen needles make it especially attractive whether outdoors in a warm country or as a tub plant for homes and hotel lobbies. The only palm on Norfolk Island is like the one in New Zealand—the nikau. Nowhere else in the world but on Norfolk do New Zealand's flax plants of the lily family grow naturally. It shares with New Zealand and Fiji and the Tonga Islands the whitey-wood trees of the violet family. Its pepper tree is of the New Zealand kind, but its *Exocarpus* is the same one as grows on New Caledonia. Yet it shows a tie to Australia by way of Lord Howe Island in having the same plants of the genus

Tropical forest grows right to the edge of the water in the Papenoo River of Tahiti, which descends from the interior mountains and then winds across coastal lowlands to the South Pacific Ocean. (Josef Muench)

Lagunaria in the mallow family (Malvaceae) as those in coastal Queensland. The land birds of Norfolk Island originally included fourteen nesting kinds, two of which are now extinct. The survivors include nine sparrowlike birds, a kingfisher that catches more insects and small crabs than fishes, a parrot, and an owl not unlike the New Zealand morepork, which is called the boobook in Australia—the only other places where it is found.

Enchanted Islands of the Broad Pacific

If a dedicated gambler had said his little prayer to Lady Luck and cast a handful of rice across a large map of the Pacific Ocean, instead of dice across a gaming table, chance might have produced the seemingly-random pattern of the far-flung Pacific islands. Actually the pattern is one of volcanic activity, following geological events that no one yet understands. Each island either has the remains of a volcano visible, or it consists of a limestone cap on the peak, produced by calcareous seaweeds and coral animals. Every stage in the evolution of these islands, from active volcano to ring-shaped atoll, is found in the area. Seeing these islands, Charles Darwin recognized quickly how they were formed and outlined it as a theory that later scientists have supported.

Generally the Pacific islands are grouped according to the people who first colonized them. Presumably the colonists spread eastward, just as they had to New Guinea from Malaya through the East Indies and perhaps from Africa. With coconuts to quench their thirst, they set out in their outrigger canoes for the unknown, and sometimes found new islands to settle.

The Polynesians, whose ancestry seems most clearly traceable to Malay stock, went farthest. They tend to be gentle people, with brown skin, large frames, black wavy hair, soft features and wide-open eyes. Their Polynesia forms an enormous triangle from New Zealand eastward to remote Easter Island, north to the Hawaiian chain; it includes Tonga, Samoa and Tokelau up the western side, Christmas Island, the Marquesas and the Tuamotu Archipelago down the eastern border, Pitcairn near the southeast corner, and the Society Islands (with Tahiti) and Cook Islands in the tropics close to the center. Polynesians reached Samoa about 500 B.C., Tahiti about 200 A.D., New Zealand 900 A.D., and Hawaii 1000 A.D.

Melanesians are small people with frizzy hair and many Negro features, the men with moderate beards. Their Melanesia includes New Guinea, the Bismarck Archipelago, the Solomon Islands, the New Hebrides, New Caledonia, the Loyalty Islands, to far Fiji. To the

north are the Micronesian people, who tend to be small too, but have straighter hair and finer features. They occupy Micronesia—islands with a total area of about 1000 square miles—from Palau nearest the Philippines into the rest of the Caroline Islands, north of which are the Marianas (with Guam the most famous); to the east are the Marshalls (with Bikini atoll) and the Gilberts, north of which is Wake Island as a distant outpost.

With the exception of New Guinea, none of these islands is on the continental shelf of any major land mass, and most of them would remain isolated even if the sea level were 1000 feet lower than it is today. The whole fauna shows progressive subtraction eastward. The last true freshwater fishes and native land turtles live on New Guinea, frogs and flightless land mammals on the Solomons, snakes on Fiji, lizards on Fiji and

The kea is a mountain parrot 17 to 19 inches in length found on the South Island in New Zealand, where it scavenges for a wide variety of foods, ranging from the forest edge through sheep pastures to the edge of the snow fields. (Michael F. Soper)

the Tongas, bats on Samoa and Hawaii, land birds on Hawaii and the farthest Tuamotus. We tend to think of many of these islands as shaded only by coconut trees (which is not quite correct), and as the home of the strange coconut crab. Known also as the robber crab, it is a hermit crab that outgrows any snail shell available to it and thereafter walks about naked. Its twisted soft abdomen becomes almost straight and hard-covered, while the crab itself clambers through the undergrowth or climbs coconut trees to reach their fruits. With its heavy pincers, it can tear away the husk and reach the nutritious "meat" inside the inner shell.

The flora of these islands in the South Pacific is richer the closer the island is to New Guinea. New Caledonia, which is fairly close and has an area of about 7200 square miles, has more than 2500 different kinds of flowering plants, perhaps four-fifths of them endemic species. In its forests is a unique bird, the large gray heronlike kagu, which is almost flightless. Raising its conspicuous crest, it dashes about at twilight, catching insects, worms and mollusks such as land snails. Introduced pigs, dogs, cats, rats and people have brought this bird close to extinction.

In the rest of Melanesia, in Micronesia and in Polynesia, the number of endemic plants and native land animals falls off rapidly. Most of the living things present belong to families that are progressively less represented in a series that begins with Malaya, not merely with New Guinea. In the whole of Polynesia except Hawaii, there seem to be fewer than a dozen endemic genera of flowering plants. One of them is the peculiar tree-like *Fitchia* of the daisy family, of which Tahiti has two different kinds among the six that are known.

The Hawaiian Islands

Diagonally across the Tropic of Cancer for 1600 miles, the Hawaiian archipelago spreads as a series of volcanic and coral islands with a combined area (6439 square miles) between that of New Caledonia and Fiji. From little Kure in the northwest, the series continues past Midway, Pearl and Hermes Reef, Lisianski, Laysan, Maro Reef, Gardner Pinnacles, French Frigate Shoal, Necker, Nihoa, Niihau and Kaula, before reaching the larger islands of Kauai, Oahu (with Honolulu), Molokai, Maui, and finally the biggest, most southeastern of all, Hawaii itself. America is another 2000 miles distant and, except for stepping-stone islands in the western Pacific, the distance to Australia or Asia is even farther.

The volcanos to which the Hawaiian islands owe their existence burst upward into air where the Pacific Ocean otherwise is about 15,000 feet deep. The water hides the vast bulk of Mauna Loa volcano, whose crest towers 13,680 feet above present sea level. It ranks among the world's highest mountains if it is measured in relation to the surrounding sea floor. Geologists regard it and the other volcanos in Hawaii as dating back only five to ten million years. Yet today, these islands are home to about 170 different kinds of ferns

and non-flowering plants, 1730 of flowering plants, 1070 of snails, nearly 4000 of insects, and 70 of land and shore birds, not counting any that have been introduced deliberately or accidentally by man.

No one needs proof that these volcanic islands had to be firmly established and cooled to a tolerable temperature before land plants and animals could settle as colonists. But for islands averaging perhaps seven million years in age to have 1000 kinds of land life today does not necessarily imply that in each thousand years during the seven million an additional kind of colonist arrived and succeeded in getting a roothold or a foothold. Of the 7000 kinds of land life, there are only about 700 distinctive types, each type a cluster of closely related species. They could have evolved from 135 kinds of ferns and non-flowering plants arriving during those seven million years; and 275 of flowering plants, 24 of land snails, 250 of insects, and 15 of land and shore birds.

Dr. Ernst Mayr, the distinguished zoogeographer, has suggested that on the Hawaiian islands other than Laysan (which is special and different), these birds probably represent just fourteen successful immigrations. The most recent would be the black-crowned night heron, which seems identical with those in America. Four with the same probable origin have evolved for long enough on Hawaii to have features earning them the status of subspecies: a gallinule, a coot, the black-necked stilt, and the short-eared owl. A hawk and a crow, probably from American ancestors, may have come at about the same time in the past. The rare Hawaiian duck, like the still-rarer Laysan teal, is different enough from the widespread mallard to be regarded as a slightly older resident of Hawaii.

America probably contributed the Hawaiian thrush and the gentle little Hawaiian goose about as far back in time as the colonization of the islands by an Old World flycatcher that apparently spread across the Polynesian part of the Pacific Ocean. The goose, called the néné (pronounced nay-nay) and chosen as the state bird, has become well adapted for life on lava fields around the big volcanos on Maui and Hawaii. Flightless at moulting time, it dwindled in numbers to forty-two individuals in 1947, all of them on Hawaii. A male and two females were caught and loaned to Commander Peter Scott for an attempt to breed the vanishing goose in captivity at the Severn Wildfowl Trust in England. His success with this enterprise and the strict protection in Hawaii have led recently to a gratifying rise in the world's population of free nénés.

Two different genera of honey eaters are represented on Hawaii by birds whose ancestors almost certainly came from New Guinea across vast stretches of open sea. A Hawaiian rail may have been there before the honey eaters, for it is now flightless; however, no way has been found to learn its origin.

Hawaii's most spectacular birds are the twenty-two different honeycreepers of the endemic family Drepanidae. On the various islands they evolved differences in size, beak form and habits that made many species into insect-, fruit- or seed-eaters, instead of nectar-sippers. Each main line in this diversification is represented by distinctive geographical forms on the principal islands of Hawaii. Some of the warbler-like birds in Central America are the most plausible representatives of the ancestral type—they have remained relatively unchanged during the same period because their isolation was much less.

On Laysan Island, the coral completely conceals all indications of the underlying volcanic peak. Unlike the other westerly islands of the Hawaiian chain, it has some freshwater and brackish ponds. These have been hospitable to the Laysan teal and allowed growth of marshy areas in which, until 1944, the flightless Laysan rail survived. Now it is extinct, along with the Laysan millerbird, an endemic Old World warbler last seen in 1923. The precariousness of animal life on these Hawaiian islands can be seen in the record of twenty-four species and subspecies of birds extinct since 1824 and another twenty-six on the list of the world's most endangered.

The land snails, particularly those of the genus *Achatinella*, and the insects tell a similar story of evolution from relatively few ancestors. No native freshwater fishes or cold-blooded land vertebrates are to be found. The only mammal is a bat so similar to the hoary bat of America that it is probably a recent colonist.

Land Plants in Isolation

Of all the floristic regions in the world, Hawaii is surely the most isolated. Fully 90 per cent of the species of flowering plants there are indigenous, known from nowhere else. The islands have representatives of seven genera of plants in the lobelia family, only one of which (the virtually cosmopolitan genus *Lobelia*) occurs elsewhere. One or two lobelia seeds reaching Hawaiian shores may have sufficed to set off these new patterns of growth that matched the ecological opportunities. On Hawaii, the plants of this family are either pollinated by honeycreepers and well adapted to attract these peculiar birds, or live where honeycreepers have been exterminated in recent years and survive by self-fertilization. Lobeliads that must manage without their

pollinators are probably degenerating, following toward extinction the birds that have disappeared, although at a slower pace.

Many of the plants on Hawaii have lost there the adaptive features that let their ancestors make the long trip over open water. The bur marigolds, for example, probably came from a seed or two that stuck in the feathers of a migrating water bird, holding on with two or three barbed prongs just as the seeds of these marigolds do elsewhere in the world. On Hawaii they evolved species in which the seeds lack prongs entirely, cannot hitch a ride, but merely drop to the ground—biased in their fall to some extent by whatever breeze is blowing. Similarly, an amazing proportion of the insects on Hawaii have short wings or none, although they belong to genera and families whose members fly well on other islands and continents around the Pacific.

Hawaii is a poor place for plants that rely upon insects for pollination. This probably explains why less than a fifth of the kinds of flowering plants are monocots, and this fifth is almost all grasses and sedges whose pollen is carried by the wind. The four-fifths include some extraordinary shrubs and trees, century-plant-like styles of growth among violets, geraniums, pinks, lobeliads, and composites that belong to genera consisting elsewhere mostly of herbs.

For every kind of flowering plant native to Hawaii but with connections in America, there are two or three kinds representing groups from Asia and Australia. Of *Coprosma* species, Hawaii has about twenty, New Zealand forty, and the remaining thirty are widely distributed from Java to New Guinea and Australia, down through the South Pacific islands to Juan Fernandez off the coast of Chile. Sandalwood forests once grew so profusely around 2600 feet elevation on the slopes of Mt. Haleakala on the island of Maui that King George IV of Britain bought an $80,000 shipment of lumber cut there in 1821 to give in exchange for Queen Cleopatra's barge—a treasure for the British Museum. The other sixteen kinds of sandalwood include six more in Hawaii, three in Australia, two in New Guinea, one (the white sandalwood, *Santalum album*) from Timor to Java in the East Indies, one in New Caledonia and the New Hebrides, one in Tahiti and adjacent islands of the South Pacific, and one on Juan Fernandez. Hawaii even has four kinds of *Metrosideros* akin to the twelve kinds of ratas on New Zealand, the thirteen on New Guinea, ten on New Caledonia, two in Australia, and one each in the East Indies as far as Sumatra, and in South Africa near the Cape of Good Hope.

Silverswords and Greenswords

The most impressive of Hawaii's plants may be the silverswords and greenswords found on higher parts of the big islands. Best known is the Haleakala silversword, which grows near the summit of Haleakala volcano on Maui and on dry lava slopes of Mauna Kea, Mauna Loa and Hualalei, all on Hawaii. Each plant develops as a tight rosette of glittering silver-gray leaves arising in a spiral from a short upright stem. The shape of the plant is roughly spherical, resembling a silver snowball on a scree of purple lava. Individual leaves are one-quarter to three-eighths of an inch wide, about four inches long on small plants and to sixteen inches on big ones. Rather thick and almost triangular in cross section, they curve upward slightly and taper to a blunt tip. Over all surfaces that the sun can reach, each leaf

The blue-faced booby, which is named for the blue-black, naked skin on its face and throat, nests in New Zealand only on the Kermadec Islands northeast of the North Island. (Michael F. Soper)

is densely covered with fine, crystal-clear, short silky hairs. These reflect the light like the glass beads on a projection screen, keeping the leaf cool without appreciable evaporation of water. They may also help the plant condense dew at night and moisture from any mist by day.

Adding a new leaf at a time for many years, each silversword seems to keep a running inventory of the amount of food stored inside its shining foliage. Sometime between the age of 20 and 200 years, the food

205

store passes a critical point and the short woody stem in the midst of the hundreds of leaves begins to extend itself. One we saw with a well-formed bud on June 10 was 36 inches tall. Daily for a week it added an inch to its height. By July 1 the bud became a towering cluster of separate flower buds 58 inches high. A few flowers had already opened, each a daisy-like head the size of a silver dollar, with golden yellow petals surrounding a purplish brown center. Sticky glandular hairs on the central flower stalk prevented crawling insects from reaching the nectar and pollen.

By August 24, when the entire flower spike stood 76 inches tall, its last flowers dropped their petals. The stalk and the silvery leaves around the base began to shrivel. By winter the seeds would be ripe, ready to blow up slope or down in the same wind that might topple the dead parent plant. We found many dry fragments of fallen patriarchs, but only a few plants in their prime of bloom.

The Haleakala greensword grows in a gap in the rim of the volcano's crater, among tree ferns and shrubby geraniums, benefiting by the almost daily mists and frequent rains from clouds formed on the outer slope

The unique silversword plant of Mount Haleakala, Hawaii, grows in the crater and on the outer rim, adding one silvery leaf after another to the large cluster, and then finally a towering display of flowers. (Lorus and Margery Milne)

of the mountain. The greensword's leaves are dull greenish gray, with fewer hairs to conceal the color of the chlorophyll within. In full bloom a greensword sometimes towers nine feet tall. Its blossoms, however, remain pendant as though to shed the rain. Each flower is covered with a fragrant resin, and never opens widely, as do those of silverswords.

Other silverswords and greenswords are limited to the second volcanic peak on Maui, and to the rainiest of all mountains in Hawaii: Mt. Waialeale, on Kauai island. Yet the nearest relatives of these extraordinary plants on any continent are the sticky-flowered tarweeds of California and adjacent states. Presumably the sticky seed of an American tarweed, clinging to the feathers or the foot of a bird, reached Maui long ago. Its descendants, diversified into strikingly new species, reward the hiker who ventures into some of the least known parts of these islands.

As we think about the strange plants and animals on New Zealand and the islands of the broad Pacific, we realize that some of the sense of discovery they offer comes from their isolation and the adaptations they show to surviving among neighbors brought together by happenstance. At least as much arises from the fact that so few naturalists reach these islands by traveling east—getting perspective by becoming acquainted first with the rich life of tropical Southeast Asia, including the Philippines. From there, through Micronesia to the Hawaiian islands, the steps across water are long—but not as long as in the opposite direction.

10
Islands
in Special
Isolation

Many of the world's islands show such close natural ties with the nearest continent that their individuality diminishes. Despite political differences and local pride, Sumatra and Java and Borneo are more like Southeast Asia than distinct from it. The living things on Ceylon are those of India, just as those of the British Isles are a sampling from Europe. Newfoundland is an extended part of the Canadian mainland. Yet currents in wind and water have so influenced the distribution of life on some other islands that, despite their geographical setting, they are especially isolated.

The plants and animals that have reached these islands in special isolation have evolved to fit strange combinations of physical environment and neighbors. On the small or recent islands, the flora and fauna are generally depauperate while still including endemic species—those found nowhere else. The larger and older islands have a richness and so many differences that the biogeographer sees endless challenges to explain what is there and what is missing.

Ownership of isolated islands is a part of recent history, sometimes relating to the nearest country and at others to conquests by naval powers in Europe that had importance in distant lands at that time but little today. The Juan Fernandez Islands off the Chilean coast and the Galápagos Islands, which are nearest to

Ecuador, are the principal curiosities of the eastern Pacific Ocean. Atlantic waters surround Bermuda (British), the Macaronesian clusters (Portuguese and Spanish), remote Ascension, St. Helena, Tristan da Cunha and Gough (all British). The big island of Madagascar and the adjacent small Comorro Islands are French, whereas the Mascarenes and Seychelles are British wards in the Indian Ocean. Far south, the Falkland Islands and dependencies (South Shetlands, South Orkneys, South Georgia and South Sandwich islands) retain a connection to Britain, remote Bouvet Island to Norway, Marion and Prince Edward islands to the Republic of South Africa, Crozet and Kerguelen to France, McDonald and Heard and McQuarrie islands to Australia, and the scattered cluster of Campbell, Auckland, Antipodes, Bounty and Chatham islands to New Zealand.

The Juan Fernandez Islands

The largest of these three small islands, and the only one to be inhabited continuously, is world-famous as the fictional site of the pioneering castaway in Daniel Defoe's story. Accordingly in 1966, its name was changed from Mas a Tierra to Isla Robinson Crusoe ("Robinson Crusoe's Island"). Here the little creeper (*Aphrastura masafuerae*) was listed in June 1967 as among the world's most endangered kinds of birds. Yet its home island is inhabited only temporarily each year by fishermen from the main island—men for whom the little creeper holds no interest.

The Galápagos Islands

In the words of Charles Darwin, who landed in this part of the world on September 17, 1835:
"This archipelago consists of ten principal islands, of which five exceed the others in size. They are situated under the Equator, and between five and six hundred miles westward of the coast of America. They are all formed of volcanic rocks . . . Nothing could be less inviting than the first appearance . . . we fancied even that the bushes smelt unpleasantly. Although I diligently tried to collect as many plants as possible, I succeeded in getting very few; and such wretched-looking little weeds would have better become an arctic than an equatorial Flora. The brushwood

Adélie penguins walk well inland from antarctic coasts to stony nestling territories (Benedict Leerburger, Jr.)

appears, from a short distance, as leafless as our trees during winter; and it was some time before I discovered that not only almost every plant was now in full leaf, but that the greater number were in flower." Botanists have explored thoroughly to widen the information Sir Joseph Hooker published from his studies of the plants that Darwin collected. About two among every five kinds of vegetation present are endemic to these islands. Yet almost all are of American ancestry, with relatives more often in Central America and Panama than along the adjacent coast of Ecuador.

No palms, not even the coconut, have established themselves. The lowlands present a tangle of thorny shrubs, poisonous manzanilla, evergreen *Maytenus* of the stafftree family (Celastraceae), aromatic balsams, crotons to represent the spurge family (Euphorbiaceae), and the strange shrub *Psidium galapageium* of the myrtle family, which is known elsewhere only from the little Revilla Gigedos Islands 420 miles west of Mexico and more than 300 from the tip of Baja California.

Inland and at higher elevation, a fog forest is carpeted and hung with mosses and lichens. Orchids thrive and so do tree ferns. The woody trees include an endemic satinwood, pisonias of the four-o'clock family (Nyctaginaceae) reminiscent of those on islands of Australia's Great Barrier Reef, and peculiar sunflower trees of the daisy family. Among the many species of sunflower trees, only six are found on more than one island; of these six, only one occurs on more than two islands; and no sunflower trees are known to be native elsewhere in the world.

The most probable route for plant immigrants and animals that could cling to rafts of floating vegetation may have been discovered in 1892 by the distinguished oceanographer, Alexander Agassiz. He measured surface currents moving as rapidly as seventy-five miles per day, sweeping south from the vicinity of Panama at least as far as tiny Cocos Island. Cocos is fully a third the way to the Galápagos and, upon occasion, might not be the end of the current Agassiz investigated. Similarly, the Galápagos are not universally the end of the northbound Humboldt Current, which brings

cold water along the Pacific coast of South America. A fortunate combination of circumstances, occurring no oftener than once in a thousand years, could lead to introductions of land plants from Middle America and account for the flora known on the archipelago.

No freshwater fishes or amphibians have made the trip. The only land mammals are rice rats—five closely related species probably derived from a single kind of ancestor—and an endemic bat similar to the migratory hoary bat of America and the single Hawaiian kind of bat.

Isles of the Giant Tortoises

Galápagos is Spanish for tortoises, island monsters that were already depleted in Darwin's day. Among these giants, thirteen subspecies have been recognized, five on the largest island (now called Isabella, but formerly known by its English name of Albemarle). Hatchling tortoises quickly grow to a weight of several hundred pounds, and then grow more slowly for an indefinite number of decades. Apparently the vagaries of weather determine which will be good vintage years for tortoises since, if a drought develops earlier than usual, the muddy ground in which the females hide their eggs becomes so hard that the hatchlings are entombed and die.

The adult tortoises and partly-grown ones plod about, eating large quantities of the vegetation they reach with their long necks. Apparently they are immune to the sharp spines of the cacti, for they prune away and munch down all of the low growth. In consequence, wherever the tortoises are numerous or the land iguanas compete for the same food, the cacti are kept pruned. They become, as the British buccaneers described the vegetation, "dildo trees and torch thistles." The branching fluted cylinders of *Cereus* cacti and the spiny pads of prickly pear (*Opuntia*) commonly rise thirty feet above the ground on remarkably sturdy trunks.

More famous than the land iguanas are the related marine iguanas, which grow to be five feet long, not merely four. The marine kind are extraordinary in feeding chiefly on seaweeds in underwater pastures along the coast. To reach this food, the big lizards dive off the lava rocks into the surging waves, hold their legs against their sides, and propel themselves by powerful sinuous movements of the body and long tail. While browsing on the lettuce-like algae some distance from shore, the reptiles may be attacked by sharks. On land, they have no native predators to fear and generally bask in the sun, often piled atop one another in social groups. Probably the ancestors of both the marine

Above left: The crested Chameleon dilepsis *uses its prehensile tail and grasping feet to hold firmly in its native Madagascar. (D. C. H. Plowes) Left: The sparrow-sized fodies of Madagascar, frequently called Madagascan cardinals from the bright color of the male, fly as far from home as Praslin Island in the Indian Ocean. (Tony Beamish)*

iguana and the land iguanas originated at least 15 million years ago in Central America, at a time when there was no land connection to either North America or South. At about that same time, similar iguanas reached the West Indies, and their descendants still survive on islets around Jamaica and elsewhere.

The Galápagos have seven different kinds of smaller iguanids known as lava lizards. They scuttle around on the lava outcroppings and also on higher land, mostly pursuing insects and small crabs. Some of the insects are large, such as the four-inch Galápagos grasshoppers, which are dark brown splashed with bright red, yellow and blue. Butterflies are numerous, but most are of kinds that are widespread: the cloudless sulfur, the gulf fritillary, and the monarch are among the commonest. A distinctive praying mantis and many kinds of spiders also lie in wait for insects of medium

Around the tundra and glaciers that cover most of Spitzbergen's 23,700 square miles, coasts attract sea birds and polar bears They depart before winter, leaving the island to ptarmigans, snow buntings, arctic foxes, small caribou and introduced muskoxen. (Lisa Gensetter)

and small size. The spectacular red crabs that sidle over the lava along the coasts are ready to scavenge for any remains they find.

On some of the islands, sites for basking and resting along the shore have been subdivided. Some areas are defended by marine iguanas, others by the twenty-inch Galápagos penguins, which are the only penguins exclusively in the tropics. On forbidding Fernandina (Narborough) Island, on Santiago (James), and parts of the big island (Isabella), the surviving fur seals haul out. There are still about 4000 of them, representing an endemic subspecies of the southern fur seal; it has relatives around Tierra del Fuego as well as on the shores of many remote islands in subantarctic seas. Because of exploitation they are far rarer than the local race of the California sea lion, whose fur coat has too many long, hard hairs to be of commercial value.

The only flightless birds on these islands are the penguins and the little Galápagos cormorants, which are unusually short in both wing and tail, allowing them to stand upright almost like penguins. Probably this infrequency of flightlessness merely shows that flying creatures on the archipelago have never been endangered by strong and continuous winds.

Where the cormorants climb out of the water to dry their feathers and rest, the endemic swallow-tailed gulls commonly settle. Often called lava gulls, they have black heads and necks, red eye rings, red legs and feet. In coves where lava pebbles and sand form a beach, small green herons dine on unwary crabs and fishes that swim too close to shore.

Darwin's Finches

The unwariness of the animals impressed Charles Darwin, and he wrote of his experiences with them at length in the first edition of his popular *Journal of Researches into the Geology and Natural History of the Various Countries Visited during the Voyage of H.M.S. Beagle round the World*. Not until the second edition (1845), however, did afterthoughts lead him to expand a mention made previously of some inconspicuous dark-colored birds now known as Darwin's finches.

"The remaining land-birds form a most singular group of finches, related to each other in the structure of their beaks, short tails, form of body, and plumage . . . Seeing this gradation and diversity of structure in one small, intimately related group of birds, one might really fancy that from an original paucity of birds in this archipelago, one species had been taken and modified for different ends."
Nowhere else in the book did Darwin so clearly forecast the direction his later investigations would take, or the new view of the earth's living things that would stimulate scientists all over the world.

Darwin's finches are the dominant bird life inland on the Galápagos. Six species are ground finches, six are tree finches, one a warbler finch because of its antics rather than for its song. A Cocos Island finch is the only close relative. More distant relatives live in South and Central America.

The warbler finch is the only one represented on all of the Galápagos Islands. Yet something in its diet of small insects has induced both the development of a more slender beak and also the formation of distinct subspecies: one different subspecies on each of five islands, a subspecies on four islands, a seventh and eighth subspecies each on two islands that have no other warbler finches. Of the ground finches and tree finches, seven kinds show no differences from island to island, but are found on only certain ones. The remaining five kinds are distributed less widely, but show differences by which a skilled ornithologist can tell from which island or islands a particular bird comes.

The distinctive features of these finches match their food habits and places where they live. Four of the ground finches have heavy finch-like beaks and feed on seeds. One of them occupies the humid zones on six islands, and has evolved into three subspecies. The other ground finches remain nearer the coasts and probably fly from one island to another, since no differences geographically have evolved. They avoid competing with one another for food by choosing for their principal diet large seeds, medium-sized seeds and small seeds. Buds, young leaves, small fruits and insects are taken in season, but again in relation to the magnitude and shape of the beaks. By contrast, the cactus-feeding ground finch has a longer and more decurved beak as well as a split tongue, adapted for probing for nectar into the flowers of the prickly pear cactus and for pecking into the soft pulp of the flat blade-like stems. The final species inhabits only three of the smallest islands and, by staying in the most arid parts while searching for small seeds and cactus flowers to yield nectar, seldom encounters the one or two other kinds of ground finches living in the same part of the Galápagos.

The bird known as the "vegetarian tree finch" spends most of its time among the higher trees inland, where it moves about in a leisurely fashion using its heavy beak much as grosbeaks do, to feed on buds, leaves, blossoms and fruits. Occasionally it flies to the ground to pick at herbaceous plants or salvage fruits that have fallen from the trees. In this way it meets no appreciable competition from related finches with thicker, longer or smaller beaks that specialize in insects and hence are insectivorous tree finches. Three different kinds of these insect-eating birds move more quickly, often hanging upside down from branches while pecking and prying into the bark for beetles and other prey. By staying inland these birds rarely encounter the mangrove tree finch, which is found along the coasts of Isabella Island and Santa Cruz (Indefatigable).

The most remarkable of these tree finches is called the "woodpecker finch." Its beak is especially long and straight, apparently in relation to an almost exclusively insectivorous diet. It not only searches among leaf clusters and in bark crevices but also bores into soft wood to reach insects half an inch or more from the surface. Lacking the barbed tongue of a woodpecker, it has evolved the habit of using a cactus spine as a tool. Plucking off a fresh spine each time, the bird probes with it into the hole until the insect is impaled and can be extracted. With no woodpeckers in the Galápagos for competition, the woodpecker finch has all of the woodborers of the islands to itself. One subspecies of this bird lives on six different islands, and another is on the easternmost island of the archipelago—San Cristóbal (Chatham) Island.

The Galápagos Islands afford such a showpiece of evolutionary change that biologists from all over the world want to visit and study their flora and fauna. Yet a more inhospitable archipelago can scarcely be imagined. Looking at the lava cliffs around Fernandina (Narborough) Island, Darwin compared them to "pitch over the rim of a pot in which it has been boiled." Introduced goats, pigs and dogs have caused great damage and exterminated many plants and animals. Yet not all of the extinctions there have been caused by human activities. A lava flow on Fernandina in 1825 all but destroyed the population of native tortoises and other kinds of life. So forbidding is this island that the existence of tortoises on it was not known until 1906, when Rollo H. Beck managed to clamber over the incredibly rough and seemingly still-fresh lava into the interior of the island and to bring out a single specimen. A new eruption in June 1968 produced shock waves recorded all over the earth, and discharged vast quantities of volcanic ash. Part of the crater rim collapsed, but no lava flowed. This may have saved the forest of sunflower trees, the land iguanas and other native living things. An expedition sent by the Smithsonian Institution to check on the survivors a few months after the eruption reported far less devastation than had been feared.

The Remote Atlantic Islands

In considering the travels of land life across barriers of open sea, distance from the nearest continent is scarcely more important than prevailing winds and ocean currents, migration routes of birds and past changes in sea level. Only the present size of an island and its distance are likely to show on a map. And if the distance is great, the location of the remote bit of land may be indicated mainly on navigational charts or representations of the whole world in which the vastness of the oceans cannot be overlooked.

Bermuda

About 600 miles east of Savannah, Georgia, the Bermuda islands offer a landfall of about nineteen square miles. Deep bore holes there reveal the remains of volcanic rock, but the strata exposed normally to view are of consolidated coral sand from dunes blown by the wind. This porous material lets the rain percolate through and out to sea, now over a coral platform whose form and depth show that the area of Bermuda exceeded fifty square miles during the last days of the Ice Ages. Beyond this is deep water.

No truly freshwater fishes have reached Bermuda or could survive there, for all of the ponds are brackish and the streams temporary. No amphibians or reptiles of the land, other than a skink, live naturally on these islands. The wading birds that haunt the marshy places, like the land birds, are those of adjacent North America. Only one of the sea birds, the Bermuda petrel (or cahow) breeds nowhere else. No land mammal—not even a bat—is a native resident.

Bermuda has about 150 kinds of flowering plants, but only about a dozen are endemic and all of these belong to genera represented on the mainland of the southeastern United States. In Paget Marsh Nature Reserve, a small forest of the sole native palm has been preserved. Under these trees and dependent upon the special ecological conditions they provide, the Bermuda sedge survives. During the 1940's, introduced scale insects attacked the endemic Bermuda cedar trees, killing most of them in just a few years. Formerly the cedar grew luxuriantly, providing the principal windbreak as well as shade, ground cover and valuable timber. Today many cedar skeletons still stand, but their role is being taken over by quick-growing introduced trees such as the Australian she-oak and the Eurasian tamarisk.

Macaronesia

At a comparable latitude on the opposite side of the Atlantic Ocean, four little clusters of islands are known collectively as Macaronesia: the Madeiras (two that are habitable, the bigger thirty-four miles long and twelve wide), the Canaries (seven islands, with Teneriffe the largest), the Azores (ten islands) and, just south of the Tropic of Cancer, the fifteen small Cape Verde Islands. All are volcanic and separated from the African mainland by deep water.

The land iguana of the Galápagos Islands seems able to eat the spiniest cacti without harm. At maturity it is about four feet long, a foot shorter than the closely-related marine iguana. (David Cavagnaro)

In each cluster of the Macaronesian islands, a correspondence might be expected between the number of distinctive plants and animals and the distance from the mainland, or the latitude as a measure of climatic suitability, or the area of the islands upon which the colonists might settle. The Canaries, in fact, seem so close—a mere sixty miles—that they hardly seem isolated at all. The tropical Cape Verdes, 280 miles from the coast and with an area of 1557 square miles compared to 2807 in the Canaries, would seem less likely to have animal immigrants and more likely to show endemic kinds of life through evolution of a few colonists to match new conditions. The Madeiras, with a total of 308 square miles and a distance of about 360 miles, and the Azores, with 888 square miles and the greatest distance (740 miles), are more obviously remote and small.

Actually, none of the Macaronesian islands has any native freshwater fishes, amphibians, turtles, snakes, or land mammals other than bats. The Cape Verdes have endemic lizards representing four separate types native to coastal Africa, whereas the Azores have none, the Madeiras one, and the Canaries three. Of the island clusters, the Azores are most isolated, yet have no endemic genera of plants; the Canaries, least isolated, have the most endemic genera.

Madeira

Among our vivid memories of strange animals seen in Madeira, the commonest large fishes in the market stand out. They are deep-sea snake mackerels (or escolars or espadas) of a small and peculiar family (Gempylidae). Their four- to six-foot bodies laid out in rows, these jet-black fishes showed formidable white teeth in their slightly gaping mouths, and big silvery eyes. They are predators commonest at the 2400-foot

The giant tortoises of remote islands in the Indian Ocean have survived only where their eggs and young escape the attention of rats, hogs, dogs and people. (Rolf Blomberg: Full Hand)

arborea, is endemic in the Madeiras and Canaries, where it grows as a handsome evergreen tree; elsewhere this family has a few representatives in temperate North America, and one endemic species in each of Japan and the Sunda Islands of the South Pacific. Yet side by side with the Madeiran white-alder may be a tree heath, a native that is apparently identical to tree heaths on the Canary Islands, the upper slopes of the Tibesti range in Chad, and the mountains of tropical East Africa. Madeiran shrubs include an endemic rowan and an elder closely allied to plants of western Europe.

Among Madeira's plant curiosities are a violet (*Viola paradoxa*) that grows as a tall shrub and a tree (*Melanoselinum decipeens*, known elsewhere only from the Azores) of the carrot family (Umbelliferae). Barberries are native to the islands, at their farthest south near the Atlantic side of the Old World; barberries live naturally farther south only in eastern Tanzania and the Andes.

The Canaries

These islands gave a name to the most popular of cage birds—the domesticated and much more golden-yellow varieties of a wild species endemic in the Canaries, the Madeiras and the Azores, where its ordinary plumage is olive-green above and dull yellow below. A local chaffinch differing only slightly from the widespread Old World bird and a handsome wood pigeon are among nearly forty different endemic birds.

"Canary seed," appropriately, was originally the grain of a grass upon which wild canaries fed. The grass itself is generally overlooked because other endemic plants of the Canaries are more spectacular. Most famous of them are the dragon trees, which grow to a height of fifty feet or more. One on the island of Teneriffe towered seventy feet tall, was fourteen feet in diameter at the base, and supposedly had an age of 6000 years when a storm blew it down in 1868. Resin from these trees, known as "dragon's blood," served as a varnish for the great Italian violins of the 18th century.

Lowlands in the Canaries, to an elevation of about 1300 feet, tend to be arid and covered by a spectacular assortment of tree-like spurges and live-forevers. In places where natural springs add moisture to the soil, small oases of the Canary Island date palm grow tall and graceful. Between 1300 and 2400 feet, almost nothing remains of the native vegetation because somewhat more rain falls and crops can be raised profitably —mostly those that would grow well in southern Italy. Between 2400 and 4000 feet, where the climate is cooler

depth, but caught a mere four miles off Madeira shores on long lines with 180 hooks, hung between 1800 and 4800 feet below the surface. All of Madeira's housewives know how to cook the fish so as to eliminate the purgative oil in its flesh.

The only endemic bird in the Madeiras is a wren. Many of the land snails are known from nowhere else. Of the moths, about a fourth are endemic, the rest like those of Mediterranean Europe and North Africa. Almost a third of the beetles are not only local but wingless, or so short-winged that they cannot fly. Those with functional wings generally correspond to kinds on Mediterranean coasts.

The plants of the Madeiras challenge the biogeographer, for some of them seem to have come from the New World. At least four kinds of ferns are of genera found elsewhere only in tropical America. Tropical America is also the principal home for white alders (family Clethraceae), of which one, *Clethra*

216

and frequently foggy with low clouds, Canary wood trees form forests as they do in the Madeiras and Azores. From 4000 to 6000 feet elevation, tree heaths are common, and tall Canary pine stands in impressive groves. The understory is chiefly laurel, a bayberry called faya, and Canary cedar whose pendulous branches bear bluish needles and orange-brown berry-like seeds; all of these shrubs are native also to the Madeiras and Azores, but endemic to Macaronesia. At higher elevations, a zone of gorse-like plants (the "genista" of florists) spreads upward to alpine meadows dotted with clumps of another member of the pea family, the dwarf broom that is common on arid highlands of North Africa.

The Azores

Birds and bats of western Europe visit these islands, and some stay to breed there before flying off again. But none of the vertebrate animals have evolved endemic subspecies. The native animals without backbones seem few too, and less varied than in other parts of Macaronesia. By comparison with the Canaries and Madeira, the Azores have few special kinds of vegetation. Apparently no trees grew on these islands until foreign ones were introduced. Tree heaths are missing, but endemic low-growing heaths include a widespread kind and a blueberry; these give a moor-like character to the uncultivated land, which seems strange so close to the tropics. A heath-like broom crowberry fits perfectly with the scene, just as it does in similar sites in Portugal; the only other member of this genus is poverty-grass of northeastern North America. Most of the other native plants in the Azores are those of western Mediterranean coasts.

The Cape Verde Islands

Most of the native trees have been cut, and the endemic vegetation greatly reduced in number and variety by the foraging of introduced goats. Yet more than 300 kinds of flowering plants have been recognized, nearly a third of them limited to these islands, and almost all more closely related to those of Mediterranean coasts than to those of Africa's Atlantic shoreline. Exceptions to this are a lavender and a bluebell, which

On many of the volcanic slopes in the Galápagos Islands, the tree-like prickly-pear cacti are the most conspicuous vegetation. Although on the Equator, the lower parts of these islands are arid. (Rolf Blomberg: Full Hand)

217

are shared with the African mainland; and a bean caper that is common from East Africa to India, and found also in South Africa and on Aldabra Island in the Indian Ocean. Today the most widespread of native plants in the Azores are guinea grass, a beardgrass and an Arabian grass, all of which are common in westernmost Africa.

Islands of the Eastern South Atlantic

The South Atlantic Ocean has still more remote islands: Ascension (38 square miles) and St. Helena (47 square miles) in the tropics, and the Tristan da Cunha group (the largest just 16 square miles) in the South Temperate Zone. Although closer to Africa than to South America, they are still between 1000 and 2000 miles offshore. All are volcanic, surrounded by deep water, but only Tristan da Cunha has had recent eruptions. Activity in the crater of the main island of Tristan in 1961 required the evacuation of the total human population of 295 people. Little Gough Island (10 square miles) is regarded as an outlier of the Tristan group, although it is 230 miles to the south-southeast. None of them has any freshwater fishes, amphibians, reptiles or native land mammals.

When first discovered, Ascension Island had vegetation only on the slopes and summit of Green Mountain (2870 feet). Grasses and shrubs were introduced so that the lower slopes would produce food for livestock, thus limiting the native plants to higher ground. Now only about eight indigenous flowering plants can be found, including a tuftybell that grows on St. Helena also, an endemic spurge and a madder.

St. Helena, famous as the place of Napoleon's exile from 1815 to 1821, is similarly arid and bare around the coast, then green with introduced plants from 400 to 1800 feet elevation. Only the central high parts of the old volcano serves as a refuge for native vegetation. A rush and a lowly plantain represent species of unknown origin, since they occur elsewhere in southern South America and in southeast tropical Africa. An alkali heath represents a little family (Frankeniaceae) with only forty-eight species, the forty-five in this genus found on coasts of the Americas, the Mediterranean, temperate Asia and Australia, and also the Cape region of South Africa. Three of the five endemic genera of flowering plants on St. Helena are tree-like members of the daisy family (Compositae): *Commidendron*, *Melanodendron*, and *Petrobium*. In many ways they resemble the sunflower trees of the distant Galápagos Islands more than the strange silverswords of Hawaii, but their development of this unusual style of growth is probably an independent adaptation to life on a volcano. Similarly, no special significance can be seen in the fact that St. Helena has one endemic bird: the small plover called the wirebird. Its relatives include the killdeer of North America and more than a dozen other shorebirds of the same genus in the Northern Hemisphere.

Of the little cluster of old volcanos comprising Tristan da Cunha, the most western is called "Inaccessible" because its borders are cliffs 1000 feet high. On the four square miles of summit, sea birds nest and so does a small flightless rail (*Atlantisia rogersi*) that is known from nowhere else. The southernmost island (Nightingale) is less than a fourth as big and inhabited only by sea birds. The main island, which is the northernmost, is roughly circular and surrounds a volcanic cone 6760 feet high. The rocks and sand of the beach are black and barren, but the cliffs 2000 feet high on all except the northwest side of the island are green with ferns, grasses and a distinctive tree of the buckthorn family (Rhamnaceae) found also on St. Helena. Its only relatives otherwise are on the Cape of Good Hope, and two species endemic to Madagascar, and one shared among Madagascar and the islands of Réunion and Mauritius in the Indian Ocean.

More than half of the forty-four kinds of native flowering plants known on Tristan da Cunha and on little Gough Island are endemics. Their ancestors seem to have come from both South America and Africa. Some are shrubs (*Nertera*) allied to *Coprosma*, with relatives in South America, Hawaii, Australia and particularly New Zealand. It has the same snowberry, called diddle-dee, as grows on the Falkland Islands, Tierra del Fuego and the Andes. It belongs to a little genus whose only other species are similarly lowgrowing evergreens in the North Temperate Zone.

The only fruit-eating birds on Tristan da Cunha and Gough are the only endemic ones: a thrush (*Nesocichla eremita*) and two finches (*Nesospiza*) on Tristan da Cunha, differing in the size of both beak and food, and one finch (*Rowettia goughensis*) on Gough. Ornitholo-

Above right: The Galápagos race of the California sea lion occupies favorite beaches on these volcanic islands. A less sociable member of the herd can be seen basking an a lava outcropping in the middle distance. Right: The endemic swallow-tailed gulls of the Galápagos Islands settle so often among the cacti on the lava outcroppings that they are commonly called lava gulls. (Both by David Cavagnaro)

gists suspect that the ancestors of *Nesospiza* came from Africa, but those of the thrush and *Rowettia* from South America. The ocean barriers are not impassable, for the South American gallinule commonly strays across 2000 miles of open sea to Tristan da Cunha, and barn swallows of the Northern Hemisphere have strayed there at least twice in the last century.

Madagascar

Off the eastern coast of Africa, opposite Mozambique, the giant island of Madagascar possesses so many distinctive kinds of life that it is believed to have been separated from the mainland for a very long time. Of the world's islands, only Greenland, New Guinea and Borneo are larger. None of these is so unique in living things.

Between Madagascar and the African coast, the channel at its narrowest is 248 miles across, which is less than the maximum width of the island itself. Geologists see no indication that the water barrier has been substantially narrower for at least 50 million years. But fossil dinosaurs and some ancient types of birds once lived on Madagascar, which suggests that no more than 200 million years ago there must have been a land bridge connecting the island to some other continent, perhaps Africa.

Flightless Birds

Madagascar's most famous birds are the extinct elephant birds, particularly the largest of them, *Aepyornis maximus*, which stood about nine feet tall and weighed nearly half a ton. Ten smaller relatives are known from fossils discovered in Madagascar and South Africa. Apparently all were flightless, but only distantly related to ostriches. Although the last of them became extinct less than 1000 years ago, no drawings or written records of their existence remain. But eggshells of the giant species have been recovered from swamps and among the boulders of stream beds on Madagascar, and dug from the debris in caves where primitive people once lived. Some of these shells are still prized as watertight bottles. One eggshell with a small hole in one end, as though it had been blown by an egg collector, was brought back from Madagascar by explorer Richard

Thermal springs and hissing steam contrast with snow on the glaciated mountain peaks of Iceland, providing a wide range of temperatures but not much stability of the kind needed for life. (Lisa Gensetter)

Archbold of the American Museum of Natural History. We had the pleasure of holding this ancient trophy and marvelling at the thickness of the shell. Its volume had been measured and found to be about two gallons, by the simple expedient of filling the shell with dry sand and pouring the sand into a calibrated container. This volume is approximately six times that of an average ostrich egg, just as the giant elephant bird was about six times as heavy as a big modern ostrich.

Flightless birds still live on Madagascar, but they are rails of three different kinds in a little family (Mesitornithidae) of their own. Known as roatelos, they inhabit the brushlands and forests, eating insects, seeds and small fruits. Birds of fifty other families live on the island, many of them restricted to it. Of those found elsewhere, a majority inhabit Africa and a minority the Oriental parts of the Old-World tropics. Yet a surprising number of common African birds are missing altogether: ostriches, secretary birds, guinea fowls, cranes, mousebirds, hoopoes, hornbills, woodpeckers, barbets, honeyguides, broadbills, African river martins, true shrikes, helmet shrikes, oxpeckers, finches, buffalo birds and widowbirds. Madagascar's birds, instead, are chiefly herons, ducks, hawks, rails, gulls, cuckoos, Old World warblers, and vanga-shrikes.

Madagascar's Mammals

We think of bats as being almost as free as birds to travel freely across open water. On Madagascar about two dozen kinds have made themselves at home. About a fourth of them represent genera encountered all over the world. A somewhat larger number have close relatives only across Africa in the one direction and as far as Australia and South Pacific islands in the other. The Madagascan fruit bat differs slightly from two other species of the same genus found in Africa. The giant fruit bats are at the western extremity for a group commonly called "flying foxes," ranging through Southeast Asia to the Philippines, Fiji and Samoa, and southward into Australia's state of Queensland. The three kinds of triple-noseleaf bats occur also on Aldabra, but their near kin are all on the African mainland. Only one kind of bat—the peculiar sucker-footed bat—represents a genus and family known from nowhere else. Ankles and wrists of this animal bear short stalks ending in suckers, with which it can support itself from smooth surfaces while resting.

The other land mammals of Madagascar are equally limited in variety and peculiar in origin. Just as Charles Darwin suggested that the finches on the Galápagos could all have evolved from a single ancestral colonist, so too the Madagascan mammals might have all come from four different types: an insectivore, a primitive primate, a cricetid rodent, and a mongoose-like carnivore of some kind. From the first would come the thirty different tenrecs (family Tenrecidae), of which some are so adapted to living conditions that they resemble opossums, moles, muskrats and mice as well as their foreign relatives, the spiny hedgehogs and shrews. One ancient primate may have given rise to the lemurs of three different families with about twenty species, of which the aye-aye is in particular danger of being exterminated as its indigenous forest habitat is cleared away. The rodents native to Madagascar now represent a dozen species in seven genera, as do the native carnivores. Yet almost none of these creatures has a name recognized in other parts of the world because all tenrecs, all lemurs, all of these genera of rodents and carnivores are endemic to the one big island.

Cold-blooded Vertebrates of a Tropical Island

A majority of the reptiles on Madagascar are of types encountered on the African continent. Yet many common African forms are missing. The Nile crocodile is there, its ancestors apparently having crossed the broad Comoro Strait, but not the monitor lizards that prey on crocodile eggs and young. Snakes are numerous, but no poisonous kinds are present; no pythons have spread to the island, but a few highly specialized constrictor snakes of the related family (Boidae) are endemic. It has small land tortoises but not the giants, such as were discovered on small islands in the Indian Ocean, although fossils of these monsters prove they used to inhabit Madagascar. About half of the world's eighty different kinds of chameleons are Madagascan, all but four of the remainder living in Africa south of the great deserts. One of the four is native to Mediterranean coasts, two occur in the southwest corner of the Arabian Peninsula, and one in India and Ceylon.

The giant among chameleons, twenty-two inches long, is a Madagascan animal showing all the versatility of its relatives. It can change color rapidly, as a whole or in patches, although more in relation to temperature, humidity and excitement than any ability to match its surroundings. Its curled, prehensile tail and grasping feet, with two toes opposed to three toes, let the chameleon cling tightly to bushes and tree branches while it examines its world through turret-like eyes. Each eye may rove independently, or the two may act binocularly while the reptile stalks prey and judges the distance it must shoot out its piston-like long sticky tongue to capture a victim. The giant can pick off and then swallow a half-grown mouse, and may occasionally catch a bird as does a big African member of the same genus. Madagascar has plenty of insects of all sizes to feed chameleons and other reptiles of lesser dimensions.

The big island has more than 150 kinds of amphibians, but no strictly freshwater fishes. The wormlike caecilians are lacking, which seems strange since tropical Africa has several genera in the east and the west, the Seychelles have some, and in Asia they are distributed from Ceylon and India to the southern Philippines, as animals closely similar to the many kinds in tropical America. Madagascan frogs are mostly endemic, with relatives in both Africa and the Orient. Toads are missing altogether. From information of this kind, the biogeographer tries to construct a mental picture of the past that will account for the present. He is well aware that ancestors of Madagascan animals may have spread from wherever they were originally to the Orient or to Africa as well as to the island, and that present distribution is no proof that they came from either the Orient or Africa to Madagascar.

The Green of Madagascar

The plant life of Madagascar relates more closely than the animals to the climate and geological features. The steep eastern slopes receive rain from the trade winds during the violent winter of the Southern Hemisphere, and sustain fast rivers suitable for generating hydroelectric power. The north and central high plateau, which consists of three massifs of pre-Cambrian and Paleozoic age, gets less moisture and at a hotter time of year—from the monsoons during November to April. In other months these regions and the western slopes, which are gentle, tend to be semiarid.

From Madagascar's wet, eastern slopes, the most famous plants are the traveler's palms of the banana family (Musaceae), from the water-holding leaf bases of which a thirsty traveler is supposed to be always able to get a nontoxic drink. (No one explains why clear water from the nearest mountain stream would not be better!) The ground on these slopes is often dense with the succulent flopper (or living-leaf plant) and other members of the same genus from whose leaves small plantlets develop, each with its own tiny leaves and roots, then drop off to grow on any bare area of wet soil. The raffia palm straggles over these hillsides, yielding fiber of many uses. Far less conspicuous are the sundews (*Drosera*) whose leaves spread out in little rosettes, each leaf with glandular reddish tentacles ready to capture insects. None of these insectivorous plants is endemic. In fact, one (*D. mada-*

gascariensis) is the second most widely distributed sundew known, being found in suitable sites all over tropical Africa.

Botanists who have examined Madagascar's plants notice that genera that are shared with South Africa are usually of smaller total range than those shared with tropical Africa. Some of the endemics, however, have been cultivated so widely in recent years that botanists now wonder whether any remain in the wild state. These include the handsome royal poinciana tree of the pea family, with reddish orange flowers and finely dissected leaves; the prickly crown-of-thorns of the spurge family; and a woody vine of the milkweed family that has been used as a source of rubber.

The Mascarenes

Between 400 and 850 miles east of Madagascar, in the tropical Indian Ocean, three small volcanic islands with a total area of 1729 square miles constitute a loose group called the Mascarenes. With distance from Africa, they dwindle in size: Réunion (969 square miles), then Mauritius (720), and finally Rodriguez (40). Yet long ago a ground feeding pigeon must have reached all three and left descendants that became flightless: the solitaire on Réunion, the famous big dodo on Mauritius, and the small white dodo on Rodriguez. When the last dodo was killed on Mauritius in 1681 and the phrase "extinct as the dodo" was coined, this island had already lost by human interference its broad-billed parrot (1638) and red rail (1675). By 1800, Réunion had lost its solitaire (1746), its parakeets and fody (*Foudia*), and Rodriguez its white dodo (1791), a flightless rail, a flightless night heron, a different parakeet and an owl.

Today almost all of the indigenous forests on the Mascarenes have been cleared away, and only a poor sample remains of the plants and animals that once lived there. On Réunion, where the volcanic peaks rise above 10,000 feet elevation, a thicket of dwarf bamboo forms a vegetative zone on the slopes beyond the cultivated fields. Extending from about 4500 to 5000 feet elevation, it is succeeded by an even denser scrubland of leafless acacia whose branches are flattened and bright green. Shrubs of the pea family (Leguminosae) include some of a genus (*Sophora*) found also half a world away on Juan Fernandez Islands, in temperate South America, on Hawaii, New Zealand, Chatham and Lord Howe Islands—the kind of distribution that makes no sense to a biogeographer. Perching lilies similar to those we saw abundant in New Zealand

The flightless cormorant of the Galápagos Islands survives only where the terrain protects it from introduced predators. (Rolf Blomberg: Full Hand)

forests grow on both Réunion and Mauritius, as well as on distant New Guinea, northern Australia, New Caledonia, Tahiti, Hawaii, and temperate South America. Réunion and Mauritius have phylicas similar to those on Madagascar and in the Cape province of South Africa. Réunion has its own flycatcher, a manioc bird and a blackbird. Only one kind of snake is known there, a harmless Oriental wolf snake.

Mauritius has two different types of boid constrictor snakes, but Rodriguez no snakes at all. All of the Mascarenes are well supplied with skinks and geckoes, but only Rodriguez has a large endemic gecko. Again the pattern is one of subtraction away from the nearest major land mass—Madagascar.

The Seychelles

Northeast of Madagascar and north of the Mascarenes lie the 92 islands of the Seychelle group, with a com-

223

The ringed, or chinstrap, penguin raising their young on islands near Antartica must stay out of the way of ponderous elephant seals. (Niall Rankin)

bined area of only 156 square miles. Mostly the water between them is less than 300 feet deep. During the Ice Ages they probably were linked by land, and to Réunion and Mauritius, if not to Rodriguez too. An outlier of the Seychelles is Aldabra, composed of two islands only 265 miles from the nearest point of Madagascar and about 500 from the African coast.

Unlike the Mascarenes, which have no amphibians of any kind, the Seychelles have half a dozen kinds of legless caecilians and three different frogs in an endemic subfamily. Fourteen kinds of endemic birds continue to survive; none are known to have become extinct. The two different insectivorous bats and the big fruit bats suggest an Oriental, rather than an African influence in the fauna. While most of the forests have been cut, seven different palms remain, all of them belonging to endemic genera. One of them is the famous double coconut, with the largest fruit in the world. Trees and shrubs include a local screwpine and a woody *Turnera*, found also on Réunion and in Indo-Malaysia, although it belongs to a tropical-America genus.

In the Seychelles and Mascarenes, the extinction of

giant tortoises repeats the story of the dodo. These amazing reptiles had evolved distinctive races on almost every island. Now only one, the race from South Aldabra Island, survives and has been introduced into other parts of the Seychelles. Individuals grow to weigh more than 400 pounds, and reproduce fairly well wherever in the tropics they are protected from dogs and hogs and rats that eat their eggs, goats that devour all of the low vegetation young tortoises can reach, and people who enjoy turtle eggs and turtle meat.

On a two-acre island in the Indian Ocean beyond Zanzibar, we spent a day studying twenty-two of these monsters where fresh water to drink and low vegetation seemed too scarce to support even this many. When we managed to pump up a few gallons of smelly rain water from the decaying cisterns of an abandoned building and poured out the liquid where the tortoises

224

could drink it, they crowded around, each trying to shove the others aside. We had to jump to get out of the way and avoid being crushed by their armor.

South Aldabra Island is now a sanctuary for animal life; its distinctive rail, ibis and dove are protected. To its lagoon each year, "blonde" hawksbill turtles and green turtles swim to reproduce, perhaps safe from workmen who come under permit and supervision to cut mangroves or to fish. Its tortoise population is increasing—the last of its kind on its native island in all of the Indian Ocean.

Antarctic Islands

Where the warm waters of the Indian Ocean extend southward, they join those of the South Pacific and the South Atlantic where no land impedes the persistent west winds and storms are more furious than anywhere at sea level. Despite the waves and because of the ocean currents, the warm waters meet the cold ones around Antarctica along a remarkably persistent line. It ranges between 45 and 65 degrees south latitude, varying only a degree or two in any year at each longitude, and maintaining a noticeable boundary between the subtemperate and the subantarctic zones.

The Coldest Continent

At the center of this almost circular boundary lies the almost circular continent of Antarctica, its 5,300,000 square miles mostly covered with ice as much as a mile thick. Larger than Europe or Australia, it has only two kinds of flowering plants: a grass (*Deschampsia antarctica*) and a pink (*Colobanthus crassifolius*). Its dominant vegetation in the few places where rocks and bare earth are snowfree in antarctic summer consists of low-growing lichens, algae and mosses, many of them plants of cosmopolitan distribution and all of them with immense tolerance for cold. A few kinds of insects and crustaceans feed on the vegetation, even in freshwater pools that contain meltwater for only a few weeks each year.

The only animals with backbones are kinds with fur or feathers and feeding habits that take them to the abundant marine resources in the seas surrounding this barren land. Four different birds and four kinds of mammals regularly rely upon Antarctica for breeding grounds: the giant emperor penguins, the little Adélie penguins, the south polar skuas that prey on Adélie chicks, and the snow petrel; the common Weddell seals and the rare Ross seals, both of which feed on fishes and squids, the solitary and voracious leopard seals

that prey on smaller seals and penguins at every opportunity, and the crabeater seal which lives almost entirely on the pelagic krill crustaceans less than two inches long. It is to eat krill that the whalebone whales approach these Antarctic coasts and the ice floes beyond the shore line.

Alone among these warm-blooded animals, the emperor penguins wait until winter to reproduce. They stand on the ice near open water, turn their backs to the bitter wind in the black of antarctic night, and take turns incubating their single eggs or brooding their chicks atop their warm feet, surrounded by folds of

The aye-aye of Madagascar, although lemur-like, is the sole species in the one genus of a unique family of primates. Emerging at night from a hollow tree or cluster of branches, it hunts insects and plant foods. (F. Vollmar: World Wildlife Fund)

225

warm feather-clad skin. The birds off duty dive into the water to find food, pursuing fishes and squids by rapid paddling with feather-clad wings, using their feet only for steering, landing on the ice again, and walking back to the emperor rookery. In 1960, at the end of the International Geophysical Year, the known rookeries of this big penguin totalled fourteen, with an estimated population greater than 150,000 birds.

Elephant seals, which breed regularly on some of the subantarctic islands, sometimes come to Antarctica —particularly on the Indian Ocean side—to haul out and gather harems. About thirty different kinds of birds fly far enough toward the South Pole to alight on these shores in summer. Some of them nest there with fair regularity, as well as on subantarctic islands. Among the commonest are the silver-gray fulmars, the antarctic petrels and Wilson's petrels and giant petrels, the antarctic whalebirds and the antarctic terns. Arctic terns that nest in the Far North and the kelp gulls of subantarctic islands come winging by in antarctic summer.

The Palmer Peninsula on Antarctica projects far into the temperate zone, close to the South Shetland and South Orkney Islands, the Falkland Islands, and Tierra del Fuego. At this point, Antarctica and South America are barely more than 600 miles apart. While the distance may not have been much less at any time in the past, unless the continents drifted from elsewhere to their present positions (a distinct possibility), living things may have spread across the intervening strait. That Antarctica once had a richer life of its own to share is shown by low-grade coal deposits and other fossils. They include a wealth of fern-like plants and cold-blooded terrestrial (or amphibious) vertebrate animals several feet in length. From the Permian period until the Triassic, at least, Antarctica was relatively warm. It did not chill beyond the tolerance of many living things until the Ice Ages, for the fossils include evidence of antarctic beech grown on these shores until the glaciers began scouring them—shoving soil and fossils beyond the coasts to pile up in deeper water.

The Windy Falklands

The Falkland Islands lie about 300 miles east of the southern tip of South America, in the cool temperate zone where winds are a major fact of life. The two main islands are similar in size, with a total area of about 4500 square miles. The relief is low, with rounded mountains rising only to about 2000 feet. Between the exposed rocks, which show an abundance of glacier scratches, great bogs of peat moss have built up and

been exploited recently for fuel. The widespread tussock grass, which grows also on Tierra del Fuego, South Georgia Island and Gough Island, suffers from overgrazing by sheep and is being replaced by heaths that the animals avoid. Except for birds, no native animals with backbones live on the treeless Falklands today. But until about 1875, an endemic fox survived, perhaps from ancestors that reached the islands from South America on drifting ice during the Ice Ages. Others of its genus live in South America, and this method of transport is still effective in arctic latitudes, bringing arctic foxes periodically to Newfoundland.

Probably the Falkland Island foxes formerly depended for food upon a mixture of plants and whatever meat they could get: insects, small land birds, sick and dead hawks and owls, sea birds and seals. Today the bird life seems scarcely changed. The endemic birds are plovers, robins, and those that swim readily out to sea as the penguins do. The flightless steamer duck uses its short wings to help itself along over the water, and finds its food chiefly at low tide when the kelp beds lie exposed in shallow bays.

The second-largest of penguins, the 38-inch king, comes to the beaches of the Falklands in early antarctic summer to stand, brooding a single egg atop its feet under a fold of belly skin, much as an emperor penguin would. Smaller 25-inch Magellanic penguins dig burrows and lay two eggs, just as their close kin, the jackass penguins, do on islets off the South African coast. Rockhopper penguins nest among the tussocks and also on the open beach in close proximity to king shags and the albatrosses known as mollymawks. Gentoo penguins form dense rookeries of their own kind on gravelly shores, often gathering branches of diddle-dee to make a nest or stealing branches from inattentive neighbors.

Skuas raid the penguin colonies for eggs and chicks. Elephant seals crush many that are in their way on the beaches. Leopard seals prey on adult penguins that are swimming in search of food. All of these hazards are matched by penguin reproduction.

From Tierra del Fuego, some of the waterfowl have spread to the Falklands without developing any distinctive differences. Yet the kinds that did not cross to the islands intrigue the biogeographer as much as those that did. The goose-like coscoroba swans, which are found from southern Brazil, Uruguay and Paraguay through suitable parts of Argentina and Chile to Tierra del Fuego, are in the Falklands too. The black-necked swans live in all of these areas except the Falkland Islands. The ashy-headed geese, the ruddy-headed geese

The elephant seals, which breed on the coast of Argentina, on Gough Island and to the South Shetland Islands, dive for fish and squids, and attain a weight of close to a ton. (E. Aubert de la Rue)

and the flying steamer ducks are alike in the Falklands and Tierra del Fuego. But the upland goose (or greater Magellanic goose) on the Falklands is an endemic sub-species, differing in several details from the one (the lesser Magellanic) on the mainland. The Falklands have also an endemic greater kelp goose closely related to the lesser kelp goose of Tierra del Fuego and the Chilean coast.

The Falkland Dependencies

Often the Falklands are regarded as continental islands, mere extensions of Tierra del Fuego despite the intervening seas. The South Shetlands and South Orkneys might similarly be considered outliers of Antarctica, relating to the curved Palmer Peninsula. Eastward from both and more isolated in the South Atlantic Ocean is South Georgia Island, which has one endemic bird—the South Georgia teal. Possibly the pigeon-like sheathbills, nonswimming birds of the little family Chionididae, hint at true relationships. One of the two species nests on South Georgia, the South Orkneys, the Palmer Peninsula itself, the Falkland Islands and some isolated bits of land in the archipelago of Tierra del Fuego. Sheathbills are bold scavengers, preying on the eggs and young of sea birds whenever possible, taking advantage of garbage dumps near human communities, and otherwise finding edible morsels among the flotsam cast ashore along the coasts.

Far South in the Indian Ocean

The lesser sheathbill, which is the only other member of this family, nests on remote islands in the southernmost Indian Ocean, where the common grass is *Poa cookii*. The strange member of the mustard family (Cruciferae) known as Kerguelen cabbage (*Pringlea antiscorbutica*) is one of the two endemic flowering plants known to this group of subantarctic islands. They include Marion Island and Prince Edward Island, almost straight south of Mozambique; the Crozet Islands far to the south of Mauritius; Kerguelen (one big island and about 300 tiny ones), McDonald Island and Heard Island south of India. The other endemic genus of seed plants is limited to Kerguelen itself. Kerguelen has one race of pintail ducks and the Crozets

227

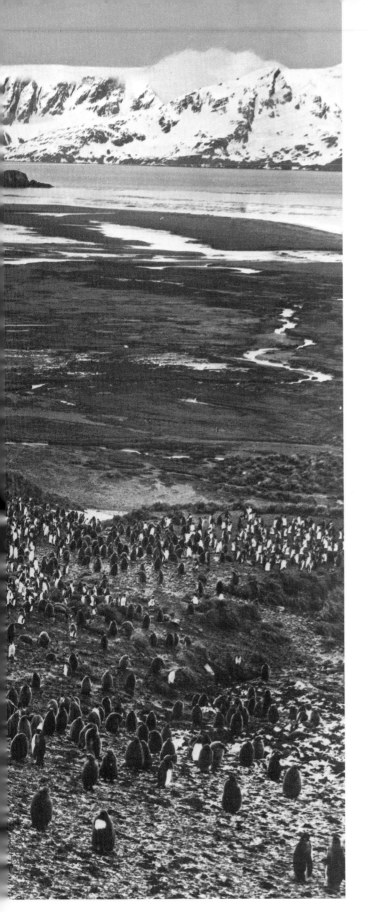

another. We wonder how they reached these islands, for all other pintails in the world nest in the northern part of the Northern Hemisphere.

South of the Tasman Sea

Farther east, south of the Tasman Sea and of New Zealand, lie Macquarie Island, Auckland Island and Campbell Island. Their tussock grasses are different again, with *Pringlea hamiltoni* on Macquarie, *P. litorosa* on Auckland and Campbell, and *P. foliosa* on all three as well as the New Zealand south island. The king penguin comes to these islands too. So do the circumpolar rockhoppers and gentoos, and more local kinds: the yellow-eyed, and the royal, which is endemic on Macquarie. Any explanation for the presence or absence of particular living things on these islands can be expected to help account for at least part of the flora and fauna on New Zealand also.

Arctic Islands

In thinking of these far-south parts of the world and of the terrestrial life inhabiting each area of land, comparisons with distribution of plants and animals in the arctic counterpart could be helpful. The northern regions have their islands too: Greenland with 840,000 square miles, Baffin Island with 183,810, Ellesmere 82,119, Victoria's ten islands totalling 81,930 square miles, and Iceland with about 40,000. All of these islands have acquired their present flora and fauna during the last 10,000 years or less, for Greenland was close to the center of the immense ice cap during the Ice Ages. It still is about 85 per cent icebound.

Greenland

The peculiar feature about Greenland's living things today is that while about 80 per cent of the 400 kinds of flowering plants came from America and the rest from Europe, the proportion among the animals is almost exactly the reverse. A few overlap, such as the American bunchberry and the Scandinavian cornel, both of which grow in Greenland just as they do in eastern Siberia. The Eurasian arctic dwarf birch forms thickets near the coast, but is at the northern limit of its range; inland the green alder of the American sub-

Second only to the emperor penguins in size, the king penguins of islands near Antarctica walk overland for long distances from the sea to their breeding grounds. (Niall Rankin)

229

arctic takes its place as a ground cover, seldom more than three feet high. Alpine blueberry carpets some of the low hills, and the sedge others; both are circumpolar plants. Often the hilltops are bright in summer with Iceland poppy and creamy white flowers of the trailing dryas of the rose family. Sandy and gravelly soil is often dense with horsetails and boggy areas with the sedge known as cottongrass.

Greenland has no native freshwater fishes or cold-blooded land vertebrates. Its land birds, other than for the circumpolar kinds (peregrine, gyrfalcon, rock ptarmigan, snowy owl, two kinds of redpolls, Lapland longspur and snow bunting) include the bulky-bodied, white-tailed eagle; the fieldfare and redwing; the widespread wheatear; the pied wagtail and the meadow pipit—all from Europe. The endemic Greenland white-fronted goose breeds only on the northwest coast, and winters in the British Isles and sometimes in eastern North America. The east coast is preferred by the pink-fronted goose, which nests also on Iceland and Spitzbergen, and by the barnacle goose, which breeds elsewhere only on Spitzbergen and northern Novaya Zemla and Kolguev. It is no wonder that the legendary origin of barnacle geese from goose barnacles on dead trees along forbidding coasts of northern Scotland, which was credited during the Middle Ages, took as many decades to disprove. Greenland is scarcely a vacation resort, even if it has a dozen kinds of mammals that have arrived from every side of the North Pole.

Iceland

Iceland is only 155 miles southeast of Greenland, but separated from Europe by a much greater width of water—water warmed by the Gulf Stream. Yet Iceland is even more an extension of the arctic area in the Old World. Its breeding land birds are all European. Its only endemic mammal is a mouse whose ancestors seem to have been the common European field mice until they crossed to Iceland and evolved slightly. Polar bears on floating ice sometimes reach the island's shores today, showing how this transport could have been accomplished.

Erosion has carved deeply into the rocky sides of the mountains in the Seychelles. A few vines clamber over the fallen rocks. (Tony Beamish)

The ice-free area on Iceland is less than a third as great as that on Greenland, and some of this consists of thermal springs and lava fields from recent and older volcanic activity. Yet Iceland has almost as many different kinds of plants as Greenland, all of them clearly of European origin. Around the hot springs, the dwarf willows and grasses grow faster and taller, the heaths and eyebright flower profusely. Some endemic mollusks, crustaceans and insects live in the warm waters. The whooper swan is equally localized in breeding on Iceland's cooler lakes and bogs. It is a strange part of the Northern Hemisphere, on the fringe of the temperate zone. It is a place where Atlantic puffins are caught to eat and the feathers from nests of common eider ducks are national resources meriting the most careful harvesting.

Polar Travelers

None of the other arctic islands shows in its native living things a comparable measure of remoteness. Although weather conditions in this northern region are similar to those around the South Pole, the opportunities for plants and animals to travel across the polar regions since the last great glaciation and to colonize have been far greater than in the cold parts of the Southern Hemisphere. To some extent this reflects the fact that the North Pole is over an Arctic Ocean, where the summer sun and the Gulf Stream contribute enough heat to produce open water every year and to allow a circulation of the ice pack. Even during winter, a herd of musk oxen and a pack of wolves might progress across the frozen ocean between the Old World and the New. Rough ice would make the going hard, but no mountains would bar their way.

By contrast the South Pole is over a major land mass with high peaks and a great central glacier more than a mile thick. To reach the opposite side of this continent, most antarctic animals must circuit the coasts. Penguins with numbered metal anklets have shown that they can indeed find their way home to their rookeries from hundreds of miles distant. But only the skua has been sighted within 500 miles of the South Pole—actually within eighty miles—as though crossing Antarctica by the most direct route. Despite all the many kinds of life that can tolerate antarctic weather, the physical fact that Antarctica is there makes other parts of their world remote.

11

Life
Between the
Continents —
in the Open Sea

Enormous as the continents are, the seas are vaster. Of the 196,938,800 square miles that comprise the earth's surface, all but about 56,500,000 are under water. Although the sea floor dips down into major trenches more than 18,000 feet deep in only about twenty places, the average depth of the oceans is about 12,600 feet—more than two and a quarter miles. If all of the world's surface, were leveled, by using the material of the continents to fill the deeps, there would be no land left anywhere. Instead, sea water would cover our planet to a uniform depth of about 8340 feet.

Encircling each continent is a relatively shallow zone, presently bounded by a contour line about 600 feet below the surface, where the sea floor becomes a cliff sloping down to about 6000 feet. Geographers refer to the shallow portion as the continental shelf, while oceanographers speak of it as the littoral zone. Much of it was exposed during the Ice Ages. It still shows evidence of erosion from former times. Yet it is being added to where rivers empty and deposit sediments. Opposite a river mouth the shelf is often forty miles wide or more. A short distance to each side of the river its width may be less than five miles.

For plants and animals that drift or swim in the open oceans, living out their lives without contact with shores or sea floor, the continents and continental shelves have scant significance. Yet most of them live in the uppermost 100 to 600 feet of water for the simple reason that daylight is available there to plants with chlorophyll. Individually these plants are almost all too small for us to see with the unaided eye. Yet they are present in fantastic numbers, constituting a sea pasture that may have a green color because they absorb less energy and reflect more in this part of the solar spectrum than in the blue-violet and yellow-orange. They filter out so much of the light reaching them that, wherever they are abundant, little solar energy passes beyond for living things at greater depths.

Even when the appearance produced by millions of these motes of plant life is familiar, the microbes themselves and their names are not. The Red Sea gets its famous color from a free-floating blue-green alga (*Trichodesmium erythraeum*) that has a bright red accessory pigment. A "sulfur-bottom" whale is just a great blue whale that has been swimming in antarctic waters for enough weeks to acquire over its white belly a flourishing coat of golden-green algae, mostly the diatom *Cocconeis ceticola*. The milky green color of surface waters in deep fjords along the Norwegian coast is often due to some of the smallest of green plants, such as *Coccolithus* and *Rhabdosphaera*, which range in size between 200 and 790 millionths of an inch in diameter. At times they reproduce in the cold water until each quart near the surface of the fjord contains five to six million of them.

Scientific knowledge of the microscopic plants and animals of the open sea began less than a century ago through the efforts of research men who participated in the prolonged cruise of the British corvette H.M.S. *Challenger* between 1872 and 1876, or who studied the many specimens collected. After crossing the Atlantic Ocean several times, the ship rounded the Cape of Good Hope and was the first to steam south of the Antarctic Circle among the remote islands. Exploration of the western Pacific was followed by passage through the Strait of Magellan, and a return northward through Atlantic waters to England. Samples were collected all along the route. A remarkable variety of nets and special bottles were used in order to keep separate the living things drifting and swimming at different levels in the ocean.

Rhythmically contracting its umbrella-shaped body, a medusa of the eastern Atlantic Ocean and Mediterranean Sea propels itself among the drifting plankton. (Alfred Schuhmacher)

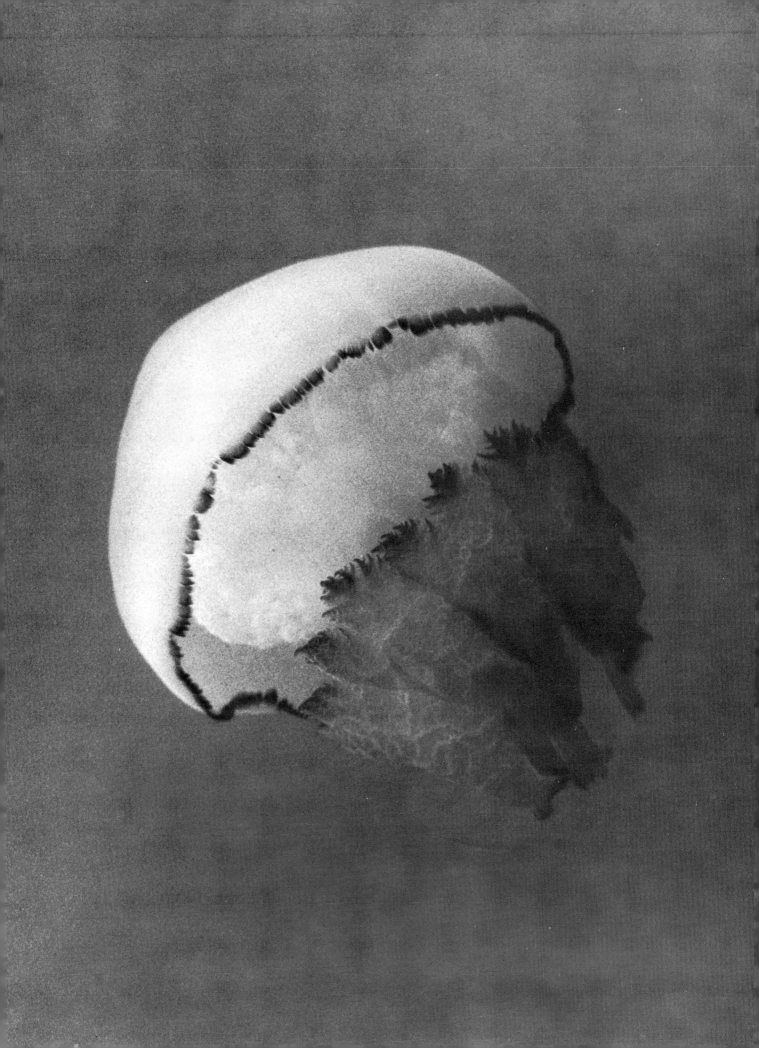

Life in Glass Pillboxes

Particularly in cold waters, a net made of fine cloth towed behind a boat gathers vast numbers of single-celled microbes called diatoms. They are golden-green algae of the phylum Chrysophyta, and the chief transformers of solar energy into stored form in the upper levels of the open oceans. The organic compounds they produce contain chemical binding energy as well as elements derived from water, dissolved carbon dioxide and mineral nutrients. The diatoms use these organic compounds for their own growth and reproduction, which continue at a rapid pace so long as water, solar energy and mineral nutrients are plentiful.

While carrying on photosynthesis, diatoms divide, each to become two in an hour or less. Each absorbs from the sea water around it enough of the scarce dissolved silica (silicon dioxide) to be able to secrete a thin silicious shell around itself. This transparent covering resembles a two-part box with overlapping parts, but with a general shape that is distinctive for each species: spindle-shaped in *Navicula*, circular like a pillbox in *Coscinodiscus*, grouped with others of its kind to form a flat ribbon in *Fragilaria*, or into a star-shaped pattern in *Asterionella*. Points and pits in the surface of the silicious shell are so exceedingly fine that they are made visible only through a compound microscope with lenses of the highest quality.

Ever since the explorations made aboard H.M.S. *Challenger*, oceanographers have realized that rapid growth of diatoms and other algae is far from uniform in any square mile of surface waters. In some places it is rapid, and they speak of an "algal bloom" where they see a local green coloration due to countless numbers of these single cells. But when so many minute plants absorb silica and other mineral substances from the sea water, they use up most of the supply, converting the elements into shells and protoplasm, oil droplets that give them buoyancy, and granules of reserve food such as starch. Without more nutrients, the diatoms cannot utilize the sunlight, the water and the carbon dioxide around them. On a microscopic scale a true famine develops. Reproduction falters, and many cells die. Bacteria gradually decompose the dead protoplasm and the food granules, releasing bacterial wastes that include mineral nutrients upon which new generations of diatoms can grow.

For some reason that has yet to be discovered, diatoms synthesize vitamin D (calciferol) and store this chemical compound in their oil droplets. When the diatoms are eaten by small animals of the plankton community, such as fragile radiolarians, the oil and vitamin D are transferred to the animal bodies with little or no change. The buoyancy conferred by the oil droplets is more important to the plankton animals than any food values they could derive in this way. When fishes eat the plankton animals, the oil and vitamin D are salvaged again and stored in the fish livers, still providing buoyancy. In cold waters of the North Atlantic, the smelt-like capelins are eaten by codfish and other scaly predators, and once more the oil and vitamin D move to a new location. Codliver oil, like the oil from livers of halibut and sharks, has therapeutic value for people and land animals suffering from vitamin-D deficiency because the products of diatoms have been maintained intact through so many steps.

The oil droplets themselves are easier to understand. Each takes up more space than an equal weight of water, and helps compensate for the weight of the diatom's silicious shell, keeping the cell in sunny levels of the ocean. But when diatoms die, the oil often remains undecomposed. Apparently it can accumulate among the sediments on the ocean bottom, and change chemically to become petroleum. Geologists know of no other large source for the world's past and present petroleum supplies. Apparently modern industry has its preferred type of fossil fuel because of the fact that countless diatoms lived and died and contributed their oil to posterity.

The silicious shells of dead diatoms offer no energy to bacteria or other agents of decay. Consequently the shells settle slowly through the ocean water. Chemically they are as inert as quartz sand, which is also silica. Some trace amounts do dissolve and renew the supply of silica from which living diatoms can absorb material for new shells. The residue sinks to the bottom, and may accumulate there as a major component of a bottom muck that oceanographers call diatomaceous ooze. It is particularly prevalent in antarctic seas. But quantities of the material have also been uplifted by seismic activity to become exposed on continents, where it is known as diatomaceous earth. It is mined in many parts of northern Eurasia and North America. Used as the inert matrix in dynamite, it reduces the likelihood that the explosive nitroglycerine will detonate spontaneously or from rough handling. Made into filters, diatomaceous earth serves to strain particles from sugar solutions and beer, making them clear instead of cloudy, and to separate bacteria from the nutrient broths upon which microbiologists culture their microbes for study.

Whirling Cells and the Most Minute Green Plants

All over the world, it seems, marine microbes known as dinoflagellates put on a wonderful display at night, emitting flashes of light when disturbed. The light puzzles us, for it seems to serve no function. Yet almost all of its energy is concentrated in parts of the radiant spectrum visible to human eyes. We met the phenomenon first in years before the danger from sharks ended our night swimming in coastal water. Suddenly we would be aware everywhere we moved of brilliant points of light flashing in the water, each one just once, not repeatedly like fireflies. This was *Noctiluca*, a tiny spherical cell that is famous for no other reason than this luminosity. Various luminous dinoflagellates off the California coast create recognizable "pools" of light where sardines are feeding and agitating the water. Fishermen have learned to make use of this phenomenon. Cruising in the dead of night by the dark of the moon, the men cautiously encircle the school of sardines with a net that can be pursed together below by tightening a line. The luminous microbes escape in the water, while a single seineful may hold 100 to 400 tons of sardines.

Dinoflagellates are commonest in warmer waters, where diatoms are fewer. Usually they move about feebly by means of two whiplash-like flagella, which is the basis for calling them flagellates. Most have one flagellum for propulsion and another for rotating the cell about its long axis. The spinning movement gives them the rest of their name, from the Greek *deinos*, which connotes whirling and terrible.

Occasionally living conditions allow certain kinds of dinoflagellates to become unbelievably abundant, until a thimbleful of sea water from the surface may contain five million of them. Under these circumstances, the production of oxygen by photosynthesis by day and the downward diffusion of this gas from the atmosphere cannot supply enough of it for the dinoflagellates and other sea creatures at night. Great numbers of fishes suffocate. Poisons from the dinoflagellates themselves can increase the devastation. It is then that the dinoflagellates are terrible as well as whirling cells, and that they cause death instead of nourishing the drifting animals among the plankton. Commonly, during this period of devastation, the water becomes red with cells and products of decay. Coastal people call it a "red tide."

The green plants called coccolithophorids, by contrast, are usually much smaller. They have two flagella of equal length, and an armor of limy plates embedded in or on the surface of their bounding membranes. So delicate are they that their skeletons remain intact only when the tiny microbes themselves are captured a few at a time in the finest nets. Yet so numerous are these tiny plants that pieces of their armor (called coccoliths) accumulate to a depth of many feet on the sea floor in certain regions. Limy material of this same type appears in sedimentary rocks all through the past half a billion years, proving how ancient the coccolithophorids are. For all those millenia, these "carriers of coccoliths" have reproduced themselves with little change, and contributed to the nourishment of sea animals.

Until the present century, the true prevalence of coccolithophorids was never appreciated. Then the German oceanographer H. Lohmann began counting and calculating the proportion of coccoliths in oceanic oozes. In a sample from the sea floor nearly 8000 feet down in the North Atlantic, he found that coccoliths accounted for about a fourth of the ooze by weight, with the limy shells of protozoans called foraminiferans making up most of the balance. But this proportion represented about 930 individual protozoans for each 4,500,000 coccolithophorids, a majority of which were of a single species (*Pontosphaera huxleyi*). Coccolith ooze of this and similar mixtures covers most of the floor of the Atlantic Ocean, the Indian Ocean, and the eastern and western Pacific Ocean wherever the depth is less than about 13,000 feet. According to Sir John Murray, the biologist who edited most of the fifty volumes of *Challenger Reports* published between 1880 and 1895, this ooze extends over about 48 million square miles—more than a third of all of the oceans' floor.

Life in a Many-Windowed House of Lime

Despite the relatively small number of foraminiferans in this calcareous ooze from the ocean floor, Sir John Murray called it *Globigerina* ooze because its most conspicuous components were the shells of *Globigerina bulloides*. *Globigerina* is just one among about twenty-eight different genera of planktonic animals of this type, which secrete tiny spiral shells consisting of many chambers. In the shell walls are window-like openings, called foramina, through which the animal extends sticky protoplasm into the surrounding water as a sensitive net with which to capture coccolithophorids, bacteria and other minute particles of food. Like an ameba, to which foraminiferans are related, the animal can shift the food it has engulfed from place to place within its protoplasm, and carry on digestion inconspicuously. Very small particles of food may be moved through the foramina into the shell, while larger ones (such as diatoms and dinoflagellates) may be digested in the open.

Globigerina ooze accumulates slowly, with about 5,600,000,000 shells settling on each square foot annually. Where foraminiferans contribute four-fifths of the weight and coccolithophorids the rest, about 6250 years would be needed for the sediments to increase an inch in thickness. Yet the consolidated sediments constitute chalk, such as the material forming the famous cliffs of Dover in England and Møns Klint along the southeast coast of Denmark. The number of individual contributors from among the plankton is unimaginably great, since these chalk strata in places are more than 200 feet thick. Many of them date from the Cretaceous period of geological time, about 150 million years in the past. But new sediments of the same form are still accumulating from foraminiferans that live and die far from land. Only where the water is too deep do their shells redissolve before reaching the bottom, because sea water under the additional pressure takes more lime into solution.

The Radiolarians

A very different style of ooze was discovered on the sea bottom in many parts of the tropical Pacific and Indian Oceans when scientists aboard H.M.S. *Challenger* began collecting and examining samples of it. It is radiolarian ooze, named for the abundance in it of silicious skeletons from single-celled planktonic animals, the radiolarians, which are kin to amebas. Their exquisite skeletons show symmetrical beauty and perfection, whether in a radial pattern like a three-dimensional sunburst, or as a microscopic latticework more intricate than the finest chandeliers.

Radiolarians from the Mediterranean Sea had been known to scientists ever since 1860, when the German zoologist Ernst Haeckel wrote and illustrated a book about these drifting microbes. His skill as an artist in depicting them led glass workers in Europe to make large-scale models for exhibit in museums. Consequently nonscientists too became aware of these creatures. It was logical that the radiolarians from the *Challenger* expedition were sent to Haeckel, and that his report on them in Volume 18 of the *Challenger Reports* extended knowledge of these animals that occur all over the open oceans.

A radiolarian's glassy skeleton serves chiefly as a support for its sticky protoplasm, spreading and extending the area on which it can capture particles of food. The small radiolarians, barely more than one five-thousandth of an inch in diameter, live chiefly on bacteria and coccolithophorids. Giant radiolarians, some as much as a quarter of an inch across, capture small crustaceans such as copepods. Many of the larger kinds seem able to control their buoyancy, rising and sinking in the sea, by changing the size of frothy vacuoles in a layer near the center of the cell. Often these same radiolarians contain minute, yellow algal cells within their protoplasm. But whether the algal companions, which seem to be dinoflagellates, are parasites or contribute to the welfare of the radiolarian remains unknown. Ordinarily, when a radiolarian dies its algae escape into the sea and reproduce on their own, while the skeleton of the radiolarian sinks slowly through the dark waters to the bottom.

The Larger Drifting Animals

Almost every phylum of multicellular animals has its representatives drifting among the plankton near the surface of the sea, feeding on the unicellular plants and animals or upon one another. A great many are temporary members of this community, remaining in it only during the early developmental stages after hatching from eggs. Others are permanent residents in the upper waters of the open ocean, reproducing without going anywhere else.

The sea surface itself is home to a few kinds of animals. It is there that the sea-going water striders (*Halobates*) prove that even an insect can live far from land. No one knows yet how they manage during storms, or whether they can lay eggs only when they find a feather floating as a raft. During fine weather they scull along atop the water film, with waxy hairs on feet and bodies keeping them dry while they scavenge for small animals as food, just as their kin do in mangrove swamps.

Other animals feeding at the surface are usually dark blue above, which makes them almost invisible to sea birds flying over the water. Only by accident is a person likely to notice on the underside of the water film a sea slug (nudibranch mollusk) half an inch to two inches long, although these creatures range widely in the

A round-browed pilot whale, or blackfish, exhales forcibly ("spouts") between sallies after squid and fish. (Willis Peterson)

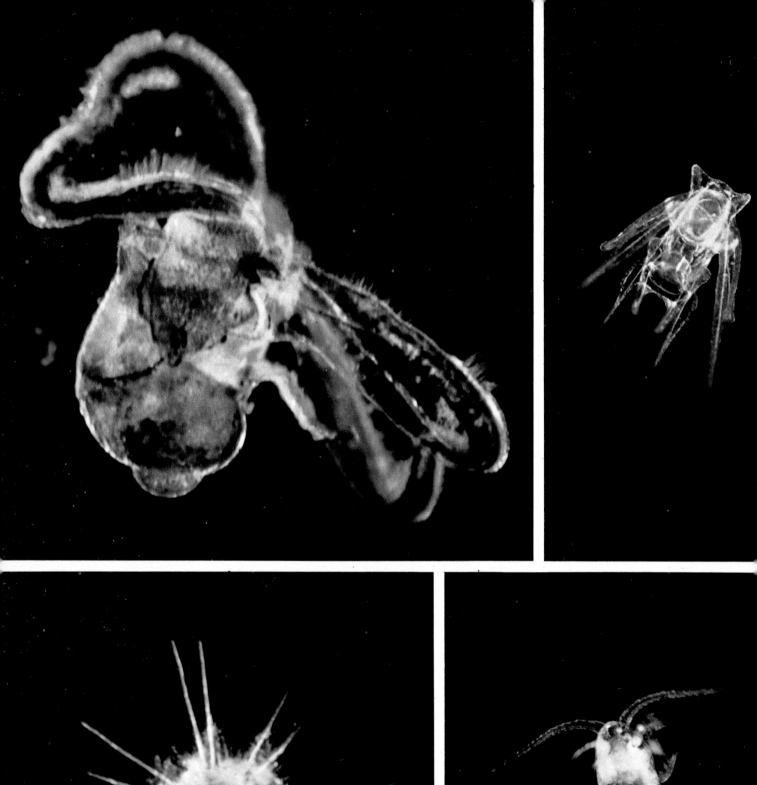
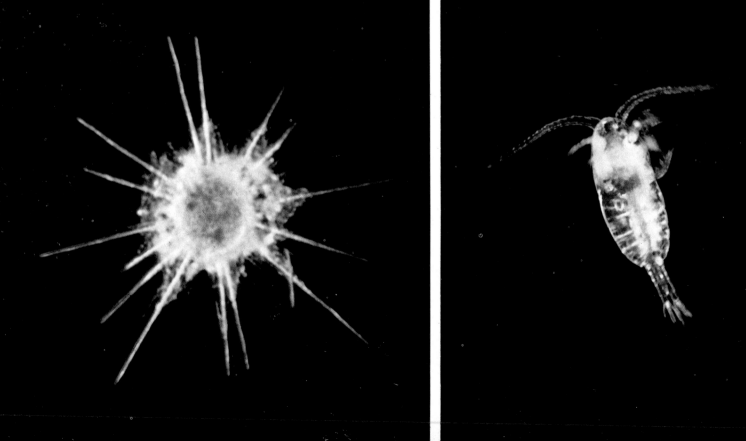

tropics. Apparently the first of them to be collected, preserved and presented to scientists arrived with no record of their color when alive. In preservative they fade to a sea green, and it is the Latin word for this hue (*Glaucus*) that has become their generic name.

Since *Glaucus* is a marine creature and not an air-breather like the sea-going water strider, it can let itself sink below the level of violent wave action during storms. The same procedure is followed by unique drifting colonies of coelenterate animals which seem to be individuals: the by-the-wind sailors (*Velella*) and Portuguese man-of-wars (*Physalia*). Both are common in warm waters far from land. Each sailor consists of dozens of tiny feeding and reproductive individuals under a stiff flat oval float with a low vane-like, diagonal sail. A colony three inches long and three-quarters of an inch broad is a big one, and harmless, whereas a Portuguese man-of-war colony may have a gas-filled float eight inches long and four inches in greatest diameter. Bathers are well advised to stay ashore when man-of-wars are drifting close to the beach, and to avoid touching those that have been stranded on the shore. The dark blue tentacles, which trail as much as sixty feet into the water below a drifting man-of-war, bear countless nettling cells with a venom that is highly toxic to human skin. Only through the glass of an aquarium are we safe to admire the flexibility of the balloon-like float, tinted blue and pink, as the member of the colony that has produced this wind-catcher contorts to rinse one side, then another, with sea water. Otherwise the only action that shows the animal nature and alertness of a man-of-war is the sudden capture of an unwary fish or crustacean that collides with one of the pendant tentacles. The victim struggles briefly before the poison takes effect. Then the man-of-war draws up its dinner for still other members of the colony to digest.

The nettling cells with which a medusa (jellyfish) captures and holds its prey differ greatly in the virulence of their poison. Small medusae called sea wasps (cubomedusae) are particularly dangerous. Off the coasts of the Philippines and Japan, healthy fishermen have been known to die in less than ten minutes after being "stung" by a few drifting sea wasps that swim in the Pacific and Indian Oceans. In Atlantic waters, the ones to avoid are the broader bells of *Dactylometra quinquecirrha*, sometimes eight inches across, and the flat saucers of any *Cyanea* in temperate seas. Cyaneas twelve inches across are common in the middle latitudes, although they seem small by comparison with the giant medusae found in the far northern Atlantic, where they grow to a diameter of eight feet, with tentacles extending downward as much as 200 feet.

The medusae jet-propel themselves, pulse after pulse, through surface waters, contracting to squirt water from under their bell-shaped bodies, seeking prey that generally is carnivorous too. They catch slender arrow-worms as much as two inches long but less than an eighth of an inch in diameter, so transparent and colorless as to go undetected in a bucketful of surface water from the sea. Shrimp-like crustaceans, known to Norwegian whalers as krill and to scientists as euphausids, are scarcely more conspicuous. The giant among them, *Euphausia superba*, attains a length of two inches in antarctic waters, and is a major attraction for whales that feed in the cold oceans. All of these creatures feed on still smaller crustaceans, whose bodies range in size from that of a common fruitfly to that of a mealmoth.

Copepods (the "oar-footed" ones) take the place of insects in the sea. They generally swim by twitching their long antennae while using their feet to comb from water the unicellular animals on which they live. Most famous is *Calanus finmarchicus* of cold northern waters, no more than a sixteenth of an inch in length, but often swarming in such numbers as to color the water and be known to fishermen as "red feed." Herring eat vast numbers of *Calanus*, and so do some of the great whales which cruise at the surface, alternately filtering the small crustaceans into the dark crevices between the whalebone plates hung from the monstrous upper jaw, and then crushing them into a nutritious paste that can be swallowed.

Daily Vertical Migrations

Generally the copepod crustaceans disappear from surface waters by day, only to reappear there as soon as night overspreads the sea. To accomplish this disappearing act, they swim vigorously downward before dawn and remain during the day near the lower limit to which appreciable amounts of solar energy penetrate —often to a depth of 500 to 600 feet. Before sunset, the copepods start up again. Their vertical migrations on so regular a schedule greatly puzzled oceanographers for years, until a plausible explanation was offered for

Active members of the drifting plankton near the surface of the sea: Above left: The veliger larva of a gastropod mollusk. Above right: The echinopluteus larva of a sea urchin. Far left: A radiolarian protozoan with glassy skeleton. Left: A copepod crustacean with long antennae. (All by J. A. L. Cooke)

Above the water, the by-the-wind sailor (Vellela) *raises a thin, clear vane while dangling tentacles below. (Charles Lane)*

the phenomenon. Because surface waters are commonly propelled several miles a day by prevailing winds, whereas deeper strata may be carried in quite different directions by steady currents, the crustaceans dive away from one area of pasture in which they have been feeding and swim up into another area. By travelling 1000 to 1200 feet under their own power each day, they sample each night a different part of the ocean pasture and have little opportunity to seriously deplete the supply of microbes upon which they depend.

Along with the copepods in their vertical migrations go fishes and many animals without backbones that feed on the crustaceans. Squids that pursue the fishes go along also. Vast populations comprising a major part of the planktonic community move up at night and down by day, as though letting the microscopic green plants carry on photosynthesis and reproduce without interference during sunny hours, then returning to harvest a share of the microbial crop as soon as darkness brings plant growth to a stop.

Pelagic Whales, Fishes and Squids

While the drifting plants and animals of the plankton shift with the currents, thriving or waning according to the amount of sunlight and of mineral nutrients available, predatory animals in the same water travel in horizontal directions on quite a different schedule. Marine fishes in the upper waters of the open ocean act as though they were aware of compass directions, and had appointments to keep on their spawning territories. The whalebone (baleen) whales surge through the plankton pastures north and south of the tropics, but always manage to arrive in the proper week at their traditional locations for calving and subsequent remating.

Squids pursue the migrating fishes, and toothed whales follow the squids. The giant is the white (or sperm) whale, known also as the cachalot. Herman Melville relied upon whalers' lore concerning this animal in writing his story of the whale Moby Dick, and in having the monster fulfil the prediction that on a definite date it would be in a particular place in the trackless ocean—just as in previous years. Modern studies of whales labeled with numbered metal darts in their blubber confirm the reality of these migrations.

Too little has been learned yet to know whether the giants among fishes cruise in a comparable way. Nor is there a satisfactory explanation for the fact that among cartilaginous fishes, the plankton-eaters are the ones that grow biggest, whereas among bony fishes, very few plankton-eaters get larger than a herring. So far, scientists have been like laymen in marveling mainly that small crustaceans with so little appeal to a man should nourish the enormous whalebone whales, the gigantic whale sharks, the basking sharks and the spectacular manta rays.

The Sharks and Rays

A length of forty-five feet is not unusual for a whale shark, which is a spotted denizen of tropical oceans. Unlike other sharks, it has a small mouth far forward on the head, just below the tip. Often, while sucking in plankton, crustaceans and little fishes, the whale shark stands vertically just below the water surface. Sometimes fishermen discover one of these giants feeding in this lethargic way right in the midst of a school of twelve- and fourteen-foot tuna, while the tuna are rushing about either in pursuit of small fishes or in trying to escape from the nets of the fishermen.

The second-largest of fishes, the basking shark, occurs mostly in temperate oceans. Like the whale shark it can process prodigious quantities of water, filtering out the edible material by means of fine combs on its gill arches. Generally a basking shark cruises along with just its two dorsal fins awash, sucking in the plankton continuously. Individuals thirty feet long, weighing more

240

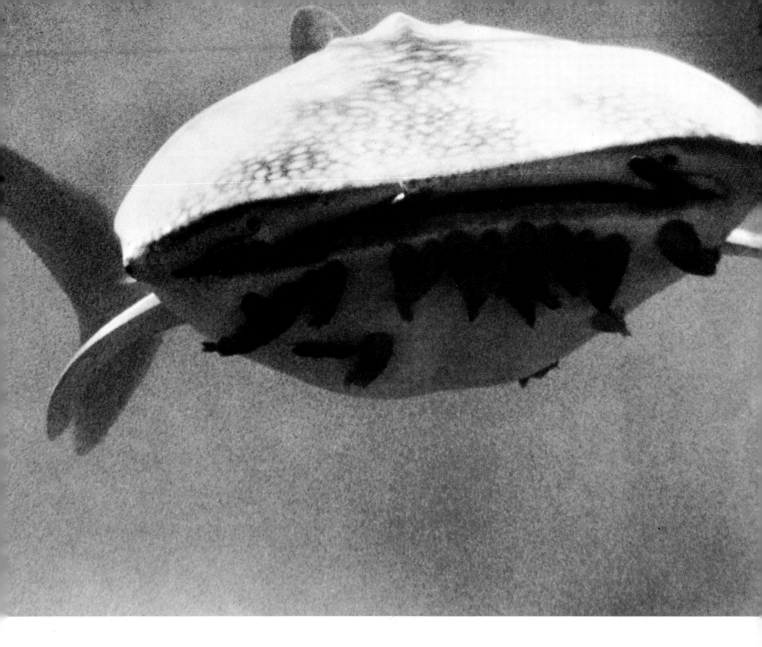

than four tons, are not unusual. The liver of such a fish is huge, amounting to at least 10 per cent of the animal's total weight, and rich with an oil that is reputed to be devoid of vitamins. From a single liver as much as 600 gallons of oil have been extracted for sale as a fuel.

Manta rays, known also as devil rays, may measure twenty feet or more from tip to tip of their widespread pectoral fins. These wing-like extensions of the body propel the ray forward while narrow vane-like projections guide surface waters and whatever food they contain into the ray's cavernous mouth. Seeing one of these monsters—even a partly-grown one—through the side windows of an oceanarium reminds us of the intake port at the front end of a jet engine on an airplane. But no fumes trail behind the cruising ray. The water from

Nearly a score of pilot fish, or remoras, hold to this 35-foot whale shark as it cruises slowly through surface waters, straining out much smaller animals as food. (Photographed off Montagu Island, New South Wales: Ben Cropp Productions)

which it has filtered its food emerges inconspicuously through gill slits below each side of the head.

Big Bony Fishes

A giant among bony fishes is the ocean sunfish, which grows to a length of about eleven feet, a height of nearly as much, and a thickness of two feet or more; its obese body may exceed a ton in weight. Mature individuals ordinarily lie side up just below the water

241

surface, propelling themselves gently by sculling movements of the tall dorsal and ventral fins, while sucking in medusae and any carrion, apparently immune to the nettling cells on any medusa's tentacles. A 200-pounder harpooned off the New Hampshire coast a few years ago surprised us, for we expected a fish that relied upon medusae for food to be almost equally soft and jelly-like. Instead the ocean sunfish proved to be a firm, coarse creature. Ocean sunfishes are not common to any specific area, but range widely over the world's tropical and subtropical oceans, occasionally riding the Gulf Stream far north in the Atlantic.

If ichthyologists applied to fishes the anthropological terms "pycnic" for heavy-set types and "asthenic" for the long, slender and less muscular ones, then the ocean sunfish would represent the pycnic extreme and the giant oarfish the asthenic. Oarfishes of this kind, thirty feet long, eight inches high and two inches thick have been captured in temperate and tropical parts of the world's oceans. Each has an unusually protrusile mouth, perhaps used for catching small squids and crustaceans near the sea surface, a pair of slender blunt-ended pectoral fins that suggest oars, and an extraordinary dorsal fin that extends from head to tail. The front part of this long fin includes rays of extra length that can be erected like a crest when the fish is excited. Oarfishes undulating along, perhaps with head and crest raised momentarily into air, may well have startled seafaring men into a firm belief in sea serpents. Until an actual mounted specimen is seen whole, almost anyone would question whether the oceans of the world could conceal so strange a creature.

Sea Turtles and Sea Birds

When an animal of any kind spends essentially all of its life over, on or in the open ocean, it surely earns a right to be called a sea animal. Male sea turtles never return to land after they escape from the egg and scramble down the beach to the edge of the waves. Females come back only to dig a nest in the sand and lay their eggs. Pelagic birds of many orders return when mature in breeding season to nest along some coast, but otherwise wander widely over the oceans to feed on fishes, squids and crustaceans. Almost certainly the remote ancestors of both sea turtles and sea birds were land animals. But these descendants have made the oceans their home now. It is there that they are most graceful and best adapted.

The cold southern oceans seem particularly hospitable to sea birds. In colder parts of the Southern Hemisphere live all of the more than a dozen kinds of penguins (order Sphenisciformes). They pursue their prey at astonishing speed by paddling underwater with their wings, often gulping down the fish or squid they catch without returning to the surface. Nine of the world's thirteen kinds of albatrosses stay mostly between the Antarctic Circle and the Tropic of Capricorn; the others—the wandering albatross, the black-browed albatross and the two different sooty albatrosses—circle completely around the world. In their scavenging habits and soaring abilities, the albatrosses are the counterparts of vultures. Their relatives in order Procellariiformes include four kinds of diving petrels in antarctic waters, which plunge and swim to catch their prey, and fifty-three of shearwaters plus twenty-two of storm petrels, which either settle on the water and reach below the surface to catch food or snatch it between wingbeats in continuous flight.

Tropic-birds (order Pelecaniformes) range far from land over warm parts of the Atlantic, Pacific and Indian oceans, hovering over schools of fish before diving to seize individual victims, then taking to the air again. Like their kin, the boobies of tropic seas and the gannets of temperate latitudes north and south, they are fast enough on the wing to snatch flying fishes that are gliding above the wave crests on outstretched fins.

Terns (or sea swallows) of thirty-nine different kinds range widely over the open seas, diving and fishing from the surface. Over the Arctic Ocean they are often accompanied by little ivory gulls, which seem able to satisfy thirst by drinking sea water without harm. Farther south the graceful kittiwakes patrol close to the waves. The only other gulls that venture so far from land are the huge skuas of the Far North and Far South, and the circumpolar jaegers of slightly smaller size in the Northern Hemisphere. Skuas and jaegers get much of their food by harassing terns and kittiwakes until these more expert fishermen drop their fish. Jaegers spend the winter months in particular far from land, in the company of other sea birds they can parasitize in this piratical way.

Because of their structure, the terns and gulls are grouped among the shorebirds in order Charadriiformes. The same classification encompasses also almost two dozen different kinds of alcids, which are the northern counterparts of the penguins. Auks and murres, guillemots, dovekies and puffins, the alcids are all heavy-bodied, short-tailed birds that dive expertly, paddle furiously with their wings under water to pursue and catch fishes and squids, but walk awkwardly on land. Only the extinct great auk (*Pinguinus impennis*)

of the North Atlantic, the last pair of which was slaughtered on their nesting site in Iceland during June 1844, had lost the ability to fly. It was the original penguin, yielding its name to the birds of the Southern Hemisphere about a century ago.

The Dark Mid-Waters

Below all of the activity and the living things in the top 600 feet of oceans are dark waters beyond the continental shelves. They are too remote from the sun to receive any energy from it directly, too far from the continents to be affected by events on land or in even the great rivers, and still well above the bottom oozes. In these midwater levels unknown monsters may be lurking. But clumsy nets have brought to light only

Ranging across the South Pacific to the vicinity of Chile, the albatross known as Buller's mollymawk returns to raise its young on Chatham Islands and others near New Zealand. (Michael F. Soper)

scavengers of many kinds and some bizarre predators.

Into these dark waters a continuous rain of food descends from upper levels. Corpses of all sizes and feces of myriad kinds sink slowly. Larger scraps of animals torn apart by sharks and other predators near the surface come less often, but are available at intervals to any scavenger that can detect and swallow them. To survive in the midwater levels, animals must either filter out the finer particles or have a way to find and engulf the bigger pieces of food.

243

To capture samples of life from the ocean below 600 feet requires special equipment and skillful work. One late May we had the good fortune to participate in collecting a netful of the deep sea from a great submarine canyon that cuts across the continental shelf toward the open Pacific Ocean near San Diego, California. For nearly an hour the captain of the Research Vessel *Stranger* maintained a slow speed forward—just enough to keep the wire cable taut on the net that was sinking into the depths. It was not to catch anything until about 4200 feet below the surface, at a level that could be predicted from the angle the cable made with the horizontal and the fact that 10,000 feet of the three-eighths-inch wire would have been released from the drum of the winch.

Once the net was down, the *Stranger's* speed increased to about three knots (nearly three and a half land miles per hour), still adjusted with the greatest care to keep that wire angle constant. At this pace a special diving vane would prevent the net from rising, and would also create a turbulent flow into its gaping mouth. Back of that opening, ten feet in diameter, a mesh cone tapered thirty feet to the actual collecting bag. At the 4200-foot depth, we knew, the overlying weight of water produces a pressure of about a ton to the square inch, compressing the water itself enough to make it more viscous than at the surface—slightly syrupy instead of free flowing. And although the ship floated on an ocean at a temperature of 68° F., the net so far below was surrounded by 38° water—a chill that varies little between the Equator and the poles.

After the net had fished for three hours, the speed of the *Stranger* was reduced as much as possible while the winch man brought in the cable at 164 feet per minute. We worried that the black and dark-colored fishes from the true abysses, below 3000 feet, would all escape as the net came up at less than two miles an hour. It would take sixteen minutes to rise from 3000 to 600 feet, through the zone of "bathypelagic" animals. During those minutes it would be fishing among creatures of almost cosmopolitan distribution, whose coloring often tell something of their habits: glittering silver if they strain the black water for whatever particles of food chanced to be there; grayish, countershaded darker above and paler below if they

A bottlenosed dolphin leaps exuberantly into the air, to splash into the warm waters of the Gulf of California. It is but one of several kinds of small, toothed whales. (Bruce Markham)

244

feed on small crustaceans and other plankton caught one at a time; and black as midnight if they lie in wait, or chase larger victims in true predatory habit.

During the final eight minutes, the net would have a chance to add to its catch a great many members of the surface pelagic fauna. We hoped we could sort these out again by recognizing them for their transparency or deep-blue color or their immaturity if they were the temporary members of the community.

When the net finally rose over the *Stranger's* stern and we could grasp the "cod" end with all the treasures, we found it solid as a stale loaf of bread and still cold from the abyssal waters. Quickly we freed the collecting can at the end, and dumped it into a pail of ice-cold sea water. We rushed it to the laboratory, anxious to examine the catch.

The pail's contents seethed like a stew in a boiling kettle, except that there were no bubbles and the temperature was little above the freezing point. Everything that could swim was trying to dive to the bottom of the pail, away from the light; only a few creatures could attain this coveted position at any one time. The translucent, hueless character of most of the contents made the exceptions stand out: jet-black fishes, blood-red crustaceans, and two eight-inch creatures of a deep purple that at first defied our efforts at identification.

Vampire Squids and Blood-red Shrimps

We reached in and seized one of the purple "things," half expecting to be stung or bitten. Instead, it was soft—almost jelly-like. With cupped hands we cradled it gently and transferred it to a shallow tray of cold sea water. Immediately we recognized it from illustrations: *Vampyroteuthis infernalis*, a famous denizen of the abysses, a vampire squid that lives out its whole life between 3600 and 8200 feet below the surface. From one end of its stocky body eight short arms arise, linked by a web that suggests the fabric of a parachute. Representatives of an additional pair of arms fit into special pockets at the base of the web; they are long, slender, sensitive feelers.

A second *Vampyroteuthis* lay limply in our white plastic pail. It must have been caught early in the deep-water fishing period, or have succumbed to being crowded by later captures. Like the first one, it was too large to be a five-inch male, and had to be a female. Each of them approximated a pint in volume, far outranking in bulk any of our other captives.

Next we picked out the half-dozen blood-red mysid shrimps, each about five inches long, with ten-inch antennae and a body as firm as that of a lobster. Their short-stalked eyes were only about an eighth of an inch across, but they reflected the daylight with a steady, orange eyeshine of great brilliance. This is characteristic of animals that hunt by faint illumination. Our red captives met these requirements, for they live principally in the abysses between 4900 and 9800 feet, where the only light is that produced by living things.

Fishes of the Dark Midwaters

The fishes now caught our eyes, and we transferred all 405 of them to dishes of iced sea water, identifying each one tentatively as we progressed. Of the five that exceeded six inches in length, two were slender snipe-eels, with beaks like those of shorebirds; both the common snipe-eel and the saw-tooth snipe-eel were represented. No one knows yet where these creatures breed in the Pacific. Those in the Atlantic are like the ordinary eel and the conger eel in that they travel all the way to the Sargasso Sea when their eggs and milt are ripe. Another of our captives was an eight-inch specimen of greater rarity: a half-grown dogface witch-eel.

Far more spectacular to us were the weirdly-shaped fishes: a slim, ten-inch, Pacific blackdragon whose snake-like lower jaw opened wide enough to swallow a fish of twice the diameter; a three-inch, half-grown fangtooth whose expansible mouth could accommodate a golf ball, and whose jaw rims were set with a murderous array of dagger-shaped teeth; a six and a half-inch Alaska blacksmelt with eyes so disproportionately large that they surpass in relative size those of any other group of vertebrate animals; and four little hatchet fishes, almost as flat and angular as though they had been run through a clothes wringer. All except the glitteringly silver hatchet fishes were charcoal black, and we knew that the exceptions had been caught on the way up. They are inhabitants of the dim twilight that rise to the surface at night and descend 500 feet or more each day.

Most abundant of the fishes captured were lantern-fishes (60 individuals of four kinds) and bristlemouths (292 of two different species), all members of the bathypelagic fauna caught by the net on its way up. Bristlemouths, rarely more than two inches long, are so fragile that they are crushed to death in almost any net that catches them. Their jaws are modified to form filters for minute particles of food. While they need no light to find their nourishment, they are bedecked with luminous spots like ocean liners with every porthole illuminated.

Possession of luminous organs is commonest among

the bathypelagic animals. Most of them direct their light downward, and from above are practically invisible. Matching this, many fishes and squids of the deeps aim their eyes upward, ready to detect food and mates as moving constellations with a distinctive pattern. Nonluminous items of diet must be no more than passing shadows that obstruct the view of twinkling animal lamps. They would be like a blimp without running lights, against a starry sky.

Small Crustaceans and the Predators That Catch Them

Most of the shrimp-like euphausid crustaceans, or krill, of the bathypelagic zone have luminous organs carried on the same stalk as the eye, as though it were a hunting light, as movable as the headlamp of a coal miner. Only one type of abyssal krill (*Bentheuphausia*) lacks both

The circumpolar fulmar is closer in resemblance to a gull than to its relatives among the tube-nosed birds, the petrels and shearwaters. After nesting on small arctic islands, it ranges southward over the open sea. (Fred Bruemmer)

luminous organs and eyes; our net had caught eight of these. Between 3000 and 600 feet depth it had captured two types with dual eyes and light production: 68 of one (*Euphausia pacifica*) and 87 of the other (*Nematoscelis difficilis*). Only the abyssal kind were more than an inch in length. In the cold water of the sorting tray all of them continued to drive themselves forward by the beating of dozens of slender feet.

Deep-sea prawns have luminous organs on sides and

246

lower surface, and wink these in ways that must confuse onlookers in the dark water below them. But if a prawn becomes alarmed, it floods its surroundings with an intesely luminous fluid, and darts away with all light extinguished. A predator would be left staring at a bright cloud devoid of nourishment, one likely to destroy for many minutes the sensitivity of eyes fitted for twilight vision.

The midwater net had captured nearly 250 prawns of sixteen kinds, ranging in length from one and a quarter to three inches, and in color from milk-white to brilliant orange red. Some bore antennae six times as long as the body—organs of touch that reached out in all directions, ready to warn their wearers of a predator's approach. No doubt every one of the sixteen different prawns had its own preferences as to depth and its own special features of life history. The thirteen individuals with sharply angled bodies (*Systellaspis*) probably adhered to the program found in close relatives near Bermuda, with females carrying their eggs to the hatching stage. By contrast, we had 196 individuals of another type (*Sergestes*) that in Atlantic waters start life as eggs hatching only after floating upward into surface waters. As they grow, young *Sergestes* pass through an amazing series of body forms, descending at the same time into the darker waters where their parents live.

With our hands in the icy water, it was easy to be impressed by the temperature change encountered during the lives of so many denizens of the deeps. Not only do they spend their youth in the plant-rich surface water the sun illuminates every day; they experience warmth too, in the 60° and 70° Fahrenheit range. Then, as they grow older and larger, they let themselves sink into permanent darkness, to temperatures near the freezing point, into water at great pressure. These habits are characteristic not only of many crustaceans but also of deep-sea eels, hatchet fishes, dragonfishes, lantern fishes and bristlemouths.

The strange deep-sea medusae follow a comparable pattern as they grow. From our pail we lifted out a number of them: eleven *Periphylla periphylla* shaped like purplish-brown coolie hats, which revealed when taken from the water a high, firm cone of clear jelly atop the hat, giving it the shape of a dunce cap; and half a dozen brownish-purple *Atolla wyvillei* the shape of English muffins, marked with a beautiful pattern of milky white and perfectly transparent. Probably the two-inch individuals of *Periphylla* and three-inch *Atollas* came from the bathypelagic fauna, while the smaller specimens of each were caught nearer the surface.

Transparency due to the presence of water-rich ge-

The Killer Whale (Orcinus orca), *is a dangerous predator with a voracious appetite. Thirteen porpoises and fourteen whole seals were found in the stomach of one.* (Helmut Heimpel)

latinous tissue is commonest among animals that move feebly through the oceans between 600 feet down and the surface. Our sorting tray was still full of such creatures after we had removed all of the fishes, the medusae, and crustaceans measuring half an inch or more in length. Bulkiest of these nearly invisible denizens were the pyrosomes, each a firm colony of animals arranged to form the walls of a slightly tapering hollow cylinder. When undisturbed, every member of the colony drives a current of water through itself into the central cavity. It captures its own microscopic food particles and oxygen while adding to a combined current out of the central cavity that drives the whole colony along slowly by jet propulsion.

The tail of a baby porpoise emerges during the moment of birth as its mother somersaults in labor pains. (Susumu Matsui: UPI)

inches in diameter had been caught at night, and the officers took turns with the scientists in amusing themselves by tracing their names with a forefinger pressed against the surface of the living colony. For some seconds each signature remained as a bright line of light.

Our big netful included fully a pint of glassily-transparent siphonophores that once had hung like grapes in a bunch below a larger swimming bell. Only a touch is needed to make a colony of this kind disintegrate. We counted 2273 separate individuals between one-half and one inch in length. Even under the microscope these creatures were too transparent to show much inner detail. In the tray each could be recognized mostly as an unoccupied region, containing in some instances a small crustacean. They were the food-catching individuals of the colony, serving their community by inhaling water along with miniature animals that could be digested.

Almost as transparent were 22 comb jellies and 7 kinds of arrowworms. Carefully we separated our many treasures, preserving each for further study in the best way we knew. Finally the sorting tray held chiefly the prime food animals of the sea—the copepod crustaceans and close kin—most of them the shape and size of a grain of wheat. A few copepods were as much as a quarter of an inch long, and distinctive enough to be picked out by color: 188 black ones (*Gaussia*) with red and white flecks; 105 of a reddish orange type (*Macrocalanus*) with a small crest on the head; 45 *Paraeuchaeta* that were red in the forward half and white otherwise; 28 plain ones (*Disseta maxima*) in which one tail was shorter than the other.

From the three-hour haul at 4200 feet depth, plus nearly fifty minutes of rising through progressively warmer water into the sunshine at four o'clock on a May afternoon, the net had brought us a remarkable sample of the deep sea. Nearly 10,000 individuals represented better than 100 kinds, one individual for each 22.6 cubic yards of water going through the mesh. The net had treated the captives fairly well, for few of them were in bad condition despite being packed in so firmly.

Even the fish scales that remained in the tray impressed us. Many of them were three-fourths of an inch across. These largest scales dropped from seven fishes caught by the net, the longest of them barely four inches. We have always thought tarpon scales to be huge. But if a six-foot tarpon had scales in proportion to those of the deep-sea bigscale, each tarpon scale would be thirteen and a half inches across instead of a mere three inches.

We had at least a quart of pyrosomes, a total of 200 specimens. The largest were five inches long and an inch in diameter. We took a few into the darkroom and touched them gently, to see them glow a greenish yellow. This was the mild counterpart of a spectacular demonstration years before aboard the H.M.S. *Challenger*. A live pyrosome about four feet long and ten

Life in the Depths

Afterward we discussed with experienced oceanographers the deep-sea animals our net did *not* catch. Where were the lancet fishes that are regarded as the wolves of the abysses, and where the slickheads, the chin-whisker fishes, the snaggle-toothed fishes, the viper-fishes, the gulper eels, the hammerjaws, and all the others we had read about? The men reminded us that oceanographic technique in the midwater depths was equivalent to towing a butterfly net ten feet wide behind a slow-flying helicopter and expecting to catch birds.

A few of the deep-sea fishes get caught on hooks and long lines set by fishermen. Others are known only from rare accidents, such as when an abyssal denizen has the misfortune to develop a bubble of gas internally, lose control of buoyancy, rise to the surface, and be picked up by a quick-eyed seafarer who recognizes its rarity and can keep the specimen in a refrigerator.

Except that the swimmers in deep water tend to be arctic, temperate-tropical or antarctic in distribution, geography seems to mean little to them. Any ocean at corresponding depths yields virtually the same kinds of animals. Yet no one has discovered for sure which kinds are rare and which abundant, or how most of the abyssal denizens live. The slender, pale-bodied fishes known as barracudinas (family Paralepidae) are believed to be common because their bodies often predominate in the stomach contents of carnivorous fishes in the dark depths. Only one genus of barracudinas (*Lestidium*) has members that migrate daily between waters more than a half-mile down and the surface, where fishermen meet them. Recently, observers in a bathyscaphe photographed deep-sea barracudinas poised vertically, head downward, as though to let food particles touch them in passing and then be seized with a minimum of effort. That no adult fishes of these abyssal kinds have been caught is taken to mean that they are too alert and fast to be captured with a net, and too specialized in their feeding habits to bite on a baited hook.

So far, the dragon-like denizens of the deep sea prove terrifying only when viewed through a strong magnifying lens. Their fearsome jaws are adapted to be amazingly expansible, to accommodate the meal of the month when it chances to come along. But their relatives, for the most part, are in surface waters. It is easy to think of their eggs as rising to levels where their ancestors once lived, then hatching, and as they grow following the downward path of evolution into the abysses.

12

Life
Along the
Sea Coasts
and Sea Floors

Ever since the world's low places filled with water and grew salty with an accumulation of dissolved minerals, the boundary between earth and sea has been an interface of change. Energy transferred from the sun to the wind to the wave is transformed for the final time, often many miles away, where the waves pound against the shore, crash against the rocks, slither up the beach, then hiss back to gather strength for the next assault. At greater depths, currents scour the bottoms, rearrange the sediments and redistribute everything that will dissolve.

Unlike the land, the oceans have always been continuous. Only a few million years have passed — a short time in the geological sense — since the waves of the Pacific Ocean washed over the present site of Panama into Atlantic waters, and the Mediterranean Sea opened broadly into the Indian Ocean. In the past, salt waters have overspread much of the modern continents, leaving among the sedimentary deposits that became rock a wealth of fossils that are unmistakeably marine. These inland arms of the seas may have been no more than 100 to 200 feet deep. Yet they reduced the total land area to far less than it is at present, and served as waterways for the spread of coastal life.

The modern expanse of continents, particularly at high northern latitudes, is extraordinarily great. Some geologists credit this exposure of land with having caused the Ice Ages, and predict recurring cycles of glaciation until once more the salt waters cover more than 80 per cent of the earth's surface.

Paradoxically, the modern distribution of coastal marine life seems more influenced by land barriers that came into existence only two or three million years ago than by previous opportunities for interchange and by the age-old continuity of oceans north, south, east and west. Probably the barriers would not have been so effective had not the Ice Ages been such trying times for living things along the continental shelves, intensifying the natural struggle for survival and isolating communities that had been interconnected.

The oceanographer recognizes definite provinces in the underwater world, each with its native kinds of plants and animals. These relate to history, and to the present barriers and ocean currents, rather than to the ways in which life is adapted to multiply on rocky shorelines, coral reefs, sandy bottoms beyond the beach, mudflats, intertidal zones and deeper water. If not for history, all of the tropical marine life would presumably be circumtropical, all of the polar life circumpolar. In actual fact, a negligibly small percentage of living things have succeeded in becoming circum-anything so broad.

With warmth and all the resources of chemical substances dissolved in sea water, circumtropical should be a most attractive distribution for coastal marine life. Perhaps coconuts from coconut palms have come close to insuring such success, if a coconut tree can be called marine, because—as tropical people claim—they grow well only where they can hear the waves. These are the coasts to which sea turtles return, the most ponderous pilgrims in the tropics. We can think of the green turtle, the loggerheads, the ridleys, the tortoise-shell turtle, and the giant leatherback, which may weigh 1200 pounds. At night, generally when the tide is high, the females haul themselves up the sloping sand beyond the drift from ordinary storms. They use their hind flippers to excavate a hole, into which they deposit their several dozen soft-shelled eggs. Covering the nest as best she can, each female heads back for the water, to roam the seas for hundreds or thousands of miles before navigating again to the same shoreline to mate, haul out and repeat the process.

When the young turtles hatch, they scramble about inside the nest, causing sand from their ceiling to fall and be trampled, raising the floor as though it were an elevator platform. If a hole opens in their roof and admits daylight, they lie low until dark. Then out they come and scramble toward the distant water. Sometime

A 7-foot leopard shark turns for another look at the underwater cameraman along the Australian east coast. (Ben Cropp Productions)

before morning the tide should be high, allowing them to swim out to sea. We remember the myriad tracks of young green turtles we have found on tropical beaches, and the few hatchlings we discovered early one morning on Heron Island at the southern end of Australia's Great Barrier Reef. They were trying to make the perilous trip after the sun rose. Only our presence prevented the circling gulls from seizing each young turtle and making a meal of it.

In the shallows off a tropical shore, we find hammerhead sharks cruising near the bottom, scavenging for whatever dead animals they can find and snapping at fishes that are too bold. The circumtropical porcupine fishes, of which the world has about fifteen different kinds, usually are too quick at inflating their spiny bodies with water to become prey. Spiny lobsters scuttle

backward into rock crevices, leaving only their long strong antennae waving in the open. Hatpin urchins occupy hollows in coral limestone, well protected by slender spines which are as jet black as their bodies. On an urchin five inches in diameter, the brittle, movable spines may be twelve inches long, each one needle-sharp, hollow and filled with a liquid that may be poisonous. Equally black sea cucumbers creep slowly over the calcareous mud, swallowing whatever of it will stick to their branching tentacles. They digest out the organic matter and discharge the cleaned mud. Their

251

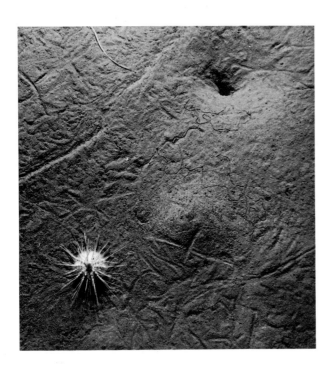

A sea urchin on the ocean floor at 6600 feet depth between Cape Cod and Bermuda (D. M. Owen: Woods Hole Oceanographic Institute)

protection is a poison so potent that tropical people have learned to pick up the sea cucumbers, hold them over tidal pools and wring them as though they were limp rags. The poison that drips into the pool causes almost every fish hiding there to rise to the water surface, so distressed that it can be caught easily by hand. When cooked, the poisoned fishes and the sea cucumbers (*trépangs*, or *bêches de mer*) become edible and harmless to man.

The Coastal Community of the Tropical Indo-West Pacific

If we wade out from a tropical shore at low tide, and find spread on the sand in shallow water the soft brown body and surrounding greenish tentacles of a solitary coral six to ten inches across, we have a clue to our location. We are somewhere between Durban, South Africa, and the big island of Hawaii. We might be as far south as Sharks Bay in the northwest corner of Australia, or Brisbane in the east, or even somewhere in Tahiti. With further exploration we should discover a few of the handsome limy products of these creatures —a convex mound four or five inches across and an inch

or two thick, bearing on its upper surface a pattern of radiating knife-thin vanes. They once fitted into corresponding grooves on the underside of the coral animal that secreted them. So attractive is the skeleton of this, the mushroom coral, that curio dealers around the world stock them for sale, either naturally white as though freshly collected or dyed some pastel hue.

Where mushroom corals live, we can find whole underwater reefs rimmed by the platform coral, which needs a continual covering of shallow water. Usually it grows only around the edge, where it is bright purple, dotted with minute feeding individuals (polyps). There should be whole forests of pink-tipped staghorn coral, and brown mounds of live brain coral. From the abundance of the staghorn we recognize additional clues as to our place in the tropical world. West of Ceylon and east of Australia's Great Barrier Reef it is less conspicuous, but seldom missing altogether.

Reefs in the South Pacific stimulated Charles Darwin to think of the corals as producing natural breakwaters, contributing to the stability of islands. According to his view, which is now widely accepted, coral animals (and coralline algae) arrive and colonize the shores of volcanic islands soon after the rock is cool. At first the reef is a fringing rim around the volcano, steep on the outside but extending downward as far as daylight penetrates. If sea level rises or the island sinks (or erodes slowly), the coral maintains its level in relation to the low-tide mark. Eventually the volcanic matter may disappear, leaving a ring-shaped atoll with a central lagoon. Test borings in the lagoons of many atolls have brought up samples of volcanic rock at various distances downward, as would be predicted from Darwin's explanation.

Other kinds of life are often spectacular on a reef. Bright green tufts known as merman's shavingbrushes catch our eye, as algae two to five inches tall growing on the bottom sand of lagoons and tide pools. Fan-shaped fronds of peacock-tail, a brown alga, seem leathery, even stiff. They are marked attractively with alternating bands of lighter and darker green, almost like some terrestrial bracket fungus. Brilliant blue sea stars six inches across glide over the coral. Brownish purple sea anemones with coarse tentacles spread over an area three feet in diameter serve as playgrounds for two-inch clown fish, which are tan with glistening white stripes. The fish are partners of the anemone among whose tentacles they dart, never harmed by its stinging cells as long as the mucous coating over their scales is intact. But larger fish pursue the conspicuous clowns and get caught by the anemone. Then the clowns can feed on parts of the large fish that have not yet been drawn into the anemone's digestive cavity.

An expert on crabs might not find the other animals of the reef so distracting. He could localize his position after a little exploration, because so many crabs are limited in their range. Of those in the Red Sea, for example, 70 per cent are local or distributed only a short distance along adjacent coasts of the Indian Ocean. The other 30 per cent include wide-ranging species, found across the tropical Indo-West Pacific to Hawaii. The only crab we recognize for sure in this coastal province is *Neptunus pelagicus*, whose larval stages are spent among the plankton. It develops slowly, and ocean currents keep it widely distributed over this broad region.

Sea Snakes

From the knowledge gained of the great wealth of animal life in the littoral region between eastern Africa and Hawaii, Tahiti and Japan, zoogeographers discern a central area as the richest: in the East Indies not far from Singapore. There we are wariest of sea snakes, for more of the fifty kinds in this family (Hydrophidae) hunt among the coral reefs there for fishes than are to be found elsewhere in the Indian or Pacific Oceans. We avoid them as much as possible because they show their ancestral relationships to cobras in their teeth and in the venom with which they subdue their prey. Fortunately they rarely bite people.

Only one kind of sea snake (*Pelamis platurus*) has been so pioneering as to spread southward along the east coast of Africa to Cape Agulhas. Apparently it cannot tolerate the cold water beyond, which prevents it from rounding the Cape of Good Hope and reaching the South Atlantic. This same species has crossed the Pacific Ocean to the coast of Central America and Panama, but arrived too late in history to find a salt-water passageway into the tropical Atlantic. If ever a sea-level canal is built across the isthmus, this poisonous snake can be expected to continue its eastward spread to the West Indies, and to join the bathers along Miami Beach.

Horseshoe Crabs and Man-trap Clams

Some of the other denizens native to waters from Singapore to New Guinea would be less surprising, and others seem much more exotic, if they could find a way to reach tropical coasts of the western Atlantic Ocean. The horseshoe crabs of Southeast Asian shallows seem little different at first glance from the one found from Yucatán to Maine. But these are the only horseshoe crabs in existence; all of their relatives have become extinct so many millions of years ago that biologists refer to the few survivors as "living fossils." Actually, they are not crabs at all but members of the ancient group of marine creatures from which the scorpions and spiders evolved. As they grow, they molt their hard shells, emerging from a slit around the front edge of the dome-shaped body. The new tail spine, which will be used to right the body when the horseshoe crab settles to the bottom on its back, is the last part of the animal to emerge as it creeps out of its old shell and swells up to a new, larger size.

Coconut crabs might add new interest for beach-combers if their distribution were broadened. Presently, these creatures range from islands of the South Pacific to the entrance of the Red Sea near Aden. For a while they live among the plankton, which accounts for their wide spread. Maturing a little, they settle to the sea floor and become hermit crabs, each one shielding its soft, twisted abdomen in the empty shell of snails. Half of a hermit crab's time, it seems, is devoted to examining every empty snail shell the animal finds, testing to see if the new one would be a better shelter than the old and still wieldy enough to carry about. Eventually a coconut crab gets too big to fit into any whelk shell. It then deserts the sea, lets its abdomen straighten out and its own shell harden. It begins climbing coconut trees and—so the natives claim—uses its powerful pincers to cut off nuts that are nearly ripe. On the ground, the crab can open a coconut and reach the nutritious meat inside. But whether this giant hermit crab, which may be more than a foot long, can perform all of the antics with which it has been credited still remains to be shown.

One of the other giants in this broad tropical province has often been called the "man-trap clam" (*Tridacna*). Some individuals of this amazing bivalve do grow to be nearly four feet long, lying hinge downward in reefs near New Caledonia. Elsewhere—west to the Indian Ocean, south to the Tropic of Capricorn in Australia's Great Barrier Reef, east to Hawaii and north to southernmost Japan—they are smaller, but live in the same way. Instead of depending, as other bivalved mollusks do, on filtering minute algae and other living particles of food from the sea, the giant clam raises algal cells in its own special "greenhouses" and uses its blood cells to harvest the crop. By day, unless exposed by a low tide, the clam spreads from the wavy gap between its shell halves a thick and colorful mantle. Within the superficial mantle tissue are large sinuses of the mollusk's circulatory system, each one an agricultural center for algae which receive light through translucent dots, which once were believed to be the clam's eyes. It has no eyes, but its nervous system keeps it well advised of any change in the neighborhood of the mantle that might be an approaching fish or hand. Every *Tridacna* that we have examined has pulled back its mantle and closed tight its shell before we could get a finger close enough to touch it. Consequently we doubt the story that is often told, which claims that an unwary person walking on a reef at low tide could thrust a whole foot into a *Tridacna's* shell and have the mollusk clamp firmly, holding the victim until the tide flooded back and caused death by drowning. South Sea islanders think nothing of diving to reach one of the bigger clams below the low-tide line, smashing a gap in its shell edge, and reaching in with a sharp knife to cut out the bivalve's closing muscle. It is a solid mass of meat more than a foot long and several inches in diameter, somewhat tougher but just as sweet as the similar meat from a scallop's shell.

The Chambered Nautiluses

In still deeper water, to as much as 1800 feet below the surface, the straits and minor seas among the islands of the East Indies conceal a style of mollusk that dates back more than 400 million years with little change. Known as chambered nautiluses, they secrete a beautiful shell in the form of a flat spiral, neither right- nor left-handed in its perfect symmetry. As the animal grows, it enlarges its shelter in the same pattern. But instead of becoming continually longer and spiral in body form, it repeatedly closes off the older, smaller parts of the shell with curved crosswise partitions and occupies only the outermost of these many chambers. All of the others it fills with gas in just the right amount to compensate with buoyancy for the weight of the shell. For this reason a live nautilus remains poised with its spiral home in the vertical plane, adding no load for it to carry, while it extends several dozen tentacles from its open doorway to feel for small animals it can overcome and eat.

Mathematicians marvel at the nautilus because the shell it secretes follows precisely the form of an equiangular spiral—a logarithmic pattern of enlargement—and because its partitions, separating the cavity into a series of chambers, are of the same logarithmic spiral form. Artists appreciate the shell for its symmetrical beauty and economy of line. Biologists admire the animals as another living fossil that has managed to escape extinction while all of its near relatives disappeared. Zoogeographers wonder what special ame-

nities of life the nautiluses find in the tropical Indo-West Pacific province. For thousands of species this region terminates near Port Elizabeth on the Indian Ocean coast of South Africa, and with the southern islands of Japan and the Hawaiian archipelago.

Coastal Communities of the American Tropics

Quite different plants and animals give character to the warm waters of the eastern Pacific Ocean from Baja California to the Equator, including the shores of the Galápagos Islands. This warm-water life is similar to that of the Atlantic side from Florida and the Bahamas around the Gulf of Mexico and through the West Indies, down the coast of South America to the latitude of Buenos Aires.

For mile after mile, these coasts are sandy or composed of the broken shells of mollusks that have not yet been broken beyond recognition. White coral sand mixes with black volcanic sand. In the shallows beyond, coral conceals the rocks and soft calcareous mud floors the bays and lagoons that are protected from wave action. Big sea cucumbers process the mud for organic matter, but no other animals may show themselves. Even the seaweeds seem inconspicuous. In the tidal gutters we find, instead, seed plants that have invaded the coastal waters—eelgrass, turtlegrass, and mermaidweed, which seldom flower except near river mouths where the salinity is less than the 3.5 per cent of the open ocean.

After a storm, the flotsam on the dry beach includes a generous sample of living things that have lost their hold in deeper water. Along the shores of Sanibel and Captiva islands, beloved by shell collectors on Florida's west coast, we find thousands of fan-shaped brittle pen shells seven to eight inches long, dead or dying from exposure. We wonder how many survivors are beyond our reach, each pen tethered to the sand by long plastic

Above right: At a depth of about 30 feet off the coast of southern California, a group of garibaldis live among the kelp beds. Attaining a length of 11 inches, these handsome members of the damselfish family choose one particular kind of red alga on which to lay their eggs. (Jim Hodges) Far right: A shell-less sea slug, or nudibranch, with distinctive soft projections feel unharmed on coral tentacles. (Barrie Poole) Right: A marine medusa, or jellyfish, propels itself by spasmodic contraction of its bell-shaped body. The animal's body consists of about 99 per cent water. (Barrie Poole)

fibers secreted by the mollusk's foot. We can appreciate the predicament in which these bivalves find themselves, for we ourselves have slept uneasily on numerous occasions inside our umbrella tent when it was staked down on sand and vulnerable to a gust of wind.

What surprises us more is the number of creatures that are cast ashore despite their ability to bury themselves deeply when waves begin to agitate their world. Heart-shaped cockles (family Cardiidae) along America's tropical coasts, like the soft-shell clams on New England shores and the dog-cockles (or bittersweets, family Glycymeridae) so common on Australian beaches, have powerful muscles and an active foot. They can pull their shelled body down, displacing the sand, surely as fast as the waves can loosen it. It is tempting to believe that they get confused when storm waves pound. Cockle harvesters along European coasts sometimes induce these edible bivalves to heave themselves into view (and easy reach) by merely stamping on the sandy bottom when ebb tide has decreased the depth of water over the shellfish beds.

Waves for Guidance

For other coastal denizens, the rhythmic signals from the waves must provide better guidance. The egg-shaped crustaceans known as sand bugs pop out into retreating waves on an ebbing tide, and promptly bury themselves at greater depth. Promptly they turn about and again filter particles of food from the water. On a rising tide, they repeat this procedure many times in the opposite direction.

Coquinas (or wedge shells, family Donacidae) let the waves propel them to new positions as the tide rises and falls. Often we stand ankle-deep to watch the abundant Florida coquinas heave their brightly colored half-inch shells out of the sand, to roll over and over in the advancing wave, then to dig in and disappear as the waters stops. Inch-long bivalves along much of the Pacific coast behave in a similar fashion, staying below the surf whatever the stage of the tide—not only on open beaches but also on the sandy floor of sea caves eroded into the coral rock or lava blocks of headlands.

The approved technique for gathering coquinas for soup is to toss shovelsful of wet sand from the wave-wet beach into a wooden frame with a flyscreen bottom.

Amid a tangle of serpent stars 60 feet below the surface off the southern California coast, a sea anemone is slowly swallowing the translucent salp it has caught. (Jim Hodges)

With a bucket of water, the sand can be washed through, leaving the coquinas clean and in full view. Other denizens of the sandy beach can be captured in this way. Slender, slithering creatures, so transparent as to appear glassy, are generally sand lances two or three inches long. Two-inch opaque, pink animals pointed at both ends of their flattened bodies are lancelets (or amphioxuses), which delight the scientist. Although lancelets develop no backbone, distinct brain or eyes, they have many other features (such as pharyngeal slits, hollow dorsal nervous system and stiffening notochord) found in animals with backbones. They probably resemble the ancestral vertebrate more than any other creature in existence. Slithering among the sand grains near the edge of the sea and filtering microscopic food from the water, somehow they have found a way to survive with fewer obvious adaptations than any fish, amphibian, reptile or warm-blooded animal.

Along these tropical American coasts, we try to be especially careful as we wade into deeper water. We know that flat-bodied skates and stingrays are likely to be lying on the bottom, the edges of their fins blending with the sand. They are too lethargic to avoid us and almost impossible to see unless they move. We fear those that lash upward and forward with a venomous spine (the "stinger") on the tail. Even a little stingray, weighing barely a pound and measuring less than twelve inches from side to side, bears a saw-edged spine capable of penetrating human flesh.

The skates and rays eat mostly mollusks and worms they find buried shallowly in the sea floor. We have only to stand or kneel motionless on the mud as the tide flows in to watch some of these hidden animals, which are alert to the slightest vibration. Before long, an undisturbed echiuroid worm (such as *Echiurus*) will slowly push out from its doorway a soft flexible hood-like proboscis and begin cleaning the mud surface in radial streaks, creating a sunburst pattern around the burrow opening. An equally unsegmented sipunculoid (such as *Dendrostoma*) will creep to one end of its U-shaped burrow and extend horizontally over the mud a whole array of threadlike tentacles, which pick up food particles and convey them by means of beating cilia to the mouth at the center of the array. Unless touched, these slender tentacles seem not to move, whereas the somewhat thicker ones from the head end of many tube-dwelling segmented worms (such as *Amphitrite*) maintain such writhing activity that they destroy any illusion of being flower petals—as might be assumed from examing a still photograph of these food-gathering and oxygen-absorbing organs.

257

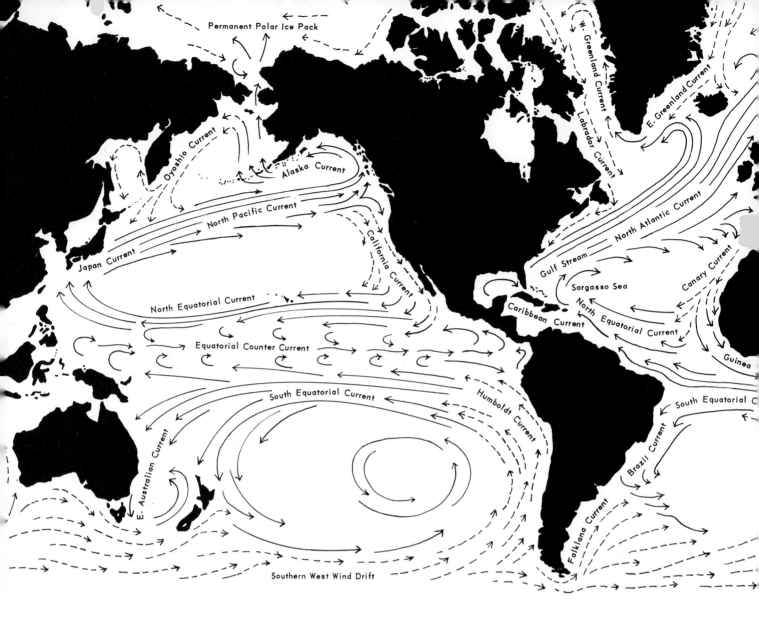

Map labels: Permanent Polar Ice Pack, Oyashio Current, Alaska Current, North Pacific Current, Japan Current, California Current, North Equatorial Current, Equatorial Counter Current, South Equatorial Current, Humboldt Current, E. Australian Current, Southern West Wind Drift, W. Greenland Current, E. Greenland Current, Labrador Current, North Atlantic Current, Gulf Stream, Canary Current, Sargasso Sea, Caribbean Current, North Equatorial Current, Guinea, South Equatorial C, Brazil Current, Falkland Current

Prevailing winds and continental barriers combine to shape the major ocean currents, which affect climate and the distribution of life.

We are generally tempted to dig out a few worms and bivalves merely to see what other kinds of animals have taken refuge in their chambers. A little goby fish may pop out of a worm tube and try to wriggle away over the wet mud. Rounded pea crabs half an inch broad in the body usually stay in place right behind a worm in its tube or actually within the mantle cavity of a big bivalve.

Animals with Stinging Cells

In deeper water, we are astonished by the number of burrowing sea anemones living in mucus-lined tubes of their own making. Each burrower is expanded at its hidden end, preventing itself from being pulled loose when it wraps its stinging tentacles at the exposed end around a vigorous but careless fish. A burrowing anemone of this kind, twelve inches long and with a flower-like display of tentacles six inches across, is not unusual. The same muddy sea floors are densely set with sea-pens with a soft anchoring stalk well hidden. The waving top is pink or tan, purple or white, which may distract us from noticing the tiny clusters of tentacles like those of a coral, with which members of the sea-pen colony capture minute animals from the surrounding water.

Sea fans and sea whips often add beauty to the sandy bottom where coral masses or volcanic outcrops channel the currents and break the force of storms. Rising upright from their holdfasts, the sea fans grow edge-on

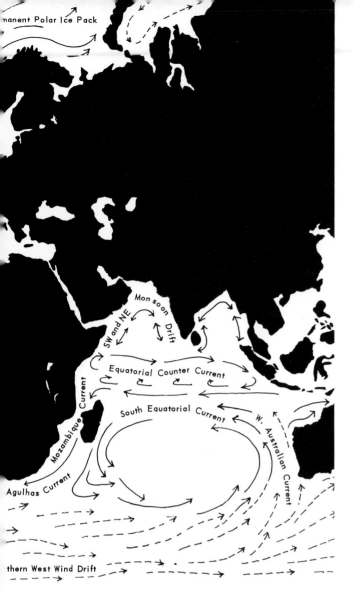

SW and NE Monsoon Drift

Equatorial Counter Current

South Equatorial Current

Mozambique Current

Agulhas Current

W. Australian Current

thern West Wind Drift

toward the current, which equalizes the feeding opportunities for the polyps on both surfaces. Sea whips, whose branches may be cylindrical or angular, also grow aligned with tidal movements of the water.

Over the surface of some sea fans and other supports that ordinarily are harmless to touch, the stinging corals sometimes spread a thin limy coating in which are embedded thousands of minute individual polyps, each with a circlet of particularly venomous tentacles. We try to avoid these hazards in the warm waters along tropical coasts, while remembering that this same stinging coral can be encountered in the shape of massive limy secretions of almost any configuration.

Stony corals, which produce respectable reefs in tropical American shallows, have six tentacles or more on each polyp, and produce their hard parts as limy secretions outside the body. Usually they are colonial,

with many small polyps in circular craters, to the rim and sides and bottom of which they add lime in a similar way. If touched, each polyp can withdraw into its shallow cavity.

Despite their nettling cells, corals and gorgonians (such as sea fans) are never safe from attack. Brilliantly-marked parrot fish use their hard-rimmed mouths to bite off pieces of coral. They swallow the coral animals and fragments of the limy skeleton too, digesting the soft bodies but discharging the lime as a contribution to the coral sand on the sea floor. Gaily patterned sea slugs (nudibranchs) creep over the sea fans and corals, sucking out the polyps one at a time. Some of these shell-less mollusks have the extraordinary ability to swallow the coelenterate without triggering the discharge of its stinging cells. Digestion is selective too, letting the white blood cells of the sea slug transfer the stinging cells of its prey to its own skin, there to serve as a repellent.

Tropical sponges may escape attack. Some along Atlantic coasts, such as Neptune's goblet (*Poterion*) and the loggerhead sponges grow to be six feet across. But they harbor a multitude of crustaceans, worms, mollusks and even small fishes, which take refuge in their water-discharge chambers or live there for extended periods. A roughly spherical loggerhead sponge measuring about twenty inches in diameter yielded more than 17,000 animals, some 16,000 of which were small shrimps. A similar assortment of wriggling and creeping creatures emerge and try to find a new refuge when commercial sponges are gathered, pounded and washed clean before drying and sale. Yet these particular sponges are chosen for their fine texture, small cavities and relative freedom from hard inclusions that might cause a finished sponge to scratch.

Recent Change

On the two sides of tropical America, the marine life differs enough to be significant. Recognizing this slightly more than a century ago, the distinguished American zoologist A. E. Verrill and the equally eminent German zoologist A. Günther made a prediction, and asked geologists to prove them right or wrong. They held that the separation of Pacific and Atlantic waters by highlands in Latin America must be recent, just long enough to let new species evolve on both coasts. Farther back, the coastal communities must have been connected to account for generic similarities. The geologists soon found that this sequence was correct.

A beachcomber might suspect this for himself from the sand dollars he can pick up after a storm. Both the keyhole sand dollar (*Melitta*) and the arrowhead sand dollar (*Encope*) live on sandy bottoms and sand bars that are exposed briefly at low tide, all the way down the Gulf of California and suitable parts of the Pacific coast to the southern boundary of Ecuador, and along eastern coasts between Georgia and Rio de Janeiro. Offshore, in deeper water, the tropical fishes include about 10 per cent that are found on both the Pacific and Atlantic sides. Among these is the sea devil or manta ray, a plankton feeder that often attains a width of nearly twenty feet. It swims near the surface by graceful movements of its great pectoral fins, as though they were wings with which to fly slowly through the water.

Experienced sportsmen who fish for snook (*Centropomus undecimalis*) from Florida southward generally claim that these sedentary swimmers grow biggest (to fifty pounds) opposite shores forested by mangroves. With a critical eye, these men may see differences in body form, markings and behavior when they catch a snook (*C. viridis*) along the Pacific side of tropical America. Similar pairs of species with significant differences can be matched up in genus after genus, not only of fishes but invertebrate animals as well. Thus at the Pacific end of the Panama Canal, the spiny lobsters are *Panulirus interruptus;* at the Atlantic end, forty miles away, they are *P. argus*.

Without some research, an oceanographer cannot rule out the possibility that time for evolution from common ancestors along the east coast and west might not have come while the fishes and various invertebrates spread northward from an original community beyond Cape Horn. Professor Sven Ekman from the University of Uppsala in Sweden decided to test this theory by examining the short-tailed crustaceans—the crabs—of the two communities. Without much difficulty he sorted out among the 750 different species present the 150 that can tolerate cold water, such as seems to have surrounded Cape Horn for millenia. Only 35 of them on the Pacific coast and 20 on the Atlantic are species ranging southward beyond the tropical provinces; their ancestors might have used this route. Of the other cold-tolerant species on the Pacific side, 20 have a distribution into cooler waters farther north (and may have come from there) and 15 to cold water at great depths. On the Atlantic side, 39 range northward and 31 downward.

Far more impressive are the tropical crabs that prove to be intolerant of cold. A total of 34 along the American coasts are circumtropical species. Four on the Atlantic side and 7 on the Pacific are known other-

wise only from the Indo-West Pacific community; somehow they managed to cross the rest of the Pacific Ocean, but not to get much farther. Two species found on both sides of Middle America and 13 living only on the Atlantic coast are represented also on the African continental shelf of the Atlantic Ocean; their ancestors must have come westward across the Atlantic. The preponderance of the crab population, however, consists of 285 species living only on the Pacific side, 270 only on the Atlantic side, and 15 on both sides, all of them known from nowhere else in the world. Professor Ekman concluded that, since most of the crab fauna came originally from elsewhere in the tropics and prior to the Pliocene period while Pacific and Atlantic waters were still connected, the diversification of other animals along Middle American coasts must have taken place under similar conditions of recent separation.

At the Equator, the Pacific Ocean extends for 8500 miles between New Guinea and Ecuador; the Atlantic Ocean is less than half as wide between the mouth of the Amazon and Africa. This alone might explain why the littoral province of the tropical eastern Pacific is poorer by far than the tropical western Atlantic in living things of so many kinds. There are eels, monk seals and sea cows in the Gulf of Mexico, or have been into the twentieth century. None of these have counterparts in the eastern tropical Pacific. Of the world's 18 different kinds of parrotfishes, 17 nip at corals in reefs along the American Atlantic coasts and the exception lives on the African side of the same ocean. To explain this distribution and the fact that related animals are to be found in the Indo-West Pacific province, the biogeographer looks still farther back in time. Cretaceous rocks give evidence that about 100 million years ago, a shrunken North Atlantic Ocean was linked across Africa north of the Tropic of Cancer to the Bay of Bengal, while a strip of land from Siberia to Middle America almost shut off the connection between the Pacific Ocean and the Caribbean Sea.

The Tropical African Atlantic Coastal Community

As we walk along the shoreline anywhere between Dakar on the Senegal coast and the mouth of the Congo River, we pick up sand dollars (*Rotula*) of a distinctive shape, which tells us where we are. They have slots or keyholes around only half of their periphery, usually three deep slots with three narrow, shallower slots between. Other animals associated with the sea floor are few, for the bottom is almost all shifting sand. It changes so continually that the few reef corals have nothing to build on, and true coral reefs are missing altogether. Nor can most seaweeds maintain themselves. Thus this coast has fewer kinds of life than any in the tropics.

Close to shore, the temperature of the water decreases rapidly downward due to oceanic currents, which thus restrict to shallows any living thing that is intolerant of cold. At 300 feet below the surface, the water is almost as cool as at this level in the Mediterranean Sea, and chillier by far than at a corresponding depth in the Gulf of Mexico. Yet the same currents cause enough upwelling of water rich in phosphates and nitrates that the microscopic algae in the plankton thrive. This makes us look for animals that can benefit, whether schools of fish or bivalves or predators that eat bivalves.

At some places along these coasts, the sea floor is firm enough for an ark shell to anchor itself by means of a massive array of byssus strands; Noah's ark shell colonizes these situations. The bivalve known as a false angel wing makes shallow burrows for itself, using its shell valves as tools just as it would if the bottom were hard clay or soft limestone. For both of these mollusks, the sand is treacherous. But enough of them succeed to support a small population of the predatory tun shells—a big snail with a thin shell—and also an assortment of rays and other fishes whose teeth and jaws are adapted to crushing mollusk shells. The curious feature about these particular mollusks, and of the spadefishes and sea breams that feed on them in tropical West African waters, is their distribution: they are animals of the American coasts of the tropical Atlantic. Perhaps a third of the fauna on the eastern coast of the Atlantic shows this dual range. Most of the remaining two-thirds are cold tolerant species from farther north, farther south, or deeper water.

Subtropical Coastal Communities

Oceanographers from the Southern Hemisphere sometimes tease their northern correspondents by referring to the Pacific, Indian and Atlantic Oceans merely as big gulfs extending from the world-circling Antarctic. Propelled by the strong, unimpeded winds in the "Roaring Forties" and fifties and sixties south latitude, the cold antarctic waters flow eastward with such vigor that the fractions diverted by the southern continents become powerful northbound currents. The Humboldt chills the west coast of South America all the way to the Equator—far into the tropics. The Benguela produces a desert in South West Africa and temperate conditions northward along the coast almost to the

mouth of the Congo River—in the tropics. A similar flow keeps the water cold along Australia's west and southern coasts. Even the east coast of South America, like the Falkland Islands, has at best a temperate climate most of the way to Buenos Aires because the Palmer Peninsula on Antarctica turns cold water north in the smaller Falkland Current.

In the North Pacific, the principal eastbound current divides off the coast of Oregon into a southern portion —the California Current—which cools the American shoreline at least to the Mexican border, and a northern portion—the Alaska Current—which first makes Alaskan waters warmer than they would normally be and then continues around in a major eddy to bring arctic water south along the Asiatic coast to the boundary of the Indo-West Pacific tropical province.

In the North Atlantic, the Gulf Stream angles across, producing temperate conditions from the English Channel northward to Iceland and the edge of the true Arctic. Much of this water returns, thoroughly refrigerated, in the Labrador Current down the west side of Greenland. Even after the icebergs melt, the water stays cool along the New England shore. It maintains temperate conditions all the way to Florida, inland from the warm Gulf Stream.

Consequently the only true subtropical coasts in the modern world extend from Brest in northwestern France down the Atlantic side of Europe and around northwestern Africa as far as Dakar; the Mediterranean Sea and the Black Sea add considerably to this subtropical coastal community. Oceanographers sometimes make a further subdivision, agreeing that the surface waters of the Mediterranean and the coasts southward along Africa's Atlantic side are subtropical, but calling the deeper portions of the Mediterranean and the Atlantic shores north to Brest "warm-temperate." They also point to the eastern Mediterranean as being distinctly saltier than the western because of evaporation, and to the Black Sea as almost brackish because of the great rivers flowing into it. Geologists look still farther back in time, to the Miocene period about 20 million years ago, when the Caspian Sea and the Black Sea were broadly joined to the Mediterranean and, for a while, the Straits of Gibraltar were closed by a land bridge between Spain and Morocco. These events have left their mark on the living things that inhabit the Mediterranean-Eastern Atlantic subtropical province today.

Mediterranean Life

These are the waters in which the sardine and the delectable anchovy are native, feeding in vast schools

among the plankton. Among the dogfishes that scavenge and prey on smaller fishes at all depths is the placental kind in which the unborn young absorb nourishment from their mother, just as Aristotle described. For 2200 years his statement was discredited, until the German zoologist Johannes Müller proved him right. Close to the sea floor, more than two dozen local kinds of gobies and one dozen of scaleless blennies travel little from place to place.

The famous ornamental coral, whose characteristic color has become a household word, was discovered first in the Mediterranean. Now it is harvested as far out into the subtropical Atlantic as the Cape Verde Islands. Like the polyps of sea fans and sea whips, those of precious coral have exactly eight feathery tentacles, and produce their hard parts internally, within the jelly

A 7-foot cub shark, known as a whaler shark in Australian waters, chewing a decapitated fish of about 15 pounds, close to the Great Barrier Reef. (Ben Cropp Productions)

layer that separates the sheer outer covering of cells from the filmy inner layer. Consequently, through the center of a coral is a natural cavity, lined in life by the digestive layer of the animal. The rock-hard pink or orange pieces found among the beach drift or obtained by diving can all be strung together easily on a thread as jewelry.

Mollusks of this subtropical province have made history too. The ancient Tyrians gathered the rock whelks and crushed the live snails in bowl-shaped cavities along the rocky coasts, to obtain a fluid they could filter and purify by secret methods. It became Tyrian or royal purple dye, one of the most handsome, permanent and colorful treatments for cloth ever invented. The essential ingredient is a fluid from the anal glands of this predatory snail, which not only creeps but occasionally swims, despite being burdened by a shell of medium weight.

Until the Middle Ages, divers around Mediterranean coasts collected golden strands from the pen shell, which uses them to hold itself in place. These byssus strands were woven into a luxury textile called cloth of gold, and made into ladies' gloves so fine that a pair could be packed without harm into an empty walnut shell. Today in museums, examples of this lost art retain their color and softness.

Modern visitors to Pozzuoli, across the bay from Naples, are often shown the work of other bivalves in the marble columns of the ruined Temple of Serapis. Both upright and fallen columns are pitted by the rock-boring mollusk *Lithodromus lithophagus*, which had access to the surfaces some time between the fourth century B.C. and the present. This silent evidence proves that, unnoticed by historians, this whole coastline sank below the waves of the Mediterranean and then rose again far above the water. There can be no question as to what made the pits, for in those that have been undisturbed, the shells of long-dead mussels are still in place. Only while submerged could each borer use its hard shell valves as tools, and possibly an acid secretion as well, to excavate a deep pit for itself. As the animal grew, it enlarged the chamber at the end of its pit and soon became a prisoner there. Reasonably secure from attack, it maintained its connection to the outside world by its long slender siphons. These brought it water, food particles and oxygen, discharged its wastes and served as avenues for reproductive cells. Today, below sea level a mile away, members of the same species continue to bore into rock and hard mud, following their family tradition.

The deeper parts of the Mediterranean and the Atlantic coast to the north of the Straits of Gibraltar harbor many kinds of animals from the boreal province beyond the English Channel. Mackerel feed on the sardines and on the abundant transparent shrimp-like krill crustaceans that Scandinavian whalers learned to recognize as whale food long before they began pursuing whalebone whales to antarctic waters. Other predatory fishes from farther north are there too: coalfish and whiting representing the genus of the cod, plaice and European flounder as flatfishes on the bottom, and the spiny dogfish as a small shark few people have found reason to fear.

Zoogeographers point to these northern fishes, to the krill and to the Norway lobster, as relics in the Mediterranean from the Ice Ages, when the climate was much cooler than subtropical. Fossils around this inland sea prove that the mahogany clam (or ocean quahog) and a scallop from far-northern waters once inhabited the Mediterranean, before the weather warmed beyond their tolerance. The largest bivalves in this subtropical province today probably evolved into a separate species in those same years, for they have relatives in European arctic waters and also down the Pacific coast of North America. The presence of European brown trout in freshwater lakes of Morocco, Algeria, Sicily, Sardinia and Corsica can similarly be interpreted as proving that sea trout once lived in the Mediterranean, where they are found no more.

The Atlantic Boreal Community

One reason that European colonists felt reasonably at home along the American Atlantic coast was that the sea life familiar to them from the English Channel north was essentially the same all the way down the shores of the New World from Labrador to Florida. Even Iceland fits into this marine coastal community, due to the warming effect of the Gulf Stream. At this longitude, the boundary of the boreal province and the Arctic comes at North Cape, the northernmost point of Norway, and skirts northern Iceland to reach the southern tip of Greenland.

Seaweeds in Cold Water

The cooler water holds more carbon dioxide in solution and favors the growth of algae. Along rocky shorelines, where the waves pound so vigorously and scour each tidal pool of anything loose, the algae reach their peak of complexity as seaweeds. They, like the clinging animals and the attached ones, are in many ways the best known marine life in the world. Some of the seaweeds are distinctive enough in form and color to bear popular names. For many, a hand lens reveals additional information. But to see the details upon which the scientific classification is based, a compound microscope is usually essential.

In this coastal province, a piece of wood wedged in a tide pool or mooring line left slack for a few weeks acquires a coating of dark green algae as limp as wet cotton batting that has been teased apart until its fibers stick out in all directions. Buoyed up by water, whether in the sea or a shallow dish, the strands separate enough to see with a simple lens. Lifted into air, they mat together into a meaningless mass. If they are unbranched, it probably is mermaid's-hair, a blue-green alga of phylum Cyanophyta. This identification is confirmed under a microscope if the cells have no distinct nuclei and the coloring material is dispersed. It would be enclosed in minute bodies called chloroplastids if the alga were a green, a brown or a red. If the thread-like filaments are unbranched or merely branched strings of long slender cells with the green color in chloroplastids, it is likely a sea beard. If their branching makes them delicate and feathery, they may be a sea-moss; no true mosses grow in salt water.

Sea lettuce is a fine name for a bright green alga in the form of a ruffled sheet or ribbon that grows as much as three feet long, holding fast to rocks in shallow water or with one end securely embedded in a mudflat. It is found in temperate waters all over the world, swaying in each wave without crumpling or tearing. On occasions when the beach drift contains sea lettuce or it is washing back and forth in the surf, we realize that a storm or unusual current has eroded the mudflat where the alga grew—a hazard against which it has no protection. On those days the beach drift is usually studded with little green spheres and egg-shaped sea-bottles and others float at the water's edge; these grape-sized plants are the largest unicellular algae in the sea.

Feathery red algae, comparable to those that utilize their red pigment in capturing light for photosynthesis 100 feet or more below the water surface in tropical waters, are much less common in the boreal province. We find some growing on other algae and on the hulls of disused boats floating where light is reflected from a sandy bottom. There we notice bright pink tufts of *Ceramium*, the branch tips of which are curved like tiny claws, and feathery *Spermothamnion*, which often seems dark red (almost brown) when attached to kelps and objects in tidal pools. In cold waters as far north as Iceland, the Irish "moss" and the common dulse grow like stiff little membranous shrubs, branching to frilled fronds. After storms have tossed them ashore, they are gathered by the ton, washed and dried by coastal people who make from them "sea moss farina," a nutritious flour useful for thickening soups, making puddings and for a wide variety of industrial purposes. Laver, which resembles a purplish-pink sea lettuce, is used in the same ways.

Brown algae have a different pigment in surface cells, concealing the green chlorophyll they hold in chloroplastids. A few are almost as feathery as any green or

red alga, but most are coarse. Particularly widespread are those that hold to rocks in shallow cool waters, and are left exposed at low tide. They form a slippery mass of flat-branched rubbery vegetation that protects from sun and dry air a host of smaller life in tide pools. When the tide returns, they float up in the quiet water, buoyed by small gas bladders. The names rockweed and bladderwrack are used for both the kind with a midrib (*Fucus*) and another with none (*Ascophyllum*). Or they are called sea whistle because their egg-shaped gas bladders can be used by children to make whistles.

Where rockweeds grow, slightly deeper water generally contains a submarine forest of coarse flat-bladed brown kelps, such as sea colander, whose single blade, up to nine feet long and twelve inches wide, has a thick midrib and abundant perforations. Various kinds of *Laminaria*, known as sea apron, sea belt, sea staff, sea tangle, sea wand, fan kelp or simply laminarias, form a mass of root-like branches clutching something solid on the bottom. They produce a solid stalk as long as five feet that ends either in a single, broad blade with no midrib or in a finger-like array of similar blades, occasionally thirty feet in length. In a cut stalk, growth rings are visible. Henware is similar, but the stalk is flattened, produces a number of small leaf-like side branches, and then continues as a midrib through the tattered ten-foot frond.

Most of the brown algae grow where they can sway with the water and remain covered by it, buoyed up at low tide. One known as devil's-shoelace forms extensive beds on both coasts of the North Atlantic Ocean, often hiding the bottom and allowing no good estimate of the true depth of the water. Its slender, cylindrical fronds may be twelve feet or more in length, with a central gas cavity. Young plants have a coating of delicate colorless hairs, but mature ones are either clean and shining or fouled by a covering of a fuzzy brown alga that sometimes grows to twelve inches.

Animals That Attach Themselves Permanently

Under the seaweeds we look for animals that gain nothing directly from the sun. Many go through early stages of development in the drifting plankton, and then settle permanently, becoming almost plant-like in their attachment. Sponges, sea squirts and barnacles have no way to move to a new location—to be animal-like—once they have settled on something solid. As inexperienced, immature larvae they choose a good site or a poor one and commit their future.

Sponges show less movement than most of the seaweeds. They grow at almost any depth, adhering to the rocks or embedded firmly in the bottom sediments where the waves are gentle. With no obvious action, they propel water and microbes into invisible pores over their surfaces and discharge the water again through larger openings. If touched, some sponges will constrict the holes through which the water emerges. Otherwise their movements seem due to growth, and become noticeable only when a photograph taken on one day is compared with one taken a week later.

Live sponges feel harsh and are inedible to most biting animals because of internal strengthening with skeletal parts: minute needles of lime in members of class Calcarea, or six-pointed spicules of glassy silica in the "glass sponges" of class Hectactinellida, or some combination of horny fibers with silicious spicules that are simple rods (or four-pointed and cross-like) in members of class Demospongiae. These details of the skeleton provide more meaningful clues than color or shape to the identity of a sponge. Other features vary according to the depth at which the sponge grows, the currents to which it is subjected, the surface to which it is attached and its age.

We recognize limy sponges by their vase-like form, their size from one to five inches long, or because we detect this underlying simplicity even when they are joined together into clusters. The smallest ones resemble the eggs of certain snails so much that we generally have to touch them to be sure they are sponges—that the small opening at the end is where the water comes out, not where a hatchling snail emerged. Urn sponges usually retain their symmetrical shape. Purse sponges become soft bags several inches long and tend to hang, collapsed and flattened into purselike form, when left unsupported by a receding tide.

Glass sponges are denizens of deep water, where living conditions remain virtually constant all year. Most abundant 3000 feet below the sea surface, they include the glass-rope sponge of the North Atlantic. The "rope" is a spirally-twisted bundle of extraordinarily long spicules that splay out in the soft ooze of the ocean floor, to anchor a bell-shaped colony a few inches tall.

Most sponges are neither limy nor glass sponges, which proves that a horny skeleton (alone or with silicious spicules) allows the greatest adaptability in sponge architecture. Some of these colonial animals inhabit almost every tide pool in the boreal Atlantic province. We most often meet the breadcrumb sponge as a pink or green or bright yellow coating on rocks, raised at intervals into little volcano-like craters where the water emerges. Equally widespread are the boring sponges, which grow over the limy shells of univalves and bivalves, dissolving sizeable pits into the shell as though never able to hold firmly enough. In shellfish beds, these sponges are pests, for they weaken the shells, making the mollusks vulnerable to predators.

Sea Squirts

Some of the sea squirts could easily be mistaken for sponges. But each individual has two openings and ejects water vigorously from both if disturbed. The openings close, and the sea squirt takes on as nearly a spherical form as its strange skeleton will permit. The skeleton is of cellulose, called a tunic, and hence the sea squirt becomes a tunicate. Internally it is mostly one big chamber, the pharynx portion of the U-shaped digestive tract, where inhaled water is propelled through a multitude of pores and cleared of both microscopic food and oxygen. Yet for a biologist these pharyngeal pores have a special meaning. They, and the larval sea squirt's small hollow dorsal nervous system and stiffening notochord, earn these strange animals a place next to the vertebrates in our own phylum, the Chordata.

So transparent is the body wall and tunic of some sea squirts that the internal organs can be seen. Apparently for this reason, scientists give the name *Ciona intestinalis* to one kind that is common on the rocks of tide pools and on wharf pilings from Boston southward on the western side of the North Atlantic and where the water is comparable in temperature on European and North African coasts. From Maine and southern Scandinavia to the boundary of the arctic province, the sea peach grows to four inches in diameter, with an almost spherical form and peach-like color that distract attention from the two small openings through which it carries on its exchange of water.

Good living conditions may let many sea squirts live side by side. Sea grapes (*Molgula*) often grow by the hundred in clusters, each individual about half an inch in diameter and brown to yellow-green in hue. A colonial habit has better organization in sea pork, which is generally pink or orange and forms irregularly rounded masses half an inch thick, through which the bright red digestive tracts of the dozens of individual colonists show indistinctly. Occasionally we find pieces of sea pork almost a foot long and five or six inches wide tossed ashore by storms. Their resemblance to salt pork is close enough to justify the name.

Barnacles

Boating enthusiasts rarely appreciate sponges, sea squirts and other animals that attach themselves permanently where food is in good supply. They call these creatures "fouling" organisms, and scrape them off at intervals from boat bottoms to reduce the frictional drag. Special attention must be given to barnacles, particularly acorn barnacles, which cement themselves so firmly and secrete hard conical limy shells.

Each barnacle starts out in life as a small swimming crustacean among the plankton. Having had an opportunity to drift far from its parents, it reaches a stage at which it must find a permanent place. Recent research, aimed at improving boat paint to make it repel barnacle larvae, shows how persistent each young barnacle is in testing surfaces during the day or two when it must make a choice. Once attached, it absorbs its swimming organs, absorbs or sheds its eyes, and quickly becomes a different kind of being. Lying thereafter on its back, it secretes and enlarges the plates of its shell, reaches out with feathery feet when the doors are open and the tide is in, to comb from the water the small animals upon which it feeds.

Where many acorn barnacles live side by side, larger ones smother smaller ones. The survivors become packed together so tightly that they can grow only longer, not broader, as they age. Their shells become elongate, resembling teeth that once fit sockets in a jawbone. Generally we find barnacles with these proportions in places where the water is rarely calm enough to let predators approach. Otherwise the dense population affords a feast to any fish or crab or mollusk that can force its way into the hard shells and remove the crustaceans from their self-made castles.

No less permanently attached, but far more flexible, are the goose barnacles that mature on the underside of a timber floating for months at sea, or of a dead branch from a tree that is exposed only when the tide is low. Each goose barnacle develops a rubbery stalk between its area of attachment and its body, benefiting from this arrangement mostly by projecting an extra inch or two into the water toward food particles. Seldom do we find a goose barnacle on a bare branch dipping into the sea without recalling the medieval superstition concerning these creatures that appear, and grow as though by magic, on the most remote boreal coasts of Europe and America. According to the story, after a goose barnacle reaches full size, it lets go from its site, grows a bird's head at the end of its stalk (which becomes a neck), changes its shell plates into wings, and becomes a barnacle goose. How else could an educated man in medieval times account for the black and creamy gray geese that came out of the foggy north to spend the winter along coasts of Ireland,

Scotland and the Baltic Sea? Not until relatively recent times did anyone discover the nests of these birds along the northeast coast of Greenland, where trees are almost nonexistent and goose barnacles are few.

The Sea Firs and Their Kin

Equally mysterious, since their microscopic larval stages are dispersed among the plankton, are the sea firs and other relatives of more mobile medusae (jellyfishes). They are the counterpart of sea fans and sea whips, which are tropical, adding interest to the coastal world of the undersea. Every one of them is a whole colony of interconnected individuals, most of which extend from around their soft mouths a ring of tentacles studded with stinging and lassoing cells. Each is equipped to capture and stuff into the common digestive cavity the small planktonic animals it catches. Few of them have a venom strong enough to affect human skin.

Sea firs and sea ferns grow on seaweed stalks, on boat bottoms and wharf pilings, and on rocky outcroppings at every depth we can explore while skin-diving. We know them to be somewhat bell-shaped as individuals, supported on separate long waving branches, or in close formation—generally frond-like and essentially two-dimensional. This arrangement lets the tentacles on both sides of a frond have unobstructed access to fishing territory. The separate feeding individuals (polyps) of *Tubularia* are on stalks two inches long; in *Bougainvillia* the branching produces a shrubby compact arrangement with no polyp far from another. Unlike these naked kinds, the members of the cosmopolitan genus *Obelia* secrete a thin and transparent horny covering to shield the soft parts of the colony, which is often a branching plume twelve inches long, bearing hundreds of medusa-producing individuals as well as thousands of feeding polyps. The medusae are the sexual stage in the life history, and disperse fertilized eggs that become larvae ready to settle elsewhere and start new colonies.

Sea ferns resemble feathers as much as fern fronds. So do sea cypresses, which are like *Obelia* in secreting a shield of transparent material over their soft parts. A few years ago we discovered some of these colonies, cleaned, dried and dyed, for sale in a novelty store as a household ornament. Sea cypress colored dark green was promoted as a paradoxical "plant" that never

Sea lions often float, basking, with head and flippers and feet exposed to the air off the California coast. (Wolf Suschitzky)

266

needed watering. The dyed skeletal remains did not grow either, but they remained amazingly flexible in the dry house for many winter months.

Particularly along rocky coastlines, in tide pools, on exposed surfaces below the high-tide mark, and on the coarse seaweeds, we find marine animals clinging and moving about at slow speed. Some are sea anemones, close kin to the sea firs and sea ferns. Others are mollusks, with or without limy shells to lend protection. Echinoderms (the spiny-skinned animals) are well represented by sea stars (starfishes) and sea urchins, often by serpent stars and sea cucumbers as well. Generally they are the sea creatures that interest the coastal explorer most because their ways of life are so different from that of any to be met upon the land.

Aristotle knew that sea anemones move to new locations. He called them sea nettles, although the stinging cells that are the badge of all members in their phylum (Coelenterata) could not be seen until compound microscopes were invented. He noted that each anemone "clings to rocks . . . but at times relaxes its hold. It has no shell, but its entire body is fleshy. It is sensitive to touch, and if you put your hand to it, it will seize and cling . . . in such a way as to make the flesh of your hand swell up." Possibly none of those he found had tentacles coarse enough to suggest petals, or brightly colored enough to suggest a flower. The commonest anemones (*Metridium*) along boreal Atlantic coasts are

Coarse ribbons of bull kelp slither past one another with each wave, and form dense tangles around rocky shores of the southern continents and islands where the water is cold. (E. Aubert de la Rue)

found in Pacific waters too; their many dozens of tentacles are slender, and their color ranges from tan to chocolate brown. When isolated in a tide pool they tend to contract into inconspicuous rubbery objects two to three inches tall, with even their tentacles hidden —inturned at the mouth. Larger and brighter anemones are denizens chiefly of somewhat deeper water. In the great depths, anemones are pale and ghostly.

In shallow waters almost anywhere along boreal coasts, sea anemones can be found in partnership with definite animals. From North Carolina southward into the West Indies, single individuals or whole groups of one species attach themselves to the snail shells that hermit crabs are carrying about. This habit gives them a good chance to capture particles of flesh swirling through the water while the crab is eating a meal. Probably the crab benefits from its burden—predators stay out of range of the stinging cells on the anemone's tentacles.

Mollusks and Echinoderms

A majority of the slugs and snails that live in rocky tide pools and along rugged coasts feed on the sea-

weeds, or on the thin film of algae they can rasp off the rocks. Like the limpets with their hat-shaped, one-piece shells, the chitons with their hinged, eight-piece armor hold like suction cups to the rock. Many show a strong territorial habit: between meals they return to the exact same place. Where wave action is vigorous, the repeated pounding of the resting mollusk against the rock may wear an elliptical groove into the rock that matches the shape of its limy armor, proving how consistently it occupied this site in one preferred position.

Carnivorous snails, such as oyster drills and moon snails, can generally be recognized by the extensible proboscis they apply to the shells of mussels, oysters, other snails and even big barnacles. Within the proboscis is the mouth with its rasping organ (radula), used in cutting a neat circular hole through the shell of the victim and then pulling out pieces of flesh from inside. A few of these snails show that they are habitual eaters of barnacles. The thorn drupes, for example, have a tooth or spine on the outer lip of the shell, which they wedge expertly between the movable doors of a barnacle's shell, preventing the barnacle from shielding itself.

Carnivorous snails, predatory worms, hungry fishes, and some of the sea stars are opportunists, able and ready to attack animals in tide pools and on exposed rocks. The sea stars (starfishes) hump their bodies over a bivalve opposite its hinge and, holding firmly with sucker-ended tubefeet to its two shell halves, exert enough force to bend the material of the shell itself. Once a crack opens, even one less than a twentieth of an inch in gape, the sea star slides its filmy stomach out through its own mouth, in through the gap in the mollusk's shell, and wraps its digestive tissue around its victim's body. After a while the mollusk dies and the sea star can open the shell to complete its meal.

However, most sea stars and other echinoderms are scavengers or grazers rather than predators. The sea urchins that cling to rocks with long slender tubefeet seem well equipped to get at either dead bodies or algae. Around its mouth each sea urchin has a set of five chisel-like teeth that can be spread apart or driven together and protruded, through the action of special muscles between limy bars in a pharyngeal organ of considerable complexity. Aristotle discovered these unique mouthparts, and commented that each set "looks like a horn lantern with the panes of horn left out." The lantern he had in mind was a type mounted on a post, with the lower end pointed and the thin panes through which the light shone slanting to let illumination reach the ground. Ever since, the sea urchin's mouthparts have been known as "Aristotle's lantern." From personal experience we can vouch for their ability to take a nip out of human flesh.

A sea urchin usually relies upon its stiff movable spines to protect itself from predators, much as a hedgehog does. Quite logically, the Germans and French who are familiar with hedgehogs refer to sea urchins as sea hedgehogs—*Seeigel* and *hérissons de mer*. Just as a hedgehog can be picked up by its spines, gulls often pluck loose a sea urchin if they find one exposed by a low tide. Still holding to a spine or two, the bird flies up over an expanse of rock (or a paved road or a parked car) and drops the sea urchin, then darts down to enjoy the soft contents of the smashed shell.

Sea cucumbers gather food in a more peculiar way. Either they display a horizontal circle of branching tentacles that would do credit to a sea anemone, and wait for microscopic food (such as diatoms) to settle on the surfaces; or they use similar tentacles to pick up organic matter among sediments at the bottom of a tide pool or over the sea floor. Often we have sat entranced at low tide, watching a six- to twelve-inch sea cucumber such as a *Cucumaria*, push several branching tentacles at a time into the surface sand, then turn them one at a time inside its mouth to strip them of adhering particles like a small boy licking jam off his fingers.

While the front end of the sea cucumber engages in feeding operations, the cloaca at the opposite end keeps opening and partly closing as though to whistle an undersea tune. Water is drawn into and exhaled from the cloaca, ventilating the branched respiratory organs concealed inside. Little rounded crabs often take shelter inside the spacious cloaca. One whole family of blenny-like fishes, called cucumber fishes (Carapidae), specialize in hiding within these cavities. Their flattened, tapering bodies lack tail fins, pelvic fins and scales, thus allowing them to slip tail first into a sea cucumber's cloaca and up one of its hollow respiratory trees.

The only way we know of easily learning which sea

Preying on smaller fishes close to the bottom over the continental shelves, North Atlantic cod swim in schools, growing to an average weight of 25 pounds. Record fish are as long as 6 feet and weigh up to 211 pounds. (Ingmar Holmåsen)

cucumber has a cucumber fish inside is to transfer several into a pan of fresh water. Cucumber fishes soon become uncomfortable and emerge to seek an escape from the low salinity. Most of them will accept a substitute home in a saltwater aquarium if the opening to a deep cylindrical hole of the right size is smeared with mucus from the right kind of sea cucumber. Probably the largest of the cucumber fishes (*Enchiodon*) from the North Atlantic, and the similar ones (*Carapus*) from cool Pacific waters, have trouble finding sea cucumbers big enough to accommodate their twelve-inch length.

The Pacific Boreal Community

Temperate coasts of the North Pacific Ocean are quite extensive, for cool southbound surface currents sustain boreal life as far south as southern California and the middle islands of Japan. The boundary with the arctic province comes along the Aleutian island chain and across to Kamchatka. Here the brown algae include phenomenal kelps with root-like holdfasts thirty to sixty feet below the water surface, serving for anchorage but not for absorption. Each holdfast remains alive because nourishment travels slowly to it down a rope-like stalk of the plant, from leaf-like blades floating at or near the surface. Ribbon kelp (or sea-otter's-cabbage) has a single float six to eight inches in diameter, from which the ribbon-like fronds spread out; because of its floats, it is sometimes called sea turnip. Often it grows in the same places as the vine kelp, which has a branching stalk and a separate, smaller float for each of its wrinkled, olive-green fronds. On rocks over which the waves break at low tide, the sea palm grows upright. Rising stoutly into air, dangling its circle of narrow dripping fronds, it yields endlessly to each surge of the sea.

Otters and Lions of the Sea

Formerly the sea otters coincided closely in distribution with the coarse kelps, from Kamchatka to California. This was not due to any definite dependence on the seaweeds, although the otters often sleep with one arm or leg holding to a kelp stalk to prevent currents from carrying them away from shore or onto the rocks. Sea otters are adapted for crushing the shells of sea urchins, mussels, snails and crustaceans, for which they dive in water preferably between ten and sixty feet deep. Lacking the usual layer of subcutaneous fat as insulation from the cold, they depend upon air held close to the body by soft hair. Unfortunately this pelt was quickly recognized as of high quality, suitable for royalty, and sea otters were hunted to near-extermination. With absolute international protection, they have increased in numbers and reoccupied about a fifth of their former range—in central California, western Alaska and the Kurile Islands. One of the joys a naturalist finds near California's Monterey peninsula is to watch these animals floating on their backs, fondling their young or smashing sea urchins on flat stones they have brought up from the bottom to serve as anvils.

So extraordinarily rich and varied is the life of the boreal Pacific coastal community that superlatives distract attention from the many less spectacular kinds. "California" sea lions, which are the trained "seals" of circuses, bark noisily from offshore rocks, where they haul out to bask and breed. They are numerous also around the Southern Sea of Japan, and are represented by an outlying colony on the Galápagos Islands. Steller's sea lions, which are the largest of the eared seals, congregate in colonies from southern California to the Aleutian Islands, and down the Asiatic coast to northern Japan. Both kinds are extremely wary of anyone approaching them on foot. Drifting close in a small boat, as we have done until we could smell the fish on their breath, can be an exciting and risky enterprise.

When the Tide is Right

At certain times of year when the tide is high, a night field trip with a flashlight along sandy beaches of California reveals shallow-water fishes called grunions making their spawning runs. Riding a wave, a female grunion and a few male consorts allow themselves to be stranded as a group. Quickly the female wriggles her tail into the wet sand until only the front half of her seven-inch body is exposed. A male or two curl close around her, discharging their milt as she expels her eggs. All of the grunions enter the next wave and wriggle back to the safety of the open water. If no storm washes out the fertilized eggs and developing embryos, and if no shorebird finds them with its probing beak, the young grunions hatch two weeks later within three minutes of the time at high tide when the next wave wets their hiding place. The little fish slither up through the sand, to ride in the water film to the wide Pacific Ocean.

By day at low tide, it used to be possible to wade out and gather abalones from the rocks to which they clung. Now only the small ones, four inches long and less—protected rigorously by law—are there. The larger abalones, including the giant red one, which is as long as ten inches and with a most delectable muscle, live in deeper water. But still the shallowest tide pool may contain surprises. Once we found about a dozen giant sea slugs (or sea hares), which normally remain submerged. They had been dislodged by a storm and dropped in intertidal waters, from which they were slowly creeping back to the kelp zone where they could feast on coarse seaweeds. The largest of these animals weigh about fifteen pounds. When we touched one, it demonstrated its defense against attack: it flooded the tide pool with a dark purple fluid, and exuded copious quantities of slime over its soft body.

From Puget Sound northward past Vancouver Island to Alaska, and down the Asiatic shore to comparable latitudes, one of the largest acorn barnacles can be found at low low tides. Its shell is often four inches across and as much as six inches high. Coastal people still follow a tradition that seemed strange when explorers first witnessed its performance among Indians: of visiting the rocky shores when the tide is right to break off these giant barnacles and eat the contents as though they were shrimp. We treasure one giant barnacle shell that came off almost intact when we tried this technique near Nanaimo, British Columbia.

A different giant grows in mudflats from northern California to Alaska, and probably along Asiatic coasts as well. It is the largest of burrowing clams, whose Indian name of "gweduk" has been converted to "geoduck" (pronounced "goo-ee-duck"). As a juvenile, this bivalve settles on the mud and digs in shallowly, with an active foot that can repeat the performance as required. Later, the clam matures as a bulky bivalve whose mantle secretes shell halves too small to enclose the body. Instead, the valves suggest the gray-white wings of a legless bird with a long headless neck. The neck is the thick siphon, almost two inches in diameter and two to four feet long, depending upon how far it must reach from the geoduck's chamber in the mud to the water above. Below the neck, the thick fleshy enlargement of the mangle edges continues to protrude as the "breast" that keeps the shell valves from coming together.

A geoduck is an epicure's delight and a sportsman's trophy. The largest have shells eight inches long and live where their sites are exposed only during a low low tide, such as comes at new moon. Apparently they attain these dimensions only where the receding waters uncover the mud above them and interrupt their feeding for just an hour or two each month.

Once, on the coast of Washington state at Gig Harbor near Tacoma, we were fortunate in seeing tide, daylight, mudflat and experienced geoduck-hunter all ready, and to be present while the hunter dug the legal limit of three giant bivalves. From hard experience, the digger had learned to recognize the little inch-high fountain of water coming through the last of the receding tide where a geoduck's neck extended to the surface of the mudflat. He carefully set a V-shaped wooden frame on the mud to deflect water draining down the beach. He pushed a small stick into the fountain, which caused a great squirt because the geoduck retracted quickly. Working rapidly with a shovel, he dug on the low side, tossing away soggy earth and the water that poured in and moving his little stick to follow more and more of the vertical tube where the extended neck had been. Finally the animal itself came into view, its neck tip dark as the mud, its siphon and breast the color of a roasted turkey's skin. Had the digger been less of a connoisseur, all of his effort might have resulted in exposing only a Pacific gaper clam with a slender neck, inconspicuous body, less meat, and ordinary flavor.

America's Boreal Coasts

The differences between the Pacific and Atlantic boreal communities along the North American continent are often stressed: the Pacific salmon (six species of *Oncorhynchus*) that make a single fatal spawning run, contrasted with the Atlantic salmon (*Salmo salar*) that spawns year after year in fresh water but returns each time to the ocean to feed and recuperate. The Atlantic has its eels (*Anguilla bostoniensis*), which spawn at a depth of 1500 feet in the same high-salinity water as is used by the European eels (*A. anguilla*); the eastern Pacific has no eels at all. But these differences may be due to physical differences. Their ascent of the big rivers along the Pacific coast to suitable spawning grounds requires immense effort on the part of the salmon there, far more than an Atlantic salmon must expend to make use of the rivers in New England and eastern Canada. The eels have a suitable rendezvous in the midst of the North Atlantic eddy—the Sargasso Sea—where the salinity is higher than anywhere else in this ocean and warm to greater depths; the eastern Pacific affords no comparable spawning grounds.

On the other hand, similarities between the coastal communities on the two sides of Canada and of the United States need an explanation. Both have their similar codfish (*Gadus callarius* in the Atlantic, *G. macrocephalus* in the Pacific), herrings (*Clupea harengus* and *C. pallasi* respectively), and halibut (*Hippoglossus hippoglossus* and *H. stenolepis*). The sand dollar

271

(*Echinarachnius parma*) and the green sea urchin with the big name (*Strongylocentrotus dröbachiensis*) seem alike on both coasts. Biogeographers suspect that a sea link across the north of Canada existed during the Pliocene or slightly earlier, and suggest that the spread of boreal life from the North Pacific coasts included a major colonization of arctic waters as well as a sharing down the eastern seaboard of North America.

Life in the Shallows of the Arctic

During the long days of arctic summer, life in the arctic shallows receives more energy from the sun than it can use. But it continues too at a surprising pace under the sea ice during the long nights of winter. Without light for photosynthesis, the green algae disappear from the plankton but their places are taken by drifting bacteria and other agents of decay. The coarse brown algae that grew twenty feet deep or more all summer decline too in the dark waters under the ice and overlying snow. Any that break free from their holdfasts and drift to shore are soon ground to pulp by the edge of the ice sheet. Yet the pulp is not lost; it spreads over the sea floor as detritus, nourishing some kinds of bivalves, sea cucumbers, sea urchins, sea stars, sea worms and snails. The plankton serves the food needs of most bivalves. Other sea stars, sea worms and snails prey on their neighbors. Fishes take their share, but many of them are caught by squids. Seals remain active under the ice, eating fishes and squids between visits to a breathing hole cut through to air. Walruses dive as deeply as 150 feet, whether beneath the ice or under open water, to get invertebrates living in or on the sea floor. Using their tusks to pry loose whatever they want, and freeing their food of hard parts by skilful movement of the bristly pads around the mouth, they generally gain weight during the winter and are in fine condition when springtime becomes breeding season.

Just a menu of walrus meals gives a good measure of the animal life under the winter ice. In Alaska, walruses prefer the arctic saxicave and the truncate soft-shell clam, both bivalves in the size range between one and three inches. Around southern Greenland, their tastes run to other circumarctic bivalves: the boreal astarte and the chalky macoma, which average about one and a half inches long. Cockle shells and whelks to eight inches long, snail shells containing hermit crabs, sea urchins of any size, sea cucumbers and sea worms are all acceptable fare.

Where walruses remain in their former abundance,
living sociably in herds of about 2000 animals, their food requirements seem prodigious. Three thousand clams daily is normal for an adult walrus. Until Eskimos got modern weapons with which to hunt walruses, there were hundreds (perhaps thousands) of herds around Arctic Ocean coasts, each herd consuming about two billion clams annually. With fewer walruses, there may now be a mollusk population explosion. Food for so many shellfish must still be circulating below the winter ice.

In spring, the long days return and the sun acts like a battery charger revitalizing the arctic seas as much as the land. The plankton becomes green, and the plankton-eaters thrive. Fishes that have been in polar waters all year mate and lay their eggs. Five-inch capelins that have escaped being eaten by cod in boreal waters of the North Atlantic and North Pacific come to rush the beaches of Iceland, Greenland, Newfoundland, Alaska and Kamchatka, starting off new generations of their kind. Even faster than grunions, they dash high with a wave, discharge their reproductive cells in the beach, and ride the same wave back to sea. A surprising number of echinoderms and mollusks give far more parental care, brooding their young until each becomes independent, in a way that has almost no equivalent in warmer waters.

Northern fur seals return from Pacific waters to breed and nurse their young around shores of the Bering and Okhotsk seas. Earless seals, such as the ringed, the ribbon, the harp and the bearded, have no need to migrate, since they stay close to the edge of the polar ice all winter and congregate on the drifting cakes after the ice packs break apart in spring. For these too the arctic shallows offer enough to eat.

The food supply in these cold waters is incredible. It would nourish a great many whales, if there were whales to swim north in season to make use of this resource. But the right whales (five kinds of *Eubalaena* and *Balaena mysticetus*) and the finbacks (which were

Above right: At a depth of about 30 feet, a colorful sculpin rests as inconspicuously as possible, ready to gulp down smaller fishes that swim close. Far right: At 60 feet down, a fringed blenny (Alloclinus holderi) chooses a vacant snail shell for a hole from which to dart out at prey or at invaders in the territory. Right: At 60 feet down, a colorful rockfish (Sebastodes serriceps) finds a large hole in the rocky sidewall of a small underwater canyon as a vantage point from which to watch for food. (All by Jim Hodges)

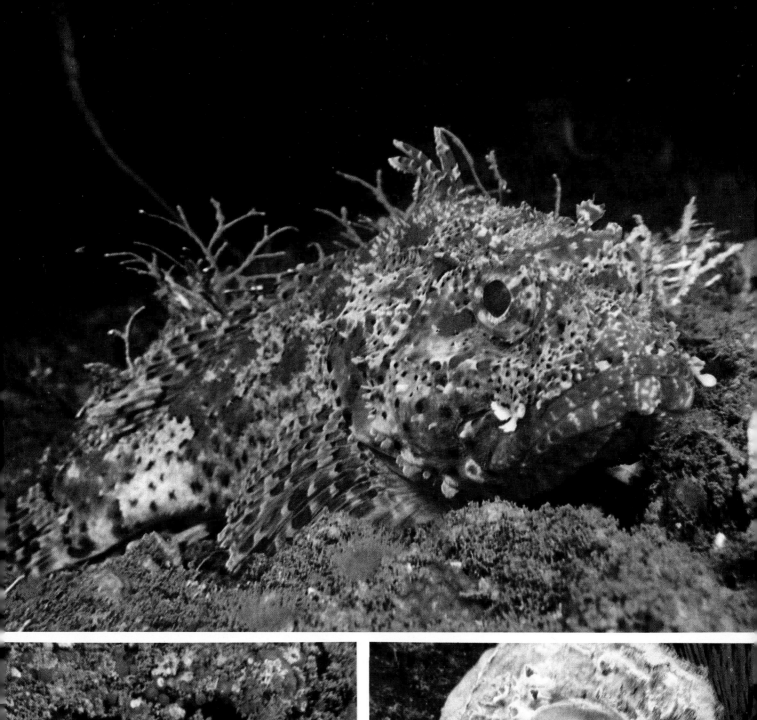

"wrong" because they sank when killed with primitive equipment) have been almost exterminated. Even the little white whales less than fifteen feet long and the spotted narwhales, of which the male has one long spiral tusk, have waned. The tusks are no longer prized as horns of the mythical unicorn, but the bodies are sought as food for Eskimo dogs.

Life Around Antarctic Coasts

The great whales and the little arctic terns evolved a migratory pattern guiding them from waters of the North Polar ocean to the vicinity of Antarctica every year, giving them more daylight hours than any other kinds of life. Now the terns are the principal travelers along this extraordinary route. Toward the southern end they approach a continent with little continental shelf, one that slopes off rapidly into the cold abysses. Deep water separates Antarctica from all other continents, whereas the coasts of the Arctic Ocean are continuous with boreal provinces along two sides of the North Pacific and two sides of the North Atlantic. Aware of these differences, biologists expected to find few kinds of plants and animals native to the narrow shallows around the southernmost continent.

At first, only the penguins seemed strictly southern. Of the seventeen different kinds, two (the emperors and Adélies) breed on Antarctica or its ice shelf, fourteen on islands and the shores of South Africa and South America, and one on the Galápagos Islands at the end of the cold Humboldt current. They constitute a separate order of birds (Sphenisciformes), differing from all others in many ways.

If the antarctic province is broadened to include the island of South Georgia, 900 miles from the tip of the Palmer Peninsula (with the South Orkney Islands in between), along with the "High Antarctic," the five families and 105 species of fishes in the Nototheniiformes become as characteristic as the penguins. The question is where to draw the boundary line around South Polar life.

As oceanographers dredged samples from the sea

Above left: A right-handed fiddler crab stands ready to defend himself on the sandy shore of Baja California in western Mexico. (Willis Peterson) Left: Bright blue sea stars remain in plain sight on the coral reefs of the tropical Pacific despite full sun and low tide that leaves them barely covered. (Photographed at Korolevŭ, Fiji: Lorus and Margery Milne)

floor and caught fishes in waters around Antarctica, the list of distinctive animals grew much faster than had been anticipated. The Antarctic continental shelf has about five times as many kinds of echinoderms as there are in arctic shallows, and about three-quarters of them live nowhere else, compared to just over one-quarter in the Far North. Primitive flatworms (turbellarians) and sea spiders (pycnogonids) prove to be extraordinarily numerous and local in distribution. And while the variety among fishes is comparable to that in the Arctic, about 90 per cent are endemic to the province, whereas only 65 per cent are similarly restricted in the North Polar shallows.

With the whaling industry closing down for lack of whales, and fishing in antarctic waters barely begun, informed people in parts of the world remote from Antarctica hear mostly of the penguins and the explorers in this Far South region. Some are aware that voracious leopard seals eight to eleven feet long prey on penguins, swallowing each bird whole. A few learn that the so-called crab-eater seals associated with drifting ice around Antarctica may be the most abundant of all seals, but rarely that the strange teeth of these animals serve as strainers for capturing their principal food, which is not crabs but krill (euphausid) crustaceans swimming among the plankton. The less abundant Weddell seals and rare Ross' seals catch crabs as a supplement to their preferred foods: fishes and cephalopod mollusks such as squids. The usual south-north range of these four kinds of seals seems small: between 1500 miles from the South Pole along Antarctic coasts and 3000 miles, to visit the very tip of South America and the remote subantarctic islands.

So special and varied a fauna needs explanation. The restricted species, genera and larger groups suggest a long isolation that is easy to appreciate, and also the arrival of occasional colonists, which the seas propelled by the strong west wind could readily pick up from South America, South Africa, Australia and islands nearer to Antarctica. The nourishment for all these animals is brought by the same steady current, and even carried under the edge of the ice shelf in winter with only the Palmer peninsula to disrupt its flow.

Apparently the Antarctic continent developed much of its present form and climate soon after the dinosaurs became extinct. For 60 million years or so, its living things have enjoyed continuity as well as isolation in which to evolve. During the Ice Ages, the main pattern of existence had no need to change although the glaciers and ice shelf probably increased. By contrast, conditions in the Arctic are mostly new. From the Miocene into the Pleistocene, the Arctic Ocean may have been virtually shut off from shallow waters farther south by land bridges between Scotland and Greenland and between Siberia and Alaska. During the Ice Ages themselves, the immense shield of opaque ice overspread the polar sea at least to the land on all sides, starving marine plants for light. Moreover, until ten to twenty thousand years ago, the Gulf Stream seems not to have begun warming arctic coasts and encouraging the plants and animals to spread north; it was then that meltwaters refilled the ocean basins and the northern continents adjusted to freedom from their previously immense burden of ice.

Temperate Coasts with Antarctic Chill

No other oceanic current in the world can compare for volume and vigor with the West Wind Drift around the Antarctic continent. Yet none of the gales and storms that help it circulate destroys the pattern of sudden change in temperature of surface waters in two irregularly concentric circles. The nearer of these to Antarctica is called the "antarctic convergence." It marks the place where water between 32° and 40° F. (0° and 4.5° C.) being blown northeastward by the prevailing winds suddenly sinks below warmer, less dense water. If used as a boundary for the antarctic province of marine life, it would group with the shores of the polar continent itself the shallows around the South Shetland Islands, the South Orkneys, South Georgia, South Sandwich, Bouvet Island, Heard Island, and Kerguelen. In general, the antarctic convergence lies between 50 and 60 degrees south latitude, coming a little farther north only in the South Atlantic and Indian Oceans.

A second boundary, which the distinguished oceanographer Sven Ekman calls the antiboreal convergence, comes where the isotherm of surface water is about 54° F. (14.5° C.) in summer and 51° F. (11° C.) in winter. Since water between 40° and 50° F. is diverted from the West Wind Drift along both sides of South America, the antiboreal convergence turns northward in the eastern South Pacific along the Humboldt Current and in the western South Atlantic along the Falkland Current, making it a ruptured boundary.

"Antiboreal" is intended only as an equivalent in the Southern Hemisphere for boreal in the Northern. The Greek myths, telling of Boreas, the god of the north wind, afford no counterpart to apply in a part of the world undreamed of by myth-makers. It is a cool temperate region or a subantarctic one, depending upon the background of the person describing it.

Within the antiboreal are the Falkland Islands, Tristan da Cunha, Gough Island, Prince Edward and Marion Islands, Crozet Island, and near New Zealand, the little clusters of Macquarie, Campbell, Auckland and the Antipodes. Chatham Island is north of the antiboreal convergence. But whether the boundary line runs between the two major islands of New Zealand, grouping the south island as antiboreal, has never been finally established. Otherwise the antiboreal convergence coincides approximately with the circle of 40 degrees south latitude.

The Humboldt Current

Along the Pacific coast of Chile and Peru, onshore winds pick up so little moisture from the cold Humboldt Current that much of the shore is a forbidding desert. Rarely have scientists ventured along it to explore the plant and animal life of the shallows. Yet the plankton there is extraordinarily rich, due to upwelling of bottom water with an abundance of phosphates and nitrates in solution. It supports incredible numbers of anchovies and other plankton-eaters, which in turn attract predators of larger size. Most famous are the Peruvian cormorants (or guanay birds), which nest by the millions on offshore islands. Pelicans and boobies depend on much the same resource, and contribute similarly to the accumulations of excrement which Peruvians mine as fertilizer.

No sooner had the future of the fertilizer industry been made secure by laws protecting the birds and their nests, than Peruvian fishermen installed refrigeration equipment in their ships and began harvesting predatory fishes: bluefin tuna (or albacore), the similar Pacific bonita with equally white meat, and great numbers of fine swordfish. All command a good price on world markets. To obtain the greatest national gain, the Peruvian government began balancing the fertilizer industry against fisheries.

Now fishermen have learned to use anchovies and other surface-feeding fishes directly, increasing efficiency by bypassing the natural predators. They net the small fishes at the surface and convert them into a desiccated fish meal. Inexpensive to handle, this protein-rich material commands a high price in the affluent nations as a nutritional supplement for livestock and poultry. With fish meal added to their food, the domestic animals respond by faster growth to marketable size and by improved quality, letting the livestock-raiser recover the cost of the fish product and increase his profit too. People with protein deficiency disease (kwashiorkor) could be helped both physically and mentally if a fish meal of acceptable quality could be added to their diets. But so far, despite wistful statements before the United Nations, no way has been found for the hungry poor to compete in buying proteins from the Humboldt Current.

Tierra del Fuego and the Falkland Current

From Chiloé island at 43 degrees south latitude on the coast of Chile, through the islands of Tierra del Fuego, north along the Atlantic shores of Argentina to a comparable latitude (hence somewhat south of Buenos Aires), the antiboreal province is marked by a coarse kelp called sea trumpet. Its long and tapering hollow stalks are often thrown by storms high on the beach, after the terminal cluster of narrow flat fronds have been ripped off and the kelp itself uprooted from the bottom. The kelp beds along the shore are the favorite habitat of the kelp geese, of which one subspecies follows the mainland while another is restricted to the Falkland Islands some 325 miles eastward in the midst of the Falkland Current. So few biologists have studied the life of offshore waters between the intertidal zone and a depth of 600 feet that no one is yet sure whether there are distinctive differences between a "Magellan" subprovince to the south, a Falkland subprovince to the east, and a Patagonian subprovince along the Atlantic coast of the continent itself.

The Benguela Current

Along the African coast, from the Cape of Good Hope to approximately the latitude of Cape Frio (18 degrees south latitude), the Benguela Current and extensive upwelling make the surface water over the continental shelf as cold as farther out at a depth of 1000 feet. It seems incredible to be technically within the tropics—a few miles north of the Tropic of Capricorn—at Walvis Bay in South-West Africa with a temperature in the intertidal pools around 52° F. (12° C.) in August and no more than 61° F. (17° C.) in February, which is southern summer. The coastal desert in South West Africa is the dreaded Namib, where so many people have died of thirst that an alternative name is the "Skeleton Coast." To the south, in South Africa, northbound rainstorms bring more moisture and the area is known as Namaqualand.

We have visited the jackass penguins nesting in burrows dug from the sand of the upper beach on an island off the Namaqualand coast. One bird of each pair tended the two eggs or chicks while the other waddled to the shore to socialize, to swim after a meal of fish and squid, or to salvage any pieces of seaweed that washed ashore. The seaweed became a crude mat in the bottom of the nest, and something for any passing penguin to steal from a nest left momentarily unattended. We remember how docile the half-grown penguin chick seemed when we picked it out of a nest unguarded by either parent, until we put the youngster back. Safe again under its sandy roof and on its two big feet, the young bird bit a piece of flesh cleanly out of the palm of the hand that had held it, clamping the sharp tip of its upper beak into a notch in the end of the lower. A parent bird in the nest will defend itself in the same way.

The conspicuous animals of the tide pools tolerate the cold on the west side of South Africa's tip and also the warmth toward the Indian Ocean, beyond Cape Agulhas. Rocks often bear a coral-like growth of limy tubes packed tight together, each tube as much as fourteen inches long and containing a segmented worm. Periwinkles and limpets abound. Where the cool water cannot escape from the pool and below the low-tide mark, green sea anemones cling side by side.

One of the world's most bizarre fishes (*Callorhynchus capensis*), well named the plow-nosed chimaera, scavenges beyond the low-tide mark around South Africa's tip and north in the antiboreal province on the Atlantic side. This same kind of fish, or one so like it that ichthyologists are uncertain of its distinctness, turns up in waters of similar temperature in antiboreal South America, and again along the south coast of Australia, around Tasmania, and New Zealand's South Island. Known only from these parts of the Southern Hemisphere, it is a creature of continental shelves, rarely descending beyond 600 feet even if the water temperature lower down is just the same. Its existence poses a question to biogeographers, one they wish the geologists could answer: were these areas of the world ever connected? Or did the West Wind Drift take from one place to the next a mass of seaweed in which were entangled some of the horny capsules that protect *Callorhynchus* embryos? These capsules are so like the "mermaid's purses" in which many sharks and young rays spend corresponding stages of development that only an expert can tell which will hatch as a chimaera.

Off Australian Coasts

Cold water fills the Great Australian Bight along the south coast, where the cliffs drop almost vertically from the flat bare gibber desert and nullarbor plan. In August, the surface temperature is 55° F. (13° C.), the same as at a depth of 600 feet. Where the water flows through Bass Strait, between the mainland and Tas-

A dead male narwhal caught close to Baffin Island and towed ashore. The one long tusk, ordinarily the left, is grown only by the male. (Fred Bruemmer)

waters ("muttonbirds"), letting the hot fat from the bird itself solidify around it inside the casing and protect it from decay. In season, the preserved birds are hung for sale in food stores, suspended like whole bologna, in dried kelp stalk segments twisted shut at each end.

Shallows and sea bottom of moderate depth along Australia's antiboreal south coast are well supplied with echinoderms, nearly four-fifths of which are known from nowhere else. The affinities of these sea urchins, sea stars, serpent stars and feather stars, however, can be traced through their genera and seen to be principally from farther north, including the tropics. Very little has arrived from either east or west.

From farther south, three kinds of penguins come to antiboreal Australia to breed: the crested, which is almost circumpolar; the gentoo, with a similar distribution; and the smallest of all penguins, the blue (or fairy penguin), which seem to imitate the shearwaters in coming to their nest burrows only under cover of darkness.

New Zealand Waters

Nearly 1500 miles of open water, much of it over 12,000 feet deep, separates New Zealand from Australia. Known as the Tasman Sea, it gives plenty of space for the West Wind Drift to veer northeast along the two flanks of the South Island. Offshore, it conflicts with a counterclockwise eddy (the East Australian warm current) which tends to bring subtropical conditions to the North Island's west, north and eastern coasts. Out of this conflict come storms and winds of almost gale force, accompanying a strong current through narrow Cook Strait between the islands, making the capital city ("Windy Wellington") often difficult to approach by sea or air.

After meeting the land plants and animals and the people of New Zealand, and seeing how different they are from their counterparts in eastern Australia, we were unprepared to find the life of tide pools, sand beaches, mudflats and offshore waters to 600 feet deep so similar to that around Sydney. The same conspicuous seaweed (*Hormosira*), whose branching fronds are bead-like and buoyant, is called Neptune's necklace along all of these rocky coasts; it supports large populations of the turban shell known as a cat's eye from the horny lid with which it closes its doorway at low tide. The same spiny sea stars with ten to twelve arms cling to the rocks, although never so tightly as the limpets and chitons, or as the abalones known to New Zealanders as paua shells. The small, dark rockfish lurks in the

mania, it may be slightly colder, matching the diversion southward. In places the water is almost choked with kelp. But it is a distinctive kind, the bull kelp, whose strap-shaped fronds align themselves like the layers of a spring, and slither with every wave to adjust their positions. Local people have learned to use the hollowed stalk with which bull kelp is tethered to its holdfast as though it were sausage casing. They cut it in lengths, and stuff it with the cooked bodies of unfledged shear-

pools, and the big pen shells known as horse mussels protrude from the mudflats. The crested and the blue penguins come equally to these two countries.

Wildlife authorities point to meaningful differences, however, distinguishing the big southern fur seals of New Zealand (*Arctocephalus forsteri*) and adjacent small islands from those of Australia (*A. doriferus*) and Tasmania. The rare New Zealand sea lions (*Neophoca hookeri*) are similarly distinct from those of Australia's south coast (*N. cinerea*) near Adelaide.

Actually, the ocean currents have brought an incredible assortment of living things to New Zealand's coasts. Nowhere in the world have we encountered citizens who appreciate more fully or who are more aware of the wealth of plants and animals around them. Yet only the world traveler is likely to recognize in New Zealand waters the brown seaweed *Ecklonia radiata* forming underwater groves as the sea trumpet of Patagonian shores. The bull kelps are the same as those in southeastern Australia. The butterfish kelp proves to be even the same species as the vine kelp that is harvested for alginates off the California coast.

Many a beachcomber from the western side of the North Atlantic, who is used to seeing fragments of *sargasso* weed (*Sargassum*) littering the shore after a storm from the east, thinks of this reddish brown alga as misplaced from the Sargasso Sea. Sargazo is a Spanish word, applied primarily to drifting seaweed of this type in the middle of the North Atlantic eddy—the windless graveyard of early sailing ships. To find this plant in tide pools and quiet channels around New Zealand is astonishing for anyone who has not explored elsewhere around the Pacific Ocean. *Sargassum* includes about 150 different species related to *Hormosira* and *Fucus*, most of them in coastal waters to 100 feet deep around Australia and Japan. As species they may be local, such as the two in New Zealand and the seventeen along the west coast of North America; as a genus they are widespread. In New Zealand they grow attached by small circular or oval holdfasts, each plant producing first several branches with large, delicate, leaf-like fronds. Later it extends a tough stalk several feet long, bearing side branches with small leaf-like parts, spherical floats on short branchlets, and egg-shaped reproductive bodies.

The Sargasso Sea—A Coastal Community Over Deep Water

For many years, biologists assumed that the currents that create the North Atlantic eddy swept floating sea-

weed into its center, creating the slowly rotating mat that covers an oval area between 25 and 31 degrees north latitude, and 40 and 70 degrees west longitude. Below this mat lies some of the world's clearest and least productive water, down to the sea floor about 15,000 feet from the surface. Actually, the sargasso weeds that float there are endemic and peculiar, for they produce no holdfasts and reproduce solely by fragmentation.

One of the most cosmopolitan of colonial moss animals (*Membranipora membranacea*, of the little phylum Bryozoa) spreads over the surface of the sargasso weed, flexing when the alga does. Empty chambers from which dead moss animals have slipped give a lacy appearance; full ones are hidden below a velour-like covering of short tentacles that trap microscopic particles as food. Goose barnacles hang below the weed, sweeping the water with comb-like feet for minute crustaceans and other animals. Compact little crabs and delicate shrimp-like crustaceans scavenge, creep and dart. Slender pipefishes propel themselves slowly, sucking particles of food into their round small mouths. Brown and tan angler fishes, with grotesque appendages making them blend inconspicuously among the weed, grow to be two inches long as the principal predators. The same species occur also among floating seaweed in the western Pacific Ocean.

The Sargasso Sea is a place of stagnation, into which currents must occasionally bring living things, but from which virtually nothing goes except down in death to dissolution. The plants and animals that continue there have gone through many generations since their ancestors arrived. But those ancestors must have come from coastal communities, most probably in the West Indies.

The Sea Floors Beyond the Continents

Oceanographers in research vessels have now mapped the sea floor in great detail. They know that beyond the continental shelves, the bottom slopes down rapidly to a depth of about 6000 feet. The slopes are the bathyal zone, and the floor from 6000 to 18,000 feet the abyss. Anywhere in the world at 6000 feet, the water temperature is about 39° or 40° F. (4° C.) and the pressure 2685 pounds to the square inch.

About 83 per cent of the sea floor is at abyssal depths. More than a third of it is covered by a firm red clay with very little organic matter and few living things— the underwater equivalent of a desert. On it, strange rounded lumps are littered about, each the size of a man's fist. They are composed of manganese oxide or iron oxide, and no one has yet discovered their origin. The clay itself is easier to explain, for it occurs mostly where the water is more than 12,000 feet deep. Before organic matter descending from the plankton gets this far, it is degraded by decomposers and dissolves. The pressure increases the solubility of lime and silica, causing the shells of dead plants and animals to disappear too.

Another third, in shallower parts of the oceans, is coated with fluffy ooze in which the limy shells of *Globigerina* and other foraminiferan unicells are mingled with the minute coccoliths from planktonic algae. Long-legged crabs stride in slow motion over the globigerina ooze. Abyssal sea cucumbers and segmented worms engulf the sediments and digest out the nutritious organic matter. Bacteria that are carrying on decomposition are among the foods the sea cucumbers and worms obtain in this blind way.

Almost a fifth of the sea floor is coated either with blue mud, which owes its color to iron sulfide released by decomposition of organic matter, or with diatomaceous ooze, which is most extensive under the cold waters south of the antarctic convergence and in some far northern parts of the Pacific Ocean. Calcareous sands from disintegrating coral reefs and radiolarian ooze littered with the silicious skeletons of minute planktonic animals cover most of the remaining sea bottom.

The Ocean Trenches

In twenty places, the sea floor dips lower into ocean trenches. Five of these in the Pacific Ocean (the Kermadec and Tonga Trenches northeast of New Zealand, the Marianas and the Philippine Trenches in the western Pacific, and the Kurile-Kamchatka Trench in the northwest) exceed 30,000 feet in depth. To the deepest of these—the Marianas—determined explorers took the bathyscaphe *Trieste* on January 23, 1960 and saw at 35,800 feet below the surface a number of animals moving slowly about.

Nearly 300 kinds of animals have been dredged up from the sea floor in the trenches. Known as ultra-abyssal or hadal creatures, most are members of genera represented at lesser depths. Colorless and blind, they include a surprising number of large (even gigantic) species, as compared to their relatives in abyssal, bathyal or littoral communities. Biologists are trying to discover whether large size has special advantages at these depths, where the pressure is as great as six tons to the square inch. It is possible too that the hadal animals include recent pioneers that emigrated down-

ward during the Ice Ages or subsequently and found conditions tolerable.

Below 18,000 feet, three-quarters of all the kinds of animals known are sea anemones, snails, bivalve mollusks, segmented worms, amphipod and isopod crustaceans, and sea cucumbers. Hadal species account for about a tenth of all recognized echiuroid worms and beard worms. Unrepresented groups may be significant too. So far, no crabs or other members of the crustacean order Decapoda have been caught farther down than 15,000 feet below sea level.

No doubt the depths of the sea will yield many surprises for years to come—animals unfamiliar to science or known only from the fossil record and still regarded as extinct. Until a beard worm was dredged up in 1900 from Indonesian waters, this whole phylum of animals (the Pogonophora) had gone unnoticed. Now more than eighty different kinds have been found, some from shallows close to shore in the Arctic. Slender-bodied and from one-third to three inches long, they inhabit tubes of their own making. Apparently they feed on bacteria caught and digested in the midst of their terminal tentacles (the "beard"). At no stage in its development does a beard worm have a digestive tract.

Living Fossils in the Sea

Until 1938, no one was aware that the world has living representatives of the lobe-finned fishes. Supposedly these creatures became extinct about 70 million years ago. Prior to that they were fairly common and they left good fossils, which showed little change for millions of years, that is, since the time when ancestors of this type diversified to produce among their descendents the first land vertebrates—the amphibians. Then a fisherman arriving at Port Elizabeth on the Natal coast of South Africa telephoned Miss Marjorie Courtenay Latimer, director of the local natural history museum, to say that he had aboard a strange large blue fish that he hoped would interest her. She took one good look at his trophy, asked him to save it carefully for her, and hurried to telephone Professor J. L. B. Smith at Rhodes University. Her description of it brought him to Port Elizabeth at top speed.

We had the privilege of inspecting the original specimen of this five-foot fish, which he named *Latimeria chalumnae*, while talking to Professor and Mrs. Smith about it at Rhodes University. It lies in preservative within a special stone container like a mummy case, with a heavy lid sealed removably with heavy grease. Almost two dozen other specimens of *Latimeria* have

280

been caught now, all from near the sea floor in the Comoro Straits between South Africa and Madagascar, at depths between 200 and 600 feet. Some have been alive and in good condition when brought to the surface. But this much change in their environment seems more than they can tolerate. So far none has survived where it could be observed carefully for more than a few hours.

In 1957, biologists of the world learned from a report in the scientific journal *Nature* that a far more ancient group of animals still lives in the deep sea. Among limpet-like mollusks dredged from the sea floor in 1956 by the Danish *Galathea* Deep-Sea Expedition were ten intact specimens and some empty shells like fossilized shells known from the period between Cambrian and Devonian times—600 million to 350 million years ago. From the soft parts of the intact individuals, confirmation came for a theoretical prediction made many years before: that ancient mollusks had evolved from the same ancestral stock as segmented worms, and should have had vestigial features in their anatomy to indicate a segmented pattern. The newly-discovered mollusk, named *Neopilina galatheae* and assigned to a new class (the Monoplacophora), has paired muscle scars and muscles inside its shell, paired excretory organs and external gills more like those of segmented worms than had previously been seen in any mollusk. Since 1957 additional specimens of this and another species have been collected, all from the sea floor of the eastern tropical Pacific Ocean at depths between 13,000 and 16,000 feet.

Glass Sponges and Sea Lilies

Improvements in technology make us hope someday to dive without aid so we can see in their native haunts other animals of the ocean floor that are now unknown. We long to explore the bottom of the western Pacific from which, for many human generations, Japanese fishermen have hauled up exquisite Venus'-flowerbasket sponges. These glass sponges may be fifteen inches tall, slightly curved, and grow at any depth between 1500 and 15,000 feet. Each is supported on a latticework that can be cleaned and displayed in a museum case as a glistening skeleton. Japanese newlyweds have long been given such a sponge as a wedding present, because ordinarily each cleaned sponge includes the remains of a mated pair of deep-sea shrimps. The shrimps enter the flowerbasket when both they and the sponge are small. The shrimps remain entrapped, and they and the sponge live out their days together.

Sea lilies—echinoderms of the class Crinoidea—fascinate us almost as much. The fact that they were not all extinct was learned first in 1873, when scientists aboard the H.M.S. *Challenger* recognized some badly-battered specimens among the trophies dredged from the abyssal sea floor. Some eighty different species are now known, a few belonging to the same genera as beautiful fossils from Cretaceous times. The only whole specimens ever examined appeared when repair men, raising a length of defective transoceanic telephone cable from great depths, let scientists have the sea lilies that had entwined their stalks around the man-made support. The habits of these animals are still a complete mystery, but one that may soon be cleared up. Recent explorers in arctic waters have found sea lilies of a few kinds in large numbers at a depth of barely fifty feet.

This discovery reminds us that the handsome little blood stars and a number of other kinds of invertebrate animals turn up in the shallows of the Arctic Ocean, in tropical abysses, and along the coasts of the antiboreal and antarctic provinces as well. They may descend no farther than 8000 feet in equatorial latitudes, but their extraordinary range is reminiscent of the living things on land that can live close to sea level in Patagonia, high in the Andes at the Equator, and low again in the Far North. By tolerating changes in pressure, they can find suitable temperature and acceptable food merely by adjusting their elevation at each latitude. North-south mountain ranges provide highways for terrestrial life between the cool climates separated by the tropics. The cool waters of the sea floor must serve as an equivalent for marine life.

Colonizing greater depths and spreading out where conditions are tolerable is part of the tradition of animals that associate with the sea floor. Apparently this is the only way that the bathyal, abyssal and hadal levels have been populated. The ancestors of deep sea life came from coastal waters. Now man is testing his ability to make himself at home in the undersea, and at progressively deeper levels.

13

The Life of Brackish and Fresh Waters

We have many reasons to bless the sun. Its energy keeps our planet habitable for all the kinds of life we know. Among the benefits it confers, few are more fundamental than the daily cycling of pure water. Each twenty-four hours as the sun illuminates the rotating earth, its warmth lifts about 210 cubic miles of this vital moisture from the oceans into sun-driven winds that blow quickly from place to place. Around 186 cubic miles of water fall back daily into the oceans as rain. We gain from the remainder, which drifts above the land, forming clouds and providing the precipitation that makes terrestrial life possible. Having served this role, the water flows back under the gentle compulsion of gravity, refilling the oceans and keeping the cycle in steady operation.

No one knows when the great water cycle began. It was far into the past, two or three billion years ago. But ever since that time, rivers have been bringing dissolved salts from the land, at first to oceans that were completely fresh. Gradually the oceans accumulated the salts, for salts cannot follow the water cycle. Only minute amounts of them blow in the form of dust from wave crests and give to sea air its salty tang.

Long ago, in a mysterious way, the salt concentration in the oceans reached a fairly steady level. Since then, all over the world—except at river mouths and where evaporation is excessive—the open ocean has apparently been 96.5 per cent water and 3.5 per cent dissolved salts. Plants and animals that are accustomed to this proportion are often unable to survive if it changes much in either direction.

Along the margins of the seas, in lagoons and salt marshes and the estuaries where the rivers joust endlessly with the tides, quick changes in the salinity do occur. It may go as high as 10 per cent salts, or down to almost zero. Those living things that managed to cope with these changes found few barriers in either brackish or fresh waters. Our ancestors were among them. For them it was the way to the land.

Rivers and evaporation affect the salinity in seas that are hundreds of miles long. The Danube, the Dnieper and the Don bring to the Black Sea more fresh water than the sun evaporates from the sea surface, and the salinity remains around 2 per cent despite the open connection through the Bosporus Strait to the Mediterranean. Few rivers empty into the Mediterranean, while the generally clear weather lets the sun evaporate a great amount of moisture. Particularly where the water is shallow near the coasts, the salinity exceeds 4 per cent. Inflow from the Atlantic Ocean past Gibraltar sustains the water level. The Red Sea is saltier still, being bordered on each side by desert and having a narrow connection to the Indian Ocean. At Eilat, where the Israelis distill water from the Gulf of Aqaba for lack of a better source, it is 4.5 per cent salt. But marine life is in no danger because no rapid change takes place.

The Baltic Sea is far more of a challenge to plants and animals from the ocean, because it becomes almost fresh water when the winter snow melts on Scandinavia, Finland and the countries along the eastern coast. In places the dissolved salts decrease in proportion until they account for only about one per cent. By autumn, the long sunny days have brought evaporative losses, increasing the salinity to above 2 per cent. Yet some of the marine fishes in the Baltic are of the same kinds as are to be found in the Mediterranean. They die if transferred quickly from one type of water to the other, partly because they are not adapted to the temperature differences. But given a few generations to adjust slowly to combinations of temperature and salinity approaching those of the other sea, they manage perfectly well.

The range of the American alligator has shrunk in recent years as a result of persecution and habitat destruction. (George Porter)

Along their northern coast, the Danes boasted to us that during cold winters, the ice spreads so firmly from their country to Sweden near Goteborg that daring men have driven across in automobiles. Ice forms regularly over the northern end of the Baltic Sea, coating the water for three to five months. These effects of cold are a measure too of the lowered salinity, for full-strength sea water must be chilled below 29° F. (–1.85° C.) before it freezes, whereas fresh water from a lake congeals just below 32° F. (–0.03° C., not 0.00° as does distilled water).

The depression of the freezing point is one of the characteristics of watery solutions that can be measured easily as a way to gauge the proportion of dissolved materials. It allows scientists to compare the aquatic environment of an animal with the animal's own blood, to see whether the two solutions match closely or differ because the animal has some way to maintain its independence of salinity. Our own blood, like that of whale and elephant, bird and desert snake, turtle, frog and bony fish (whether freshwater or marine), has a freezing point between –0.55° and –0.74° C. Thus, we are "warm-blooded" creatures. That of sharks and their kin, and of marine crustaceans, closely matches the freezing point of the sea water in which they live. They are known as "cold-blooded".

This contrast between modern animals stimulated the Canadian biochemist A. B. Macallum at the beginning of the twentieth century to offer a beguiling idea: that far back in time, when the ocean's salt concentration was such as to give the water a freezing point around –0.60° C., the ancestors of modern bony fishes, land vertebrates, and whales migrated from marine situations into fresh water. They could leave the ocean because they had evolved a way to isolate their internal environment (the blood bathing their organs) from the external environment. He marveled only that during the intervening years no major change had occurred in the balance of dissolved substances in the blood of all these animals. By contrast, the sharks and marine crustaceans had adjusted continuously to the rising salinity of the seas. Other scientists admitted that living things would have found invasion of fresh water easier from oceans with a freezing point of –0.60° C., and hence a salinity of about one per cent, than from the higher concentrations that are now so widespread. But so little information can be obtained from the fossil record as to the salt content of the seas during Ordovician and Silurian times—when the first big invasions of fresh waters actually occurred—that Macallum's attractive idea languished for lack of factual support.

Waters in Conflict

Viewed from the ocean, an estuary is a deep inlet where the sea water is diluted by fresh flow from a river. Considered from the opposite direction, it is the place where the river water becomes brackish and then saline, according to its admixture with the salty solution beyond. Irregular fluctuations in the volume of water that the river brings in wet seasons and dry, and in the regular pattern of tidal rise and fall, make any estuary subject to a wide range of changing conditions. When large rivers such as the Amazon and Orinoco are in flood, the fresh water pushes out far beyond the estuary into the ocean itself. In normal times, whenever the moon is new or full, the "spring" tides alternately rise higher than usual, pushing salty water far up the estuary, and then fall lower than usual, letting fresh water replace the saline. In dry season, spring tides cause an increase in salinity farther upstream than at any other time of year. Small rivers, such as the Swan River in western Australia, flow so little in most months that the estuary may become almost as salty as the open sea.

Rarely can we walk along the shore of an estuary and say confidently, "Here is where the mixing takes place, where the fresh water meets the salt." Generally the two kinds maintain some degree of separation, the denser, salty water sliding into the estuary along the bottom, below the lighter, outgoing flow of fresh water from the river. Changing too with tide and season, these conditions challenge living things to find a suitable location. The fishes and squids and swimming crabs can move. But many other animals are like the larger plants in being attached, and in vanishing if the aquatic environment varies beyond the limits of their tolerances. Many of them survive by withdrawing into the intertidal muds and sands, where the salinity of the interstitial water changes little from one high tide to the next.

By looking very closely at the surface of a mudflat when the tide is part way in or out by day, covering the mud with ten inches or less of water, we see a greenish coating. It consists of a film of single-celled plants, such as diatoms and euglenas. They migrate to the surface and back to a depth of about a quarter of an inch on a schedule that matches the tidal rhythm and the arrival of light for photosynthesis under water. Tiny snails and creeping worms browse on them, and so do small flattened crustaceans known as scuds.

The conspicuous plants that the tide immerses regularly include seaweeds anchored in place only by holdfasts, and flowering plants with firm roots. The seaweeds have spread in from shores with full salinity. The flowering plants represent lines of evolution from terrestrial ancestors that developed enough tolerance for salt to compete for space in brackish water. Water milfoil is clearly an extension from populations in the fresh water of the rivers and ponds. Eelgrass and ditch-grass are members of the pondweed family (Potamogetonaceae) that thrive best where the salt content is between one and two per cent; they are estuary plants now, having lost their tolerance for fresh water without gaining the ability to grow where the salinity is 3 per cent or more.

While the tide is in, we see these plants buoyed up by the water, swaying with it. They tend to hide from our eyes the large numbers of shrimp that enter the estuary, staying within the layer of salt water and feeding close to the bottom. There turbulence stirs up some of the microscopic plants and makes them available as particles of nourishment. Predatory fishes and squids follow the shrimp. Small fishes in the shallows pursue the scuds, and often come within striking range of a heron standing motionless near the water's edge. Larger fishes, such as mullet as much as two feet long, come in schools far into brackish water, using their strong jaws and teeth to crush the vegetation they find in greater variety than in the open sea.

We tend to forget about the hidden worms, burrowing crustaceans and mollusks when the tide is out and the sloping sides and broad mudflats of the estuary lie exposed to air. Then curlews, plovers and other shore-birds probe with their long beaks into the soft mud to reach these morsels. Judging by the persistence of these birds, we feel convinced they find a rich supply of food. Where the mudflats are studded with stones, along coasts from Japan to New Zealand, Iceland to South Africa, New Jersey to Argentina, and California to Chile, black-and-white oystercatchers stride solemnly about on blood-red feet, using their strong flat beaks to open bivalves including oysters, and to chisel limpets from the rocks.

Sand Bars, Salt Marshes and Mangrove Swamps

Slack water appears regularly twice each day when the flow of the tide and the current of a river match each other. Before and after that moment is an hour or so when the sediments the river carries have time to settle. If the particles are coarse sand, they are likely to form a bar across the river mouth, one exposed to the air whenever the tide is low. Silt sinks to the bottom gradually, and does so most where the drag of vegetation slows the movement of shallow water. As the silt builds up, it favors the growth of plants that extend their leaves into air. In most of the temperate zones, the silty shallows transform into a salt marsh with grasses and herbaceous vegetation. In warmer regions, the plants are mostly mangrove trees, which form dense swamps along protected shores.

The sand bar we have explored most often links Bar Island to the principal part of Mount Desert Island on the coast of Maine. In most places it is densely covered by a "scalp" of live mussels, each holding to its neighbors by means of strong thread-like strands of brown secretion from the foot. When covered by the tide, these bivalves filter out a wealth of nourishment, grow three to four inches long, but are limited in number by the amount of space. Under the mussels, however, the sand is coated by a thick layer of oyster shells, showing that these were earlier colonists of the bar—ones the mussels smothered to form the present scalp. Already we can see where the mussels are becoming fouled by barnacles, which may eventually inherit the food supply. Yet this change causes the cycle to renew itself if no more sand is added and the tides continue to sweep across. When the mussels die, their byssus strands break; the scalp disintegrates, exposing the oyster shells again.

Salt marshes seem more stable. With practice we can learn from the kinds of plants we see what the living conditions are along the edge of the salty and brackish water. The lowest ground is generally bare because the tidal current sweeps by too vigorously each time it ebbs from full. Next comes a pastel green zone of succulent little leafless plants called glasswort (or samphire) because their cylindrical jointed stems snap when bent sharply, and often produce a crunching sound when stepped on. Glasswort excels in tolerating salt water, but it roots too shallowly in the mud to resist a current. Cordgrass penetrates the mud more deeply and grows as much as three feet tall where the water moves considerably; but its presence shows that seeping springs or the adjacent river have reduced the salinity somewhat and taken away glasswort's advantage. Still less salt lets the taller reeds grow densely, sometimes intermixed with tussocks of bulrush, which, being a kind of sedge, has stiff stems that are triangular in section.

Trees with Knees

Mangroves, which are shrubs and low trees with a characteristic way and place of growth rather than one

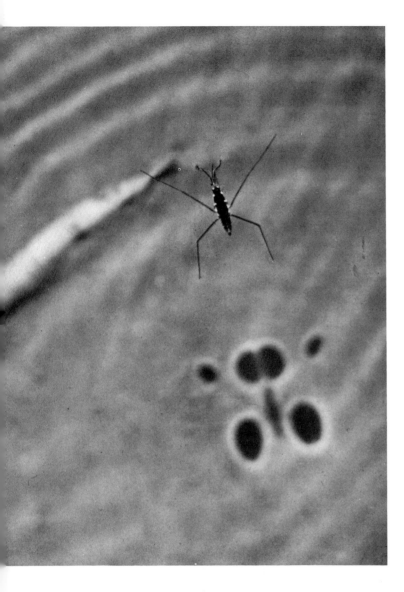

The insect known as a water strider stands dry-shod upon the water film, rowing along by means of its middle pair of legs. The foot and hind legs press large dimples in the film and create big shadows on the sandy bottom. (Lorus and Margery Milne)

directions from the main trunk, the mangrove sends down slender arching adventitious roots. These branch at high-tide level, producing a tangle referred to as "mangrove knees." For a while these may support dozens of oysters and barnacles, as well as active snails and invertebrate animals of many other kinds. Crabs scuttle among the roots, and fishes swim there while the tide is full. Gradually, however, the fallen leaves and debris left by the tide accumulate and form a soil. Red mangrove has changed its own environment from wet to almost dry, and made it suitable for other kinds of vegetation. At the same time, from among the glossy paired leaves, it has been flowering, fruiting, but having its seedlings sprout within the fruit while still suspended from the tree. Each seedling forms a stiff sharp pendant root as much as eight inches long before breaking free. It can fall like a javelin and stab into the mudflat, or float some distance on the tide before being stranded and taking hold.

Black mangrove begins to invade where the water is shallower than twelve inches, only to be displaced by button mangrove and then by trees of tropical, moist soils. From Malaya to the Philippines and Australia, the mangrove community has members of still other families of plants with a similar habit of growth. Often it is tangled further by branching stems of the trunkless nipa palm. All of these thickets attract birds for roosting and nesting, sometimes in such abundance that their accumulated wastes poison the trees. If the trees die, grasses may complete the succession that the mangrove seedling began.

Regular Travelers between Salt and Fresh

Through the world's estuaries twice a year, certain fishes travel on a regular schedule in which mankind has often sought to find a meaning. Salmon of the North Atlantic that have attained maturity head for the coasts and cease their feeding. Loaded with maturing eggs or astronomical numbers of sperms in a dense milt, they drive their streamlined bodies up the rivers to find spawning grounds in shallow riffles. There these big fish prepare a nest, mate and start off a new generation, then casually swim down to the sea and resume their normal lives. Pacific salmon fight the current in far more precipitous rivers, but do so only once at the end of a long life spent cruising as much as 6000 miles from the home stream to which they return. After spawning in fresh water, these salmon die.

Other members of the trout family (Salmonidae) show less spectacular eagerness to reach freshwater

genus or family of plants, contribute similarly to extending the land along estuaries and sea coasts protected from storms in the tropics and subtropics. Red mangrove along both sides of the Atlantic Ocean is the best known. Its seedlings rarely establish themselves where the water is less than a foot deep at high tide, yet they flourish despite wave action and form dense thickets. From outstretched limbs extending in all

shallows for reproducing. Alewives (or sawbellies) of the herring family and smelts in a family of their own struggle if they can through the tidewaters and lay their eggs in the freshest water they can reach. Fisheries scientists with a knowledge of Greek use the word "anadromous" (*ana* = up, *dromous* = channel) for these marine fish that spawn in fresh water, to distinguish them from eels and others that migrate in the opposite direction (the "catadromous" fishes).

Actually, eels (Anguilla) are only slightly more than halfway to being catadromous. Male eels travel no farther than the brackish water of the estuaries, and there grow to maturity. Females wriggle far upstream and, on wet nights, travel across grassy fields to reach isolated ponds in some mysterious way. When they have attained a length of about five feet, the females wriggle back the way they came, attract a coterie of males in the estuary, and set out through the ocean to the abyss in which they hatched. There the adult eels mate, lay eggs and die. Their young, which are flat and resemble transparent willow leaves, drift upward into the plankton and are carried by surface currents. Off continental coasts, they may be able to detect the flavor of fresh water from rivers. Somehow they are stimulated to swim into the estuaries and there to transform into cylindrical transparent "glass eels" scarcely thicker than the lead in a pencil. Slowly they grow accustomed to the lower salinity, and as slowly darken and become opaque, in the shape and color of maturing eels although still far smaller.

Not until 1922 did the scientific world learn of these extraordinary travels of eels, from a thorough report of persistent detective work. Most of the research was done by the Danish fisheries officer E. Johannes Schmidt aboard a special Norwegian oceanographic ship, the *Michael Sars*. He showed that American eels (*A. rostrata*, with fewer vertebrae) and European eels (*A. anguilla*, with more) both spawn at the bottom of the Sargasso Sea. Their young rise into the Gulf Stream and sort themselves out, the European eels remaining in the current until it comes close to their home shores. More recent studies have identified similar abysses in the Indian and western Pacific oceans where eels of the Southern Hemisphere reproduce in an almost identical pattern.

In waters from Indonesia to New Zealand, a few other fishes are known to enter rivers for the principal part of their feeding period, then return to the sea for reproduction and the first long portion of larval development. Perhaps the choice of anadromous fishes to spawn in fresh water has no deeper meaning than that their ancestors in the marine world found some freedom from parasites or predators by using their tolerance for near-zero salinity. Spectacular as these habits are, they may not date from as long ago as Ordovician or Silurian times, when ancestral vertebrates entered the streams and lakes, there to evolve fishes with jaws and the various classes that spread onto the land.

How Did Plant Life Reach Fresh Waters?

The number of ponds, lakes, streams and rivers in the world is beyond counting. They seem isolated, at least those in one river system from those in another. Yet almost anywhere in the world, once we row in a small boat on a slow reach of a river, around the margin of a lake or explore the waters of a pond, we recognize familiar living things met in many other places. A water lily or a bladderwort, a pond scum or a water moss is likely to belong to the same genus whether we encounter it in polar regions, the temperate zone or the tropics. The characteristic marsh plants called cattails in North America, reed mace in parts of Europe, are known in New Zealand by their Polynesian (Maori) name as raupo. The plants tend to be like those in a similar habitat on any other continent, rather than like those along the seacoast between one river system and the next.

It is true that the single-celled plants, such as diatoms and euglenas, seem little different in fresh water from their kin in the salty seas. But the strings of blue-green algae, such as *Nostoc* in its slippery jelly and *Oscillatoria* with its slowly waving thread-like filaments, are unlike anything in the salty seas without showing any special features fitting them for fresh water. Tiny green spheres of *Volvox*, with hundreds of coordinated cells arranged in the surface, roll along in ponds under the compulsive beat of microscopic whiplashes. Unbranched strands of *Spirogyra* and *Zygnema* mat into bright green floating blankets often called simply "pond scum." Under a microscope their cells are regular, bizarre, exceedingly beautiful. Yet they too show no features that can be identified as pertaining to fresh waters rather than to a marine environment.

For years after we knew stonewort (*Chara*) by name, we remained unaware that it is an alga rather than some type of freshwater moss, or perhaps an aquatic horsetail since its stalks feel rough and bear whorls of branchlets. It is a cosmopolitan denizen of ponds and streams that have a high content of dissolved lime. Upon its decay it releases some of the sulfurous odor we associate with stagnant ponds and also a contribution of limy marl that may accumulate to considerable depth upon the bottom. Botanists regard *Chara*, and its more delicate but equally cosmopolitan relative, *Nitella*, with special interest because these green plants bear their eggs amid a swirl of slender branchlets and their sperms in multicellular sex organs. This, and the pattern of elongated cells in a *Chara* stalk, seems a half-way step toward mosses and other higher plants. To qualify as a moss, each egg would have to develop inside a multicellular sex organ too, and go through embryonic stages after fertilization, which *Chara* eggs do not.

Water Mosses

Botanists agree unanimously that the mosses and their relatives, the liverworts and hornworts, came originally from some type of algal ancestor. So too, it is believed, did the ferns and their allies, which include almost all the plants upon the land. But today, virtually the only mosses that live immersed in fresh water are the makers of acid bogs—the peat mosses (*Sphagnum*)—and denizens of swift streams, such as the fountain mosses and water hypnums that grow attached to rocks and beyond the water where spray strikes frequently. We cannot be sure that the ancestors of either have been aquatic all the way back in time. They may have reentered the water through new adaptations after millions of years on land.

Peat mosses seem amphibious wherever we find them, and the water mosses only slightly less so. In autumn we can sometimes walk dryshod over the quaking mat of *Sphagnum* that roofs a small landlocked lake, such as a glacial pothole. Between us and the dark water are pale, sunbleached crowns of the branching moss, then a layer that is moist and pastel green, next a soggy mass that still retains its buoyancy because so many special cells in the leaf-like branches are air-filled and resistant to crushing. Under the floating mass are slowly-rotting portions of the moss, releasing organic acids into the water below and inhibiting the agents of decay. When the lake fills with undecomposed debris or the water drains away and leaves the dead moss, the residue is peat. It has a fuel value about the same as that of an equal weight of wood, but is several times greater in bulk.

The first time we wondered whether mosses evolved out of water to land or in the opposite direction, we stood beside a slim cascade on the eastern flank of Mount Whitney in the Sierra Nevada of California. Through our field glasses we watched a dipper (or water ouzel) as it hunted for insects and other prey in the coating of wet moss over a steep rock. A breeze blew the falling water like a curtain, often to drench the moss and bird, sometimes to leave them glistening in the sunlight on the mountain. As far out on the rock as the misty spray blew, the moss grew luxuriantly. It coated the rocks in the stream above and below the falls, and the dipper hunted there too—holding its breath while going under the swift cold water. The bird's lungs proved to us that dippers had terrestrial ancestors, despite their adaptations to feed on aquatic life. But the only clue we could think of as pointing in the same direction for the plant was its production of dust-sized spores to be carried by the wind. Peat moss reproduces in the same way, just as terrestrial mosses do. A true water plant, like the ancestors assumed for all of the vegetation on land, would probably rely on the water to distribute its spores.

Floating liverworts in the temperate zones, floating ferns where frost is no hazard, and the drifting duck-weeds almost everywhere can grow in air atop wet mud. They also rise on the water film after rains or the spring melt of snow, and drift in all directions. Duckweeds become loaded with starch in autumn, and sink to the bottom. This keeps them below any ice that forms over ponds in winter. They survive on their reserve food and become lighter, able to rise to the surface again in spring.

The duckweeds are flowering plants, which appeared on earth only toward the last of the Age of Reptiles less than 130 million years ago. As far as the fossil record shows, all flowering plants originated on land or at least in freshwater swamps. By its inconspicuous flowers and fruits, a duckweed shows that it is a monocot. None of this important subdivision of flowering plants is known prior to the Oligocene period, which began about 36 million years ago. Despite their wonderful adaptations to life on and in fresh waters, the duckweeds are relative newcomers in the world and may still be evolving rapidly.

Immersed Plants

When we stop to think about it, we realize how few families of higher plants have members that have spread from the land into fresh water far enough for their foliage as well as their roots to be immersed continuously. One (Potamogetonaceae) has done so and invaded brackish water too, for it includes the eel-grasses upon which the brent geese and other coastal waterfowl depend so much for food; and also the pondweeds in fresh water, which are favorite foods of ducks and geese. Sportsmen in small boats generally know *Potamogeton* simply as a "weed", respect its role in the nourishment of waterfowl, but avoid it if they can since it gets caught easily on rotating propellers. Some members of the frog's-bit family (Hydrocharitaceae) have the same effect, and often choke the waterways by midsummer. Water celery (or tapegrass) of Eurasia and water-weed of temperate and tropical America are important fish foods too; they contribute to the oxygen supply in slow-moving water as well as furnishing concealment for many kinds of life.

Without examining their flowers, it would be hard to tell that the members of the water-milfoil family (Haloragaceae) and the bladderwort family (Lentibulariaceae) were dicot plants, for their leaves are equally adapted to complete immersion. Feathery water-milfoil itself is cosmopolitan, and often sought as an aquarium plant. We find it, and the related mare's-tail, growing submerged in shallow parts of ponds and slow streams of temperate zones, both north and south of the Equator. Bladderwort usually seems stiff, almost wiry, and is easy to recognize by the green bladders about an eighth inch in diameter, each with a trapdoor and able to catch small crustaceans or other minute pond creatures. Ranging from the Arctic to the tropics, bladderwort is commonest and most varied in South America and the East Indies. In Brazil and West Africa, we have to look more closely to distinguish it from the similar and related "lobster-pot" plants (*Genlisea*), which have larger traps of a similar design.

Not all members of these two dicot families are aquatic, although they grow regularly where there is plenty of water. Butterwort of the bladderwort family spreads over wet mud its rosettes of shiny sticky leaves to capture insects. We discover it occasionally around bogs, among the peat moss and next to other insectivorous plants such as the unrelated sundews and pitcher plants. By obtaining the nitrogenous food they need from their prey, these strange types of vegetation can let their diminutive roots serve only as anchorage. Those in bogs absorb almost nothing from the acid water, but get their moisture from the humid air. In dry air they soon die. The vine-like gunneras of the water-milfoil family, found in humid shady valleys of tropical America, Hawaii, and from New Zealand to Java, are generally close to streams from which they can draw fresh water; at least some species have on their roots conspicuous nodules containing bacterial partners that make nitrogenous nutrients from the nitrogen in air.

Most of the plants that we associate with freshwater marshes and swamps reach into air. At least the upper surface of a water-lily leaf is dry, and the plant opens its flowers above water for flying insects to visit. The cosmopolitan cattails (or reed-maces) grow in dense clumps, their narrow leaves and terminal spikes of flowers rising as much as ten feet above water level. Lower vegetation of the marsh borders is generally equally distinctive: the rounded, soft leaves of marsh marigold in dense rosettes; the arrowheads with each leaf on a separate leaf-stalk long enough to raise it into air; and the pickerelweeds with many leaves also shaped like arrowheads but spreading out from a vertical stalk which the leaf-stalks clasp. More grass-like leaves mark the burreeds, the rushes, the arrow-grasses, the various sedges (family Cyperaceae) and the true grasses (family Gramineae) that grow in shallow water and along the wet margins. It is there that Asiatic people found the grass they adopted as cultivated rice and that waterfowl and North American Indians discovered wild rice.

The trees of freshwater margins and swamps, such as the many willows and alders, the bald-cypress of the southern United States and the larches of bogs on all sides of the North Pole, have most of their relatives on land and almost certainly had ancestors there. In fact, the more we learn of the plant life in fresh waters, the more convinced we become that only the algae spread to these locations directly from the sea. None of the higher plants, after evolving from algae in freshwater swamps (probably in the tropics), stayed there from the Devonian to the present. Instead, they became terrestrial. Those we find today in ponds and streams arrived there in relatively recent times and tell us nothing of the steps that let plants move out of the oceans.

The Animals in Fresh Waters

Just as the vegetation in rivers and ponds is so similar all over the world, the animals of fresh waters show a striking uniformity. Yet for the simpler kinds, the adaptations that are distinctive to their living in this environment are related to the formation of ice in winter and the drying up of fresh waters in a late summer, rather than to the miniscule amounts of dissolved salts. The single-celled animals (protozoans), the wheel animalcules (rotifers) of similarly minute size, and the many kinds of roundworms (nematodes) in fresh water have stages in their lives when they can shrivel to dust and be blown anywhere on earth by winds. If they developed this feature while living in the sea, they could hardly help being introduced to streams and ponds; many of them live in the soil moisture as terrestrial animals.

289

Several other phyla are represented in fresh waters by just a few kinds that manage to travel from one wet place to another in mud on the feet of birds. This type of transport is believed to account for freshwater sponges, moss animals, and coelenterates such as the little hydras, a colonial hydroid and some peculiar medusae, of which one (*Craspedacusta*) is almost cosmopolitan. *Craspedacusta* achieved scientific fame first by appearing in the indoor water-lily tank at Kew Gardens near London. Some years later, we encountered it in a private swimming pool outdoors in Virginia. A friend discovered it in South Africa, in a small river where the current was slow. It turns up almost everywhere, and in the most unlikely places.

The animals that show the greatest variety in fresh water are the mollusks, the arthropods and the vertebrates. Any good explanation for their move into flowing and standing waters should apply too to the fewer kinds of free-living flatworms (turbellarians) and segmented worms (annelids) found in streams and ponds, and in or on moist soil. Their ancestors almost certainly arrived in fresh waters as a consequence of their own locomotion and of adaptations matching the more variable conditions they met away from the seas.

Among the many types of mollusks, only a few kinds of bivalves and of creeping snails managed to spread into fresh waters. Surprisingly, the differences they show from their nearest relatives in the marine environment seem related to neither the low concentration of sodium chloride in fresh water nor the scarcity of dissolved calcium needed for their limy shells. In tolerating low salinity and in accumulating calcium despite its scarcity, their ancestors must have been superbly adapted prior to Ordovician times, almost half a billion years ago. We cannot forget how efficient the large bivalves are in absorbing calcium from waters of the Mississippi River, because for many years a pearl-button industry has depended upon harvesting shells of freshwater mussels from which discs could be cut and pierced, each button with one face showing the mother-of-pearl which previously lined the bivalve's shell.

The adaptations unique to freshwater bivalves show in the form and behavior of the young they release, after their eggs hatch within the mother's pouch-like gills. The immature bivalves, called glochidia, are fitted for catching onto the gills or skin of passing fishes, and being carried about as temporary parasites. Because the fish ordinarily swims against any current that might sweep it toward the sea, the glochidia are still in fresh water—the river or the lake—when ready to transform into little bivalves and continue life on their own.

Freshwater snails are adapted to fresh water mainly by having a lung-like organ with which they breathe atmospheric air. This feature could have evolved in an estuary, where the oxygen content of the water is often astonishingly low. With a lung, the snails needed few additional adaptations to let them live on land. Apparently only a few kinds—those now living in fresh waters—failed to make this move long ago. All land snails and shell-less slugs have a lung for breathing.

Arthropods in Fresh Water

We used to believe that the arthropods fitted neatly into categories of marine, freshwater and terrestrial. Then we began meeting the exceptions: crayfishes of fresh waters, resembling marine lobsters closely enough to have common ancestors, but migrating by the thousands across rain-wet roads at night; large freshwater shrimps in rivers and streams; scuds high above the ground in the rainwater held by plants perching on trees of tropical rain forests; ghost crabs hibernating in the coastal sand dunes of New Jersey, and robber crabs climbing coconut trees in the South Pacific; salt-marsh mosquitoes whose wrigglers (larvae) tolerate 2 per cent salinity, and caddisflies whose caterpillar-like caddisworms build portable cases out of seaweed; moths with caterpillars that live in ponds. Now we expect arthropods of some kind to fit almost anywhere. But we still want to know how their ancestors came out of the sea.

The mists and humid air from a waterfall support the growth of many delicate mosses. (Martin Litton)

Overleaf: Above left: The floating leaves of the giant waterlily of northeastern South America are often 5 feet across. Underneath they are strengthened by heavy, radiating veins and armed with sharp thorns. (Robert Perron) Below left: Three male ornate chorus frogs waiting for mates to come to the breeding ponds in the American Southeast. (Robert S. Simmons) Center: Chorus frogs, native only to North America, call loudly from among the floating duckweed. (Edward Degginger) Above right: The black-necked grebe of southern Eurasia is called the eared grebe in North America. It builds a floating nest of vegetation among the marsh plants along some channel. (J.-F. and M. Terrasse) Center right: The brightly-colored purple gallinule stalks among the marsh plants or runs over the lily pads. (William J. Bolte) Below right: Water fleas seem to dance in shallow water, reproducing by virgin birth through much of the summer (J. A. L. Cooke)

290

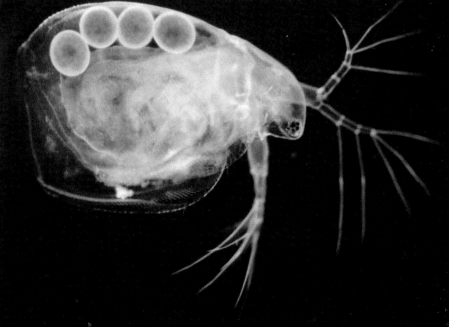

Ever since Cambrian times at least, arthropods have had an inherited advantage and one built-in handicap. Each of these animals lives inside a jointed armor, a cuticle that it secretes completely around itself before hatching from the egg. Except at the joints, this armor is waterproof, modifiable in the extreme to match different living habits while the individual grows, of minimal weight for the animal to carry about, and giving maximum support and protection. It seems an ideal covering for animals ranging in size from that of a small dog downward to that of the largest single-celled protozoans. Its one drawback is that to allow growth, it must be discarded periodically. For a few hours the arthropod is left unprotected, while its new cuticle expands and hardens.

Only through the membranous joints in the cuticular armor can water soak into an arthropod's blood. Water will enter faster from an environment where the water concentration is close to 100 per cent than in the ocean where the water is only 96.5 per cent because of the dissolved salts. The excess can be voided in urine if the animal has efficient kidney-like organs. Apparently the ancestors of the larger crustaceans in fresh water were efficient enough to survive in fresh water as they took the estuary route out of the oceans. Their descendants are the crayfishes, the freshwater shrimps and scuds, which resemble their relatives in the sea even to the extent that they retain their gills and continue to breathe in the same way.

These larger crustaceans did have to make one major change in their way of life to reproduce safely in fresh water, for they could no longer risk going through early developmental stages as members of the drifting plankton, as lobsters do in ocean water. They avoid the danger of being carried down to the sea by rivers during their development by remaining unhatched for a longer period of time. The eggs are carried about by the mother, until the young are more nearly mature and can emerge as miniature facsimiles of their parents. The extra days spent within the eggshell are made possible by additions to both the genetic instructions for growth and the amount of yolk in the egg. The matching changes place a minimum of strain on the mother, since she simply produces fewer eggs, dividing

The roseate spoonbill chooses nest sites in tropical or subtropical swamps of the New World, but to find food seeks shallows where it can swish its peculiar beak from side to side and capture tiny animals. (William J. Bolte)

up her supply of yolk into larger units. This extends the period of parental protection and keeps the young in fresh waters, to which they become progressively better adapted.

Lakes and ponds do have plankton, including microscopic plants and crustaceans that are generally less than an eighth of an inch long when fully grown. The adaptations of these tiny arthropods seem centered on exploiting the food resource and reproducing with the greatest possible speed before calamity strikes. All of the millions we have encountered have been females, reproducing without need for males. Like plant lice (aphids) among insects on the land, they change to sexual activity only at the end of their resources, such as when a pond threatens to dry up. Then a brood of mixed males and females appears. They mate and produce fertilized eggs with a heavy shell that resists damage by frost or desiccation. Such an egg can remain dry for years and blow as dust. Eventually a few eggs drop into a new lake or pond, hatch within hours and start new generations of tiny crustaceans into full activity. This is the way of life of ostracods, such as *Cypris*, with an oval, bivalved shell through the gape of which tiny legs extend, smoothly propelling the animal through the water or helping it creep over the bottom silt and submerged plants. It is the habit too of cladocerans, such as the cosmopolitan *Daphnia*, which dances in the water by twitching its branched antennae. Other adherents are copepods, such as little *Cyclops*, whose pear-shaped body progresses in jerky movements due to the rowing action of a pair of long, unbranched antennae. Generally these plankton crustaceans can browse undisturbed on the microscopic cells of green algae, with no vertebrate animals around to prey on them. This may be their biggest gain from living in temporary fresh waters. Probably their ancestors reached this environment from dry mud flats around the sea coast, not through the wet estuaries but by way of a gusty, dusty wind.

Invaders from the Land

The other arthropods in fresh waters have clearly come there from the land. Almost all of them are insects. A few are aquatic mites, relatives of spiders. Their ancestors came ashore out of tropical swamps after journeying up the estuaries of late Devonian times. Among the fossils from this period and the succeeding Carboniferous are many remains of ancient insects, of spiders and mites, and of early scorpions that seem equally fitted for life in fresh water and on land. Since the very beginning, the spiders, the mites and the scorpions have been predators. The insects on land have been their traditional prey. Seemingly when the number of predators on land increased and the fresh waters were no longer more hazardous, some of the insects and mites evolved additional adaptations that let them reinvade the aquatic environment. Like stone-flies, mayflies, dragonflies and caddisflies, they laid their eggs where their young would hatch in water and grow there until one more molt—one last shedding of the outer cuticle—would make them mature. Then out into the air they came, generally to fly in search of a mate and to repeat the process. Some of the true bugs (order Heteroptera) and beetles found advantage in swimming as adults too, carrying into the underwater world a bubble of air from which to breathe; we call them diving insects, and realize that most of them can emerge from the water and fly.

Vertebrate Animals in Fresh Waters

Of all the emigrants from the coastal seas into estuaries and then fresh waters, the ones in which people have the greatest personal interest are their own ancestors, including fishes among the vertebrate animals. The fossil evidence leads to widespread belief that the ancient fishes of Ordovician or Silurian times made this move because fresh water posed no threat. They had evolved a stiff covering of dermal bones, which prevented water from diffusing into their blood at a rate faster than their kidneys could match by excretion. Feeding on detritus they could suck up from the bottom of rivers and lakes, these creatures began to diversify in fresh water. Most of them evolved paired fins, biting jaws, and more efficient kidneys—organs able to cope with larger volumes of water. The dermal plates, which hampered movement, were replaced by overlapping scales or lost entirely while the skin thickened and became a more flexible barrier, isolating the fish's blood from the watery environment. Many fishes with these new advantages spread down through the estuaries into the oceans and left descendants that are the marine fishes of modern times. Paleontologists question whether any of the fishes that remained in fresh water are ancestral to kinds we find there now. With all the changes in continental form and drainage systems, these ancient fishes in fresh water may well have vanished and their places been taken in later times by new immigrants from the sea.

Unlike the invertebrates of streams and ponds, which so often have ways to be carried from one suitable habitat to another, fishes rarely travel overland either passively or actively. Female eels, the climbing perch of Malaya and the East Indies, the mud skippers of tropical Asia and the Siamese catfishes accidentally introduced into southern Florida and now spreading there, all seem exceptions to the general rule. Even these fishes have used their abilities on land very little to extend their range for hundreds or thousands of miles.

Freshwater fishes generally have had to wait until an accident let them swim to another body of fresh water. For this reason, their geographic distribution helps significantly in understanding the past connections between continents, or from a continent to an isolated island off shore. It generally supports the theory that freshwater fishes originated at many separate times, with the last large invasion from the sea probably during the Cretaceous. This period apparently produced the dominant freshwater fishes of modern times, the members of the order Ostariophysi, which include the characins of warm Africa and America, the eel-like electric fishes of Central and South America, the carps and minnows of all continents except South America and Australia, and the catfishes. Of the approximately 5000 different kinds of ostariophysid fishes, only a few catfishes with marine ancestors have succeeded in entering fresh waters of Australia and Madagascar. Elsewhere the success of this one group seems to have doomed many of the older kinds of fishes in fresh waters, reducing their variety to a mere token of that in earlier times.

This replacement of old-style fishes seems complete in tropical Asia and most of temperate Eurasia, which suggests that the ostariophysids made their first explosive invasion from the oceans up rivers in Southeast Asia, and then spread onward over land bridges at each opportunity. Sturgeons (*Acipenser*) still represent their old distinctive order in rivers flowing into the Black Sea. The bichirs of another order and the lung-fishes of order Dipnoi continue to hold their places in Africa despite the many ostariophysids that have arrived. Among the relics of old-style fishes we can count a paddlefish in rivers of central and northern China, another in the Mississippi, a sturgeon in north-eastern North America, the bowfin in the east-central part of the continent, and gar pikes from the Great Lakes to Lake Nicaragua. They have held out despite the spread of newer fishes across Asia northeastward

The snapping turtle of North American ponds is a powerful predator, darting out its head on its long neck to seize fish, frog or duckling. (Jack Dermid)

into North America and south all the way to Tierra del Fuego. South America still has its lungfish. So does Australia, to which few ostariophysids have yet made their way unassisted. These relics account for five of the ten orders of fishes that have most or many of their members in fresh water.

Four more orders of bony fishes have their representatives in rivers and lakes, but their headquarters in the oceans. Probably members of each of these orders emigrated up the estuaries into fresh water independently. Briefly at spawning time or for permanent residence, various of the order Isospondyli make this change; they include the salmon and trout of the Northern Hemisphere, the graylings in lakes left by the Pleistocene glaciers, the galaxiids of cool fresh waters in the Southern Hemisphere, and such peculiar fishes of Africa as the mormyrids and gymnarchids, which find their way about and locate prey in turbid water by means of a keen sensitivity to electrical fields of their own production. Eels of the order Apodes spend most of their lives in fresh water but end up as they started out, in the depths of the sea. Many members of order Mesichthyes retain their tolerance of brackish and salt water, even after hundreds of generations in lakes and streams; among their various kinds are the mud minnows of the North Temperate Zone, the sticklebacks and the pikes and pickerels of cool waters in the Northern Hemisphere, some cave fishes (amblyopsids) of North America, and the topminnows (or killifishes) of all tropical and warm temperate regions except Australia and New Zealand. Of the final order, the Acanthopterygii, only a small number of repre-

sentatives enter fresh water, although some of these are popular: the freshwater basses and sunfishes of North America, the true perches of eastern North America, Europe and northern Asia, the peculiar cichlids in the tropics of South America and Africa (often appreciated for their beauty in tropical aquaria), and the silversides of Tasmania and South Australia, which have close relatives in lakes of southern South America.

Except for the old-style lungfishes, which must now be regarded as living fossils, these fishes of fresh waters are all remote from the evolutionary path that led to amphibians, to the land, to reptiles and warm-blooded animals, including ourselves. Indeed, the closest survivor to the line our ancestors took is the marine lobe-finned fish of the Comorro Deeps between Madagascar and Mozambique. Its kin are fossilized in other parts of the world from the Devonian to the Cretaceous, always in fresh waters until the Triassic, when some members of the order emigrated to the sea.

The lobe-fin fishes (order Crossopterygii) of the late Devonian seem to have been large, powerful, fast-swimming predators. Generally scientists have assumed that the temporary nature of the swamp waters in which these fishes lived put a premium value on their being able to travel overland from a swamp that was drying up or filling in to one that still had open water. The ability to gulp air into one or two lungs, as a lungfish does today, became more important when mutations rearranged the bony support in the fleshy fins into the pattern familiar in arms and legs. But the same adaptations could help a young individual, while still too small to defend itself from a big hungry crossopterygian or other fish, to get out on land beyond reach and thereby have another chance to grow. There was major survival value for these earliest amphibians to transform from being a swimming, vulnerable tadpole to being a terrestrial creeping animal—no matter how stubby its legs were and how much its slippery body had to be hauled along, belly against the ground.

These first terrestrial vertebrates had a long way yet to evolve after their appearance in the late Devonian. Seven of the ten known orders became extinct many millions of years ago. The legless caecilians (order Apoda) are almost all tropical, burrowing like earthworms in the moist forest soil and feeding on small invertebrates. *Typhlonectes* in northern South America seems exceptional in remaining aquatic for life and in producing active young instead of laying eggs. Salamanders (order Urodela) retain their tails permanently and usually have four legs of about equal size with which to swim or drag their bodies overland; they live in and near waters of eastern and western parts of both North America and Eurasia, but not in the central portions. Frogs and toads (order Anura) are the most successful and familiar amphibians today, hopping along as adults or swimming in fresh waters virtually all over the land world.

Few groups of aquatic animals show the intolerance of dissolved salts in their environment as much as the amphibians. Except for the toads, they are in constant danger of death from desiccation. Yet in the last 100 million years or so they have traveled widely. During this time deserts that were a barrier have become moist land, and vice versa. Mountains that were too cold at the passes for an amphibian to cross have eroded away, and new ones have risen. Land bridges have appeared and vanished. And rafts with a few stowaways have chanced to travel across most salty straits in the right direction. Within those millions of years, rain and meltwater from the great natural cycle have fulfilled the requirements of amphibians many times over, including temporary ponds in which their tadpoles could swim.

As we think about the living things in brackish and fresh waters, we tend to divide them into two groups, a small and a large. The small assembly came from the parent oceans and got no farther. The larger category consists of plants and animals that returned from the land because, for them, the dry continents could not supply the necessities of life as well as the estuaries, the rivers and lakes, down to the smallest ponds. Those that returned to the fresh waters seem almost to be resuming the childhood of their long evolution, except that they bring with them to the fresh waters a heritage of features their ancestors added on land. When we search through the surface film of these waters for the life below, we reach a boundary they went through twice. They have gone part way back to the seas from which we and they started out so long ago.

Wood ibises in the Florida Everglades are at the northern edge of a range that extends far into South America. (Patricia Caulfield)

14

The Spread of the Cultured Primate

Less than two decades have passed since we personally took our first opportunity to travel beyond the limits of Anglo-America and to see how close to nature some people were living in distant lands. We cherish those experiences, for many of them cannot be repeated now. Western clothes, outboard motors, transistor radios, cigarettes and antibiotics have spread at amazing speed into cultures barely out of the Stone Age. In far less than a generation of mankind these products of the "outside world" have often upset the slow evolution of human customs that fitted people to their natural environment. In many places, the native people now scarcely fit; they are unwilling and increasingly unable to survive as their ancestors did.

We arrived in time to appreciate the system of values that had evolved among the Choco Indians of eastern Panama, the San Blas Islanders on the Caribbean coast of the same isthmus, the Djuka people (Bush Negros) of Surinam, the Pygmies of the Congo's Ituri Forest, the Masai herdsmen on Tanzania's dry plains, the Zulus on their reserves in South Africa, the Bedouins along the fringes of the North African deserts, the Aborigines in Australia's Outback, the friendly Polynesians on Fiji, and others who now face a need to change from reliable old ways to strange new ones.

As people and places become increasingly homoge-

nized, a heritage of hard-won experience is being lost, along with due respect for the adaptations in culture and body differences that served mankind to the present day. It becomes harder to shift our thinking from a highly industrialized, artificial society back to the conditions under which our species evolved, diversified, and developed its many cultures.

Amid the vast variety of living plants and animals, the human organism is certainly unique in many ways. We can find only one family—our own (the Hominidae) —with its single genus and solitary species, that shows the extraordinary tolerance for diverse environmental conditions, which permits the wide geographic distribution so characteristic of the human primate, self-styled *Homo sapiens*. Plenty of other families include only a single genus. Very many genera have only a single species anywhere on earth today. Commonly these are rather local creatures. The widespread types of life usually belong to big families, composed of dozens of genera, many of the genera huge—consisting of hundreds of species.

Fossils discovered in the present century give a consistent view of our family tree. There have been species of *Homo* for at least two million years, since the beginning of the Pleistocene. At that time there were perhaps three different species in the genus, and the family Hominidae included also the survivors of a genus of near-men (*Australopithecus*) with two more species. The skeletal remains and the artifacts accompanying them suggest that one of the near-men (*A. africanus*) knew how to make and tend a fire, and probably was by choice a meat-eater living in Africa. The other of the near-men (*A. robustus*) seems to have been a vegetarian who spread out in search of food from Africa as far into the East Indies as the island of Java. One of the true men (*H. erectus*) achieved a wider distribution before becoming extinct: he lived from northern Europe across to Northeast Asia, as well as in Africa to south of the Limpopo River.

Our ancestors must have competed with these other members of the family Hominidae, and displaced them completely. Little groups of *Homo sapiens* subsisted as hunters and food-gatherers, alert for any animal that could be overcome and eaten, for any plant with edible

Indian elephants, caught and trained to work for man, are here piling teak logs in Southeast Asia. Domestic elephants are extremely intelligent. However, they do not reproduce readily in captivity and the supply of wild elephants is dwindling. (M. Krishnan)

parts, and for any beast that might be dangerous. With a vulnerable childhood lasting at least ten years and a mature body unarmed and weak by comparison with the larger predatory animals, a human individual had to cooperate as part of a group to survive. No one knows now whether members of our species won out in the struggle with others because of better cooperation or for some different reason.

The Great Ice Ages

These contests were being won and lost under some of the most peculiar weather conditions the earth has undergone. At the beginning of the Pleistocene a wintry chill in summer let snow accumulate, reflecting away the sun's warmth until the glaciers spread southward from the Arctic and downslope from each high mountain. Technically it became an Ice Age when the frozen mass reached continental size and began calving icebergs, loaded with debris, into the oceans on each side.

The first of the four successive extensions of the thick ice is known as the Nebraskan period. It lasted until about 1,725,000 years ago, when abruptly the climate warmed and the glaciers retreated. For almost 400,000 years an interglacial period let plants and animals spread northward and up the mountain slopes. Then came the great Kansan glacial period, which ended about 900,000 years before the present. An interglacial period with a climate somewhat cooler than the previous one continued until some 550,000 years ago, when the Illinoian glacial period commenced. This time the polar cold was only about 175,000 years long. The following interglacial was shorter too. About 175,000 years ago, the latest (the Wisconsin) glacial period began, showed a brief recovery toward more normal temperatures, and then continued until approximately 15,000 years before modern times.

A continuous record of these changes was kept, as though for scientists later to decipher, in sites about as far away from emerging mankind as could be imagined. While the glaciers scoured the northern continents and erosion carried off the unconsolidated debris exposed in interglacial periods, tangible evidence collected quietly on the bottom of the Atlantic Ocean. There the nature of the deep-sea sediments varied in relation to the evolution and the coiling of the shells of minute foraminiferans amongst the plankton in surface waters. The extinction of members of the family Discoasteridae and the appearance of a new species known as *Globorotalia truncatulinoides* are taken to mark the end of the Pliocene and the beginning of the Pleistocene. This

Globorotalia still produces a shell that coils clockwise in times of warm climate, but counterclockwise when the weather chills. A fair measure of the average temperature of the sea surface can be read from the proportion of shells with counterclockwise coiling, as they accumulated in successive levels among the sedimentary deposits. After the end of the second extension of the glaciers, another group of foraminiferans evolved and thrived in each interglacial period; the coiling direction of their shells provides a still more sensitive measure of the temperature of surface waters.

So far, the fossil record on the continents has yielded no comparable information about the timing of man's spread over Africa from the area of his origin. We do not know when little groups of people traveled eastward through the Near East to Europe, to Asia north of the Himalayas, and across the Indian subcontinent into the tropical East Indies and Australia. Not until less than 11,000 years ago did any of them leave the remains of campfires in Alaska, indicating that they had entered the New World. By 8000 years before the present, people were living in Mexico. Some day an exploring scientist may discover when mankind reached Tierra del Fuego and settled there because no further habitable parts of the world lay beyond.

Diversification of Mankind

If not for the climatic and geological changes during the Pleistocene, it is doubtful that *Homo sapiens* would have evolved so quickly. Anthropologists sometimes regard the Australian aborigines as least changed. But despite a body build on the small side of average, an aborigine has hands the size of other modern men and makes his few tools to match. Aboriginal culture is scarcely as high as that of the Neanderthal people in Europe during the last interglacial period. Yet in knowing their way around in the arid heartland of Australia, in tracking edible animals and judging uses for plants, they are superb. In their home environment, it does not matter that their sloping foreheads and narrow skulls allow an average of only 1200 cubic centimeters for the brain.

The nearest kin of the Australian aborigines may well be the low-caste Veddah people who are the aborigines of Ceylon, the diminutive Pygmies of the Congo forests, the Bushmen of South African deserts,

On mountain slopes in Crete, cultivation of orchards and row crops hastens erosive loss of the thin soil. (Helen Buttfield)

302

and more distantly (with negroid admixture) the Hottentots of South Africa.

The three other racial groups that evolved during the Ice Ages all show a forehead rising more vertically between the eyes, almost no indication of brow ridges, a much larger brain and a more complex civilization. The caucasoid people who evolved in North Africa, Europe, Asia Minor and India also ranged eastward to Japan, where they include the Ainus—hairiest of all mankind. Their hair form may be wavy or straight, dark brown to red and palest blond, their eyes dark brown to blue, and their skin from dark brown to pink. The skull has complex sutures; the forehead is high and prominent, the chin well developed but not projecting. The root of the nose is high, and from it a large sharp nasal spine extends, supporting the cartilage that gives form to the narrow nose. The lips are usually thin and the upper rim of the upper lip tends to form a complex bow. These features are found among the Berbers of North Africa, the Hamitic people of Somalia, Ethiopia and westward near the Equator—including the Masai and the tall Watussi—the Arabs, the Afghans and Dravidians of India, the Mediterranean peoples, the Alpine groups in Europe and eastward, and the Nordics far into Scandinavia.

The mongoloid races appear to have evolved north of the Himalayas in Asia, probably while cut off from the rest of mankind by the long Kansan glacial period. Surviving under chronic conditions of long cold winters with high winds, they became noticeably short in arms and legs, conserving their body heat. The skull, with relatively complex sutures, developed the widest cheekbones of any modern people. Over it, fatty deposits formed around the eyes, protecting them from cold and perhaps from glaring light reflected from snow and ice. An epicanthic fold over the inner corner of each upper eyelid gives the eyes an oblique appearance. Head hair is coarse, straight and black, but body hair is scanty; the skin ranges from yellowish brown to yellow. These features apply to the Mongols and Tatars of Asia, Oriental people farther south into Malaya, the Indonesians and Polynesians of the East Indies and Pacific Islands, the Eskimos and the American Indians in both hemispheres of the New World.

Members of the negroid races are generally best adapted for life in the tropics, with a body build contrasting sharply with that of the average mongoloid

The American bison, now extinct in the wild, thrives in the sanctuary of national parks. (James Simon)

person. The bones of the forearms and lower legs are usually much larger in proportion to the body, increasing the surface through which heat can be lost; the heels project more than in other races. The broad, flaring nostrils and short noses waste little space on heating inhaled air or humidifying it. This nasal form is partially due to the low root of the nose, the very wide angle at which the bones that comprise the bridge of the nose meet one another, and to the smallness and roundness of the nasal spine. The skull sutures are mostly simple and, while the chin is less prominent than that of other races, the face has a forward slant and projecting jaws. The everted lips are less ape-like than those of any other primate; the bow of the upper lip's margin is generally simple. The hair is always black, that on the head being woolly or kinky or even so scanty as to be arranged in clusters called peppercorns; body hair is far less prominent than among caucasoids. Negroid skin is dark brown to black, and the irises of the eyes are always black. These characteristics are found among the Negroes of Africa, the Negritos of the Philippines to New Guinea, and the Melanesians of the Australasian islands.

As long as the members of these races depended upon foods that grow naturally, about two square miles of fertile territory were needed on the average for each person. Since the entire earth offered the equivalent of only about 20 million square miles of land of this quality, the human species had to remain sparse merely to survive. The total population may have been about ten million between 12,000 and 7000 years before the present, when members of *Homo sapiens* reached essentially all parts of the easily-habitable world. These roving hunters, fishermen and persistent collectors of edible parts from plants knew how to use fire to keep warm and to cook food. This knowledge gave them extra tolerance for intemperate climates and increased the variety of foods they could digest. It seems to be the only real advantage primitive man had over other tool-using animals, such as chimpanzees.

Recent success in delicately measuring the proportions of radioactive substances and the products of their decay in fossil remains and human artifacts from the last 12,000 years has confirmed one suspected point: that many of the large meat animals survived through the Ice Ages, but became extinct shortly after human hunters came as immigrants. Whether the mammoths of North America, the giant ground sloths of Hispaniola, the moas of New Zealand and the flightless roc of Madagascar were already waning may never be learned. They vanished when man arrived.

Improving the Environment

The first major change that indicates human intelligence came in Africa, Europe and Asia, later in America as well, when men and women began improving their environment by raising animals in confinement and sowing the seeds of useful plants. Time and labor were expended far more efficiently than in random gathering of food, by biasing the products of the soil toward an increase in living things that could serve as food, fiber, fuel and medicine. Suddenly the old limitation of two square miles per person no longer applied. The human species could settle on the land in social groups of increased size, and test other practical ways to make life easier.

The tribal group with the most productive crops could become stronger in health and numbers. By force it might take land from less efficient neighbors. Raising better grains and meat animals had survival value. Husbandry took on real significance. Advantage lay in choosing the most productive plants and animals as breeding stock, and in eating the second- or third-rate. Through selection under domestication, mankind sped the evolution of useful kinds of life and directed the changes toward human benefit.

Animals whose natural fear of man had been eliminated by selective breeding, and plants that produced abnormally large edible parts, found special favor. It did not matter that the animals man herded could no longer compete under wild conditions, nor could the chosen plants propagate themselves successfully for many generations without human care. A new and many-sided symbiosis was being established; it was an association between cultivated species dependent upon man, and man dependent upon having these products of the land abundantly available.

Simple partnerships are fairly common among non-human life. The corporate lichen can live and grow and reproduce where neither its component fungus or alga can be active or hold a place. The pine tree and the orchid owe much of their success to the fungus strands that serve them in place of root hairs. The legume and the alder have no lack of nitrogenous nutrients so long as the bacteria in their root nodules cooperate. The

Whooping cranes have been exterminated over most of the large area in North America where they once nested. Their territory is now crop land. A few survive, traveling between marginal sanctuaries. (Lorus and Margery Milne)

giant clam of the South Pacific reefs gets the food it needs quite easily from the algal cells it raises in its mantle greenhouses. Mutual benefit in each symbiotic relationship leads to evolution of adaptations that improve the cooperative venture. For mankind, the adaptation is called culture or civilization; it evolved as soon as cultivated plants and animals provided a reliable resource. The domesticated kinds of life can be regarded as generators of culture—cultigens. With their help, the human population could double repeatedly, and people could cluster into semipermanent communities.

Cultigens let urban people grow ignorant of nature and unconcerned about events upon the land. They busied themselves, instead, with making themselves more comfortable and in subdividing tasks until each person became a specialist. Beyond the communities, the fields of crop plants and herds of docile livestock proved an irresistible attraction to wild animals, both herbivore and predator. Farmers and herdsmen became protectionists, whose living depended upon having produce to market. They had to fight off nomadic people, too, who continued to regard all land as public property and anything useful on it as belonging to the taker. Isolated on their productive acres, the rural citizens had to remain versatile. Those far from town had to contend with trader as well as raider, while raising food for people who could forget man's ultimate dependence upon the soil and sun.

Proliferating Man

Until we see with our own eyes the artifacts of artist and artisan left in stone by people of bygone cultures, we rarely appreciate what could be accomplished with manual labor, with tools scarcely more complex than the lever and the wheel, with help at times from strong oxen, horses, camels, goats, elephants and water buffaloes. The Egyptian pyramids, the sphinx and the royal tombs are scarcely more impressive than the temples of the Incas at Macchu Pichu, the ruins at Angkor Wat, the remains of an astronomical computer at Stonehenge. Based upon food and fiber, fuel and medicines prepared from cultigens, and fresh water conducted to the cities by skilfully-engineered canal and aqueduct, the human species was able to double in population about five and three-quarter times by the year 1650 A.D. Each doubling took more than a millennium, and showed the increased exploitation of the land made possible by conquest, forest felling, marsh draining, irrigation, construction of roads, and the cheap labor of people still living at the subsistence level.

The reconstructed history of civilizations that rose and crumbled tells far more of man's spread and successes than the dates of battles, the names of generals and rulers, good and bad. Often we can recognize, as earlier generations of people could not, how the logistics of urban life failed through no single act of any military leader or other notable. Around each city, the forests vanished as woodcutters felled the trees to make wooden ships, wooden buildings, wooden utensils, and fuel, or to clear the land for food production. Exposed to the elements, to the teeth and feet of sheep and goats, the soil eroded until it could sustain few useful kinds of life; food had to be hauled from progressively greater distances. With fewer forests to trap the rain and dole out the moisture through dry seasons, the springs became unreliable; rivers varied from flooding to a mere series of stagnant pools. Aqueducts had to be extended repeatedly. These troubles at home doomed the city-states where culture centered, while historians recorded the rebellions in outlying parts of each empire and then the decisive battle with which some other civilization claimed control.

If from 1650 to 1970, the human population of the world had continued to increase at its previous rate, the total would have gone from 545 million to about 665 million. Much of the gain could be expected because colonists—herders and farmers in America, South Africa, Australia and New Zealand—displaced people who had not yet learned the gains from cultigens. Native plants and animals would have been displaced too, and to the list of ten kinds of birds and ten of mammals that vanished from the earth between the beginning of the Christian Era and 1650 another one or two might have been added as newly extinct. The rate of loss was about one species per 165 years, due partly to persecution but mostly to destruction of essential forests, marshes and useful grasslands.

The rate of growth of the human population and of its spread over the lands of all continents except Antarctica did change sharply in the seventeenth century. So did the rate at which nonhuman kinds of life lost their place in the world. By 1750 A.D. there were already 728 million people according to the best estimates. The 1975 total is expected to be 4500 million. Mankind is doubling several times each century, not just once in each 1100 years. Birds and mammals, for which a tally of species is most reliable, are becoming extinct now at the rate of one per each eight months or less.

The change in the number of people for whom food

307

and space could be found can be accounted for in terms of better tools for agriculture, improved means for transport of crops, and greater buying power among city dwellers. All of these related to the rapid introduction of a new source for energy with which to give man's enterprises power, heat and illumination. It came only when Europe's forests were largely depleted and the need for a substitute for wood as fuel had grown acute. The iron industry had been the greatest user, relying upon charcoal for heat. This led the English Royalist John Evelyn to philosophize in his book *Silva: A Discourse of Forest Trees*. Noting that although smelters usually were located close to ore pits, the land around most of them had originally been forested, he wrote in 1686: "Nature has thought fit to produce this wasting ore more plentifully in wood-land than in any other ground, and to enrich our forests to their own destruction."

Late in the 1600's, ways were found to use fossil fuels, freeing energy that had been stored in the earth from sunlight millions of years ago. First coal and coal gas, later petroleum oils, replaced wood as the fuel for industry. These fuels made it worthwhile to invent the steam engine, the steamship, the steam locomotive. The cost of metal dropped while the quality of the material improved. The extra power at man's disposal made possible a vast range of new inventions.

The seventeenth century saw the first permanent settlements of Europeans in North America and South America. For mankind it was an age of promise. But for a great many kinds of wild plants and animals, it was the beginning of the end.

Food by Mass Production

The change in man's use of the land came gradually when measured in terms of human generations, which is almost overnight on the scale of geological time. New and more efficient tools replaced the homemade, wooden implements. Successful farmers bought adjacent land and increased the size of their fields, getting better use from their expensive machines. They plowed deeper, bought fertilizer to improve the soil, filled in gullies and learned to reduce erosion to a minimum. The internal combustion engine became the motive power for the tractor and the harvester that replaced teams of horses and people with simple flails. Agriculture and animal

Domestic goats climb into Moroccan argan trees to reach the olive-like fruits and edible foliage. (Gerda Bohm: UNESCO)

husbandry became business operations, with growing investments in equipment of many kinds.

To make a living from the land, producing food and fiber in quantities sufficing for his family and 50 or 100 others, a farmer wars constantly with nature. His goal of monoculture—one kind of crop from this field and a different one from the next—is completely unnatural, for it cannot tolerate living things of other kinds. A field is to be an outdoor factory. Yet without tight walls, it is open to plants and animals from every side. They enter his fields to share the space and the resources made possible by the sun. Nature favors the maximum variety of life the climate will allow, each kind in modest numbers. Neither modesty nor tolerance will support a modern civilization. Human culture is based upon intolerant exploitation, never upon taking a reasonable proportion. Crops grow larger faster when competitors are eliminated. Who wants to share the rain and sun with a weed, or an apple with a worm?

We favor crops that can be mass produced, and take them to foreign lands as a reliable base for our modern way of life. Local resources may be overlooked as being scarcely worth the trouble to convert them into cultigens. We want wheat and sheep from the Caucasus in the Americas and Australia, potatoes and tomatoes from South America in northern America and Europe, maize from the Americas in Africa, rice from the Orient in Latin America, cattle from the Nile valley and from India almost everywhere.

In Panama once we made a point of inquiring from the managers of grocery stores why their offerings included so rarely any of the papayas, passion fruits, bananas, pineapples, custard apples and mangoes that could be bought in the native markets. Instead, the display shelves were heavy with fresh bananas from Ecuador, apples from Washington state, citrus from Florida, lettuce from California, pineapples from Hawaii, and canned fruit cocktail from Australia. The answer was uniform: the managers could count on delivery of perfect fruit in the amount ordered on the date specified from the distant suppliers, but only produce of uncertain quality and quantity from local growers. The native people were still living too close to nature to compete on a wholesale basis.

Today, in the most technologically advanced countries, we are healthier, longer-lived, better fed, and more widely informed than people have ever been. Yet we deceive ourselves about the logistics of urban existence. It is true that farming produces record crops of food, and that the green plants from which we get our nourishment constitute a renewable resource. Cultigens are living things. But the food as it reaches our table contains less energy than has to be expended from fossil fuels in order to get it there. This energy from the ground is needed to make and propel the machines that prepare the fields, sow the crops, spread the pesticides, harvest the grain, carry it to the nearest market, haul it to the corner grocery or shopping center, and deliver it to us in edible form. While eating our meals we devour the earth, using up the coal from the Carboniferous, the petroleum from the Cretaceous, and the power in atoms that we cannot replace. This is part of the price for city living and the efficient agriculture that underlies it.

Displaced Plants and Animals

Until the present century, few people concerned themselves about the native plants and animals that were displaced by human populations and cultigens. We hardly know the names of plants that vanished as the forests were cut, the marshes drained, the tall-grass prairies plowed and planted to foreign crops. Those that could not become weeds or survive in neglected corners stood a good chance of vanishing forever.

The record for large animals is better documented. It shows that the rate at which species have become extinct has risen parallel to the increase in the human population. Many that are now threatened have undergone a progressive constriction in the range they occupied. Between 1650 and 1850, while the number of people in the world went from 545 to 1171 million, about 40 species of warm-blooded animals vanished. This is one in each five-year period, not one in each 165 years as in earlier times. About half of the 40 were birds, including the dodo of Mauritius and the flightless great auk of North Atlantic coasts. Between 1850 and 1900, while the human population rose to 1608 million, another 64 species of mammals and birds vanished, at a rate of one species each nine and a half months. Among them was the Labrador duck.

Since 1900, extermination has continued even more rapidly. Even in North America, where the citizens have invested heavily in protecting wildlife, conspicuous birds have disappeared: the passenger pigeon (1914), the Carolina parakeet (1920), and the heath hen (1933). The future of the ivory-billed woodpecker, the California condor, the Bermuda petrel and others are in doubt. Only public sentiment saved the trumpeter swan and the whooping crane from vanishing. The American bison, like the European one (the wisent), is gone in the wild, although perpetuated in parks. Public zoos may become the last refuge for some of the world's wildlife.

Two international organizations centered at Morges, near Lausanne in Switzerland, have accomplished much since their origin in 1950, alerting people around the world to the endangered state of various kinds of animals and plants. The International Union for Conservation of Nature and Natural Resources (IUCN), assisted by the United Nations Educational, Scientific and Cultural Organization (UNESCO), began publishing in 1964 lists of threatened species and subspecies and a library of Red Data Books in looseleaf form that could be kept current. Presently the list of endangered mammals includes 258 kinds, and of birds 334. A total of 162 kinds of birds are regarded as having become extinct between 1600 and 1968. Meanwhile, the associated World Wildlife Fund (WWF) is serving as a charitable foundation in support of emergency programs to save dwindling species and of ecological studies to learn what needs to be done to safeguard endangered animals and plants.

Taking Stock

Only by careful planning for multiple use can a place be left for nonhuman animals and plants with no economic importance. Now that mankind's frontiers have come to be the sea, the polar regions and outer space, every bit of land must be used carefully. The total area on the continents and islands is about 57 million square miles. But of these, about one-fourth (14.0 million) are under glaciers—including 5.3 million square miles in Antarctica and another 0.7 million in Greenland—or consist of barren tundra, jagged mountains, or deserts that produce almost nothing valuable to man unless irrigated with fresh water from far away. A slightly larger area (15.2 million square miles) produces almost nothing man can eat and supports few kinds of animals because it is forest land; nearly half of it is submarginal and of little present worth, although, with care and at considerable expense, it might be made useful. A third quarter of the world's land is livestock range and pasture, yielding valuable meat and dairy products but only about a fiftieth as much energy in food per square mile as would come from good fields of grain. Of this area for domestic animals, more than half (8.2 million square miles) was formerly forested; the rest (6.1 million) was grassland. These are the parts of the world where wild species can have a place with a minimum of conflict with human interests. The final quarter of the land receives too intensive use: 9.2 million square miles of expanding cities, industrial parks, airports and garbage dumps (which produce almost no food) and 4.3 million planted with useful crops.

Out of industrial growth and medical advances in the twentieth century have come financial plans that allow people weeks of annual vacation and years of health after retirement from gainful work. Millions of these people choose to spend their free time and money hiking and camping, visiting national parks and wilderness areas—preferably those where wild animals are visible. Smaller numbers of hunters, fishermen, boating enthusiasts and water skiers, invest even more heavily in their outdoor sport. The outcome is a new use for land and shallow waters, one that is often consistent with a place for wild animals and plants of many kinds. A paradox of modern times is that, although the number of people and the need for food for them are greater than ever, the demand for recreational areas that produce almost no food has grown too.

Wilderness areas and facilities for visitors are being developed in many parts of the world. Some of the newly-independent African states, for example, have recognized, realistically, that no product of the land or of industries that could be established in the near future can earn as much foreign exchange as well-managed national parks with native animals and plants. Since 1963, Kenya has established several new wildlife reserves, including two marine parks along the coast where coral reefs of special interest are being preserved. Publicity outside the country attracts attention to these regions, while within the country school children from all over are being brought by bus to the national parks. There experienced native ranger-naturalists acquaint the youngsters with their living heritage. A reclassification of land is being undertaken, to learn its most effective use and to schedule concentration of people close to soils of potentially high yield, while converting unproductive areas into tourist attractions.

Man's Rearrangement of Life

By introducing cultigens, pets, animals to hunt, and decorative plants in distant lands, the human species has provided uncounted opportunities for nonhuman life. Horses have escaped and become wild in the western United States, where they are called mustangs and resented because they compete with range cattle for scarce drinking water. Edible giant snails (*Achatina fulica*) from East Africa, introduced first in India as a possible food for starving people (who refused to eat them) and later through much of the South Pacific island area and to Hawaii by Japanese (who ate them and regarded them as medicine), have multiplied enormously and caused tremendous damage to vegeta-

Bald eagles find nest sites but also tainted food in the Florida Everglades swampland. (Patricia Caulfield)

tion. European rabbits, released in Australia and New Zealand as game animals, competed disastrously with native herbivores and introduced livestock until brought under incomplete control by means of a disease (myxomatosis) from South America.

Carrots escaped from cultivation in North America to become a widespread, familiar weed known as Queen Anne's lace. Water hyacinth from South America, discarded by fanciers of tropical fishes from their aquarium tanks, has thrived in the canals of Florida and in the

lake behind Africa's Kariba dam, interfering with navigation and shading the native water plants to death, starving water animals that could not tolerate a steady diet of water hyacinth.

In a new area, local fungi, insects and animal parasites may greatly prefer a cultigen to the hosts upon which their ancestors evolved. Until potatoes were planted in Colorado, having been transferred from South America to Europe and then back to the New World, a black-and-yellow insect of the Rocky Mountain slopes was just a leaf beetle with ten stripes upon its back. Abruptly it became a pest, famous as the Colorado potato beetle, and to control it the potato fields had to be sprayed with poison. Nor was treating the crop plants an adequate protection as long as the same kind of insect lived also on local weeds. The weeds it could

Wastes from a sugar-beet processing plant polluted the stream and killed the fish. (W. E. Seibel: County Fish & Game Association)

eat must go too. Often this man-made war proves endless, because the other kinds of insects that eat the same types of weeds often turn to man's crop plants as second best as soon as their favorite foods have been destroyed. Moreover, the pest species can generally evolve fast enough to avoid being exterminated completely by any chemical treatment that lets the crop grow well.

For many years now a stalemate between resilient nature and resourceful men has existed in Africa, from Gambia on the west coast through the southern Sudan to Somalia on the east, to northern Mozambique and Angola. Over much of this huge territory and particularly along lakes and streams, the blood-sucking tsetse flies transfer infections of blood parasites from one vertebrate animal to another. Native animals, whether antelope or owl or crocodile, are virtually immune to the infection. But it causes a deadly sleeping sickness in people, and a similar disease called nagana in cattle. To open up this land to human colonists with cattle as cultigens, all sorts of drastic measures have been tried. Over tremendous areas every bush and tree has been cut and burned, so that tsetse flies would have nowhere to hide from the intense sun. Every native vertebrate animal that was suspected of harboring the disease has been killed to eliminate the reservoir of infection. Yet as soon as the land recovered enough to be useful to people and cattle, the tsetse flies and the diseases returned. So far the web of food relations has proved too complex and tough to be upset without destroying the territory for mankind too.

Tomorrow's Challenge

With increasing numbers of people in need of food and other renewable resources from the land, plans are often made no farther ahead than the next crop. Yet experience has proved repeatedly that the yield diminishes if the methods of cultivation harm the soil or the fresh waters. In each region the green plants determine the carrying capacity for animal life, be they native wild plants well adapted to the soil and climate or various introduced cultigens. Unless the plants can recover from being grazed and browsed upon, can grow and propagate for their own benefit, they die and leave the place a desert. Sheep-made deserts are well known in Spain and elsewhere around the Mediterranean. In arid and semiarid lands, the damage done in a single summer may take decades or centuries to repair. Generally the reclamation work requires skilled labor, costing more than the price of the livestock that caused the damage.

Perhaps it is characteristic of human independence that the ways to care for soil have to be invented over and over. We think of this whenever we encounter terracing for agriculture. South of Tripoli in Libya, we found people on the steep slope of the high desert still benefiting from stonework engineered by the Romans before the Christian Era. It holds soil and moisture around the roots of gnarled olive trees that may have been bearing fruit for 2000 years. In the Philippines north of Baguio, the primitive mountain people tend rice paddies their ancestors built spectacularly tier upon tier to get the maximum benefit from rain. In the western mountains of Guatemala, where Indians developed a similar tradition for the benefit of crops of maize, we have watched small children hurry out during a rainstorm. By hand they molded the earthen terraces on the slanting fields, protecting the soil they will inherit. In industrialized nations, adult labor minimizes erosion with similar procedures. Surveyors establish the pattern along which tractors are to be driven, dragging the cultivating equipment one level after another.

Often we feel impatient over the reluctance of people in underdeveloped countries to change from customs that formerly matched hazards in their environment. Democratically we believe that the gains from science and technology should be made available everywhere without delay. No longer is there a reason to fear predatory animals, or most parasites and infections. Many hazards can be minimized by public health measures and marvellously effective medical techniques. Improved varieties of cultigens are on hand, and knowledge of how to speed their growth or protect them from competition. More could be gained quickly by applying what we know than by waiting for further scientific progress.

Spectacular as the gains from research and development have been, letting men explore the greatest depths of the oceans and travel through outer space, they have not freed us from sharing with virtually all animals a requirement for oxygen produced by green plants. With no sudden change in temperature, we exhale a molecule of carbon dioxide and one of water for each molecule of oxygen we use in respiration. The same exchange occurs in combustion, whether of organic matter from recent growth or of fossil fuels from long ago. Combustion is our competitor for the useful one-fifth of air. But unlike respiration, it raises the temperature significantly, spreading heat into the surroundings and making inert chemical compounds both volatile and reactive.

Not since warm-blooded animals appeared on earth has so much oxygen been removed from the atmosphere and so much carbon dioxide added. Each of these changes caused by the large-scale combustion of fossil fuels has a different and potentially profound effect on living things. In the upper atmosphere, oxygen atoms separate from their normal pairs in molecules and recombine in threes as the gas ozone. Ozone acts as a filter, absorbing much of the ultraviolet radiation from sunlight. It reduces the intensity of these dangerous rays before they reach places where plants and animals live. With less oxygen and less ozone, more ultraviolet

313

will pass through and cause radiation damage. Carbon dioxide high above the earth acts as a blanket, capturing infrared radiation like the glass in a greenhouse, gentling the change from day to night or summer to winter. With more carbon dioxide in the atmosphere, warmer weather can be expected—even a return to conditions that let plants flourish as they did in Carboniferous times, when decomposition could not keep up and coal accumulated.

Waste Heat and Chemicals

Geologists fear that a rise in the average temperature by only a degree or two will cause the gigantic glaciers to melt on Greenland and Antarctica. Any major shrinkage in the area of these ice fields would let the earth absorb heat from the sun at an increasing rate. Like toggle action, the change might be quick, perhaps complete in a decade or less.

Until the middle of the twentieth century, the heat added to rivers and lakes by industrial enterprises was scarcely recognized to be an important kind of pollution. The waste water carried much more obvious poisonous and malodorous compounds. But when these are removed, as government agencies now insist, the heat becomes apparent. Any significant rise in the temperature of surrounding water is generally harmful to aquatic plants and animals. Warm water holds less carbon dioxide in solution for green plants to use in photosynthesis. This limits plant growth, and the food available to animals. Less oxygen is dissolved too. Yet the warmth spurs on cold-blooded animals to greater activity, increasing their need for oxygen in respiration and for food. As the temperature rises, fewer kinds of life can survive.

Today the aim is for industrial users to return whatever water they remove from lakes and rivers at least as clean and cool as it was to begin with. Changing to thermonuclear reactors increases the difficulty because of the large amounts of waste heat. Most of us overlook the heat, thinking only of the radioactive byproducts that are so obviously dangerous.

We generally think of fertilizer and chemicals designed to protect cultigens as beneficial. But fully two-thirds of the fertilizer, the insecticides, the fungicides and the weed-killers ("selective herbicides") miss their targets. They blow away as dust, or are leached out of the soil by percolating water, or they are washed by rain into the nearest stream.

When fertilizer reaches rivers and lakes, it certainly enriches the water. It changes the environment for aquatic plants from one with few dissolved nutrients to one with an abundance. Called eutrophication, the enrichment supports growth of plants at rates faster than herbivorous animals can match. Soon the plants clog the water, interfering with its movements and shading one another. Death and decay take the place of plant-eaters in disposing of the excess. The decomposers, however, compete for oxygen so successfully that insects, fishes and other animals die of suffocation. As though it were a poison, the additional fertilizer actually decreases the wealth of life in the freshwater realm, and must be regarded as a pollutant.

Truly poisonous materials, designed for use in killing insects and fungi that attack crops and in killing weeds that compete for space and moisture, are pesticides. Almost all of those applied in quantity were devised in the years since 1945. Most are effective in minute quantities, and will kill 50 per cent of the individuals of an unwanted species in a few years. But twice this standard lethal dose, called the LD 50, does not eliminate 100 per cent in a week or a year. Usually the survivors have progeny that are more resistant. The dose must be increased, or a different pesticide of greater potency employed.

We know that for discovering the insecticidal properties of DDT, the Swiss biochemist Paul Mueller received the 1948 Nobel Prize in Medicine and Physiology. Not until 1962 were the often disastrous side effects of this and other agricultural poisons drawn forcefully to public attention, as Rachel Carson did in her book *Silent Spring*. Nor did anyone expect that by 1969 all further use of DDT and kindred chlorinated hydrocarbon pesticides would be banned in many states and countries. These compounds and their poisonous derivatives are the worst offenders because they resist natural decomposition and accumulate in soil and water. Many kinds of life concentrate these pesticides until they become biocides—agents of death for the unintended. Sensitive tests for DDT and its derivatives prove how these poisons have been dispersed around the world, far from the fields and forests to which they were applied. Walruses in the Arctic, penguins in Antarctica, and people everywhere have chlorinated hydrocarbons in their fat, stored there for lack of a way to excrete them. Thousands of species are endangered by agricultural poisons that escape.

Following a report in the scientific journal *Science* that linked the persistent pesticides to a steady decline in numbers of Bermuda's endangered petrel and predicted extinction of this sea bird by 1978, we visited the islands to learn about the situation firsthand. These petrels come to land only to make nest burrows in

Small, white whales of arctic waters are hunted regularly from Churchill, Manitoba, in Hudson Bay (Authenticated News & Information Service)

coastal cliffs, incubate their single eggs, and tend the chicks. Otherwise they range widely over the ocean, eating small fishes and squids from the surface. From this food gathered far from land they become contaminated with DDT and similar poisons, for none of these insecticides may be used on Bermuda. In recent years the proportion of petrel eggs that fail to hatch has increased. Often the chicks that do emerge are so abnormal in behavior that their parents kill or desert them. The unhatched eggs and dead chicks contain the toxic substances; surviving chicks must contend with extra poison brought to them unwittingly by the parent birds which feed them by regurgitation. The concentration of pesticides in the petrel eggs and young matches well that found in the eggs and dead offspring of ospreys and eagles along American coasts.

By being at the top of the pyramid of food relations, these birds are particularly vulnerable. It is their role to skim off the sick and unwary from the populations of their prey. Normally they eliminate the old, the maimed and the infected individuals that otherwise would compete for food (always a limited resource) with the healthy, wary members of the prey population. But when the predator eats prey that is sick with poisons, it accumulates dose after dose of the chemicals until its own future is endangered. On urban and suburban lawns, robins meet the same fate when they act

315

as predators, eating earthworms tainted with DDT. The earthworms obtain the pesticide from fallen leaves which, while green, were sprayed with the long-lasting poison.

We notice these changes in the numbers of familiar birds more often than in populations of other animals because our feathered neighbors are mostly active by day, large enough to recognize easily, and subject to many of the hazards that affect us. People too get small amounts of biocides in almost everything they eat, in most beverages (including water), and in the air they breathe. If cow's milk contained as much DDT as human milk now often does, it would be illegal to ship it across state lines. No one knows yet what the consequences will be, but a growing number of responsible citizens prefer to take no chances with a continued accumulation of poisons in the environment we share with all other kinds of life.

A Shortage of Time

Every living thing is now involved to some extent because of the attempts of the cultured primate to support an ever-increasing population. A great many old residents are endangered and some have become extinct because of sudden competition from new neighbors that our species has introduced accidentally or deliberately, carrying them past barriers they had not been able to hurdle on their own. Mostly we have set back the succession among living things toward pioneer conditions, with fewer but more aggressive species in a drier, less productive environment. Given time, a fresh state of relative equilibrium and greater richness will develop, although we may not like the outcome. It is sure to include the hardy, prolific pests and weeds. They have evolved immunities to the harshest measures yet devised.

The real question is which other kinds of life have time to enrich the world until mankind reaches some upper limit in numbers and settles harmoniously into a benevolent pattern. It is too late to restore the planet to the conditions that produced the human species, that contributed to the development of civilizations, and that led to the distribution of geological and biological features among which our ancestors spread out. It is too soon to know how the earth will respond to our release of heat from nonrenewable fuels and to the changes already made in the chemical and living nature of the environment. Psychologically it seems too early, but ecologically it may be too late, to decide on the maximum number of people the world should be expected to support—and then to keep the total population within that goal as the self-made, self-appointed custodians of all life.

Deliberately we can make that stride, letting our species and other living things renew an orderly evolution of adaptedness. The encouragement of diversity would be a move toward higher quality as well. By comparison, any other use to which we put our intelligence is inconsequential. The frantic fluctuations we see in the man-altered landscape are marks of a chaotic community, one that has lost many of its important components and is in danger of destruction before new stabilizers can evolve. Now is the time to act upon our discoveries about the inner workings of the living world, letting us be proud to be a cultured primate—a unique species whose existence will pose no threat to the rest of creation.

Index to Animals and Plants